T0181455

Springer Proceedings in Mathematics & Statistics

Volume 270

Springer Proceedings in Mathematics & Statistics

This book series features volumes composed of selected contributions from workshops and conferences in all areas of current research in mathematics and statistics, including operation research and optimization. In addition to an overall evaluation of the interest, scientific quality, and timeliness of each proposal at the hands of the publisher, individual contributions are all refereed to the high quality standards of leading journals in the field. Thus, this series provides the research community with well-edited, authoritative reports on developments in the most exciting areas of mathematical and statistical research today.

More information about this series at http://www.springer.com/series/10533

Daniela Cadamuro · Maximilian Duell
Wojciech Dybalski · Sergio Simonella
Editors

Macroscopic Limits of Quantum Systems

Munich, Germany, March 30 - April 1, 2017

 Springer

Editors
Daniela Cadamuro
Institut für Theoretische
Physik Universität Leipzig
Leipzig, Germany

Wojciech Dybalski
Zentrum Mathematik, M5
Technische Universität München
Garching, Germany

Maximilian Duell
Zentrum Mathematik, M5
Technische Universität München
Garching, Germany

Sergio Simonella
UMPA UMR 5669 CNRS
École normale supérieure
Lyon, France

ISSN 2194-1009   ISSN 2194-1017   (electronic)
Springer Proceedings in Mathematics & Statistics
ISBN 978-3-030-13187-6   ISBN 978-3-030-01602-9   (eBook)
https://doi.org/10.1007/978-3-030-01602-9

Mathematics Subject Classification (2010): 82C10, 35Q82, 81Qxx, 32A70, 11F72, 49-XX

Preface

"It is amazing how different the world appears on different scales." Moving from the subatomic up to the intergalactic world, we cross several physical levels apparently ruled by very diverse laws. The connection between steps of this hierarchy is a never-ending source of discoveries and mathematical challenges.

Our *incipit* is borrowed from the book by Herbert Spohn on the large-scale dynamics of interacting particles, which is an outstanding reference in the field for almost 30 years. Nowadays, research activity is intensively focused on quantum systems. The present volume collects contributions from a number of experts and aims to be a survey of current directions of investigation. Most of the authors presented their research during the 3-day workshop "Macroscopic Limits of Quantum Systems", held in spring 2017 at the Technical University of Munich, which was attended by nearly 60 people from several institutions and countries. The conference was also an occasion to celebrate the achievements of Herbert and his reception of the Max Planck Medal.

The main subject of the workshop was the mathematical investigation of collective phenomena emerging from quantum theory at observable scales. On the side of bosonic systems, the problem of Bose–Einstein condensation, ground-state properties of Bogoliubov Hamiltonians and the Gross–Pitaevskii limit were covered. Concerning fermionic systems, the rigorous validation of Hartree–Fock equations and the motion of tracer particles were discussed. Moreover, superconductivity and mathematical aspects of the BCS theory were presented, as well as the statistical mechanics of the uniform electron gas and its relation to Jellium. Further important topics were in the scope of the workshop: the justification and properties of quantum Boltzmann-type equations, the theory of indirect measurements, the strong-coupling limit for the polaron model, Peierls substitution in perturbed periodic systems, and the adiabatic theorem for many-body systems.

This proceedings volume is a selection of works from leading scientists including a large part of (regrettably, not all) the speakers of the workshop, and proposing an overview of fields in which the current research is most active. The book will serve both advanced graduate students as an introduction to research, and specialists in mathematical physics, as it presents technical details in an accessible way. As is well known, to prove rigorous results in quantum physics requires the development of powerful tools within abstract mathematical theories. Here, we encounter fruitful connections to functional analysis, spectral theory, statistical mechanics, renormalization group techniques, and variational calculus. This book is also an account of that remarkable effort.

Leipzig, Germany   Daniela Cadamuro
Garching, Germany   Maximilian Duell
Garching, Germany   Wojciech Dybalski
Lyon, France   Sergio Simonella
July 2018

Contents

Kinetic Theory and Thermalization of Weakly Interacting Fermions

Jani Lukkarinen

Abstract Weakly interacting quantum fluids allow for a natural kinetic theory description which takes into account the fermionic or bosonic nature of the interacting particles. In the simplest cases, one arrives at the Boltzmann–Nordheim equations for the reduced density matrix of the fluid. We discuss here two related topics: the kinetic theory of the fermionic Hubbard model, in which conservation of total spin results in an additional Vlasov-type term in the Boltzmann equation, and the relation between kinetic theory and thermalization.

1 Introduction

Kinetic theory describes motion which is *transport dominated* in the sense that typically the solutions to the kinetic equations correspond to constant velocity, i.e. *ballistic*, motion intercepted by *collisions* whose frequency is order one on the kinetic space–time scales. Weakly interacting quantum fluids provide one such example system, as discussed in detail in [1].

We focus here on one particular case of a weakly interacting quantum fluid, the case of weakly interacting fermions hopping on a lattice. Such a model would arise physically as a description of a fluid of electrons in a crystal background potential. For our purposes, this model has also other attractive properties, namely, it has interesting non-trivial kinetic theory with relatively few technical and mathematical difficulties. Much of the discussion below can be straightforwardly adapted to bosonic lattice systems, at least for initial data which exclude formation of Bose–Einstein condensate. For more details about such extensions, we refer to [1]; for instance, Remark 2.3 summarises the changes and new properties which arise for fermions and bosons moving not on a lattice, but in the continuum $\mathbb{R}^3$.

J. Lukkarinen (✉)
Department of Mathematics and Statistics, University of Helsinki, P.O. Box 68,
00014 Helsingin yliopisto, Finland
e-mail: jani.lukkarinen@helsinki.fi

© Springer Nature Switzerland AG 2018   1
D. Cadamuro et al. (eds.), *Macroscopic Limits of Quantum Systems*,
Springer Proceedings in Mathematics & Statistics 270,
https://doi.org/10.1007/978-3-030-01602-9_1

The purpose of this contribution is not to provide a comprehensive review of literature on kinetic theory and properties of fermionic systems. Instead, we focus on building a bridge between mathematically rigorous results and the physics of fermionic systems. To this end, we begin with a fairly detailed Sect. 2 on the definition of finite system of fermions hopping on a periodic lattice with pair interactions between the particles, from the point of view of both a fixed particle number Hilbert space and the full antisymmetric Fock space. In Sect. 3 we recall how the probabilistic concepts of classical particle systems can be generalised into systems of fermions, namely, the definition of quasifree states, reduced density matrices and truncated correlation functions. The weak coupling limit of the first reduced density matrix of a translation invariant system and its approximation by the appropriate spatially homogeneous Boltzmann equation is reviewed in Sect. 4. As a conclusion, we discuss in Sects. 5 and 6 the implications of the properties of the solutions to the Boltzmann equation on thermalization in the original fermionic lattice system. Acknowledgements and references can be found at the end of the text.

2 Dynamics of Lattice Fermions

We recall in this section the mathematical description of fermions, possibly with spin, which are hopping on a finite periodic lattice of length $L \gg 1$. The particles move on a lattice whose points are labelled by $\Lambda := \mathbb{Z}^d/(L\mathbb{Z}^d)$ which we parametrize by a square centred at the origin. For instance, if L is even, we use the parametrisation

$$\Lambda = \left\{ -\frac{L}{2} + 1, \ldots, \frac{L}{2} - 1, \frac{L}{2} \right\}^d .$$

In particular, all arithmetic on Λ is performed 'modulo L': if x, y are in the above parametrisation of $\Lambda \subset \mathbb{Z}^d$, then $x + y \in \mathbb{Z}^d$ needs to be identified with its counterpart in the parametrisation. Explicitly, '$x + y$' is equal to $x + y - Lm \in \Lambda$ where $m \in \mathbb{Z}^d$ is the unique vector for which $x + y - Lm \in \Lambda$.

To describe Fourier transforms, we employ the corresponding discrete dual lattice $\Lambda^* := \Lambda/L = (L^{-1}\mathbb{Z}^d)/\mathbb{Z}^d$. If needed, we use the parametrisation implied by the above notation; for instance for even L, we use $\Lambda^* = \left\{ -\frac{1}{2} + \frac{1}{L}, \ldots, \frac{1}{2} - \frac{1}{L}, \frac{1}{2} \right\}^d$. The arithmetic on Λ^* is then performed modulo 1, i.e. using the arithmetic inherited from the d-torus $\mathbb{T}^d = \mathbb{R}^d/\mathbb{Z}^d \supset \Lambda^*$.

Such periodic arithmetic is particularly well adapted for use of discrete Fourier transforms. For a function $f : \Lambda \to \mathbb{C}$ we take its Fourier transform to be the function $\widehat{f} : \Lambda^* \to \mathbb{C}$ defined by the formula

$$\widehat{f}(k) := \sum_{x \in \Lambda} e^{-i2\pi k \cdot x} f(x), \qquad k \in \Lambda^* .$$

The inverse transform of $g : \Lambda^* \to \mathbb{C}$ is then given by $\tilde{g} : \Lambda \to \mathbb{C}$ defined by

$$\tilde{g}(x) := \int_{\Lambda^*} dk\, e^{i2\pi k \cdot x} g(k) = \frac{1}{|\Lambda|} \sum_{k \in \Lambda^*} e^{i2\pi k \cdot x} g(k), \qquad x \in \Lambda.$$

Here and in the following, we use the shorthand notation

$$\int_{\Lambda^*} dk \cdots = \frac{1}{|\Lambda|} \sum_{k \in \Lambda^*} \cdots.$$

On a finite lattice, the discrete Fourier transform is always pointwise invertible, i.e. for all $x \in \Lambda$, $k \in \Lambda^*$, $(\hat{f})\check{}(x) = f(x)$, $(\tilde{g})\hat{}(k) = g(k)$.

We assume that the dominant free evolution is defined by giving the dispersion relation $\omega : \mathbb{T}^d \to \mathbb{R}$ corresponding to free evolution after a thermodynamic limit $L \to \infty$ has been taken. More precisely, we let the periodic lattice hopping potential $\alpha : \Lambda \to \mathbb{R}$ be defined by the inverse Fourier transform of the map $\omega|_{\Lambda^*}$,

$$\alpha(x; L) := \int_{\Lambda^*} dk\, e^{i2\pi k \cdot x} \omega(k), \qquad x \in \Lambda. \tag{1}$$

The function α determines the free n-particle Hamiltonian $H_0^{(n)}$ by its action on n-particle wave vectors $\psi : \Lambda^n \to \mathbb{C}$,

$$H_0^{(n)} \psi(x_1, \ldots, x_n) = \sum_{j=1}^{n} \sum_{y \in \Lambda} \alpha(x_j - y) \psi(x_1, \ldots, y, \ldots, x_n).$$

The above construction allows an L-independent diagonalisation of $H_0^{(n)}$ by taking the discrete Fourier transform:

$$(H_0^{(n)} \psi)\hat{}(k_1, \ldots, k_n) = \sum_{j=1}^{n} \omega(k_j)\, \hat{\psi}(k_1, \ldots, k_n).$$

We assume that the dispersion relation is *smooth* and *symmetric*, $\omega(-k) = \omega(k)$. Then $\alpha(x; L)$ is always real, and denoting the inverse Fourier transform (i.e. the Fourier series) of ω by α, we then have $\alpha(x; L) \to \alpha(x)$ for each fixed $x \in \mathbb{Z}^d$ as $L \to \infty$. In addition, the range of α is finite, in the sense that $|\alpha(x)|$ decreases faster than any power as $|x| \to \infty$.

An explicit often considered example case is nearest neighbour hopping. This corresponds to

$$\omega(k) = c - \sum_{\nu=1}^{d} \cos(2\pi k_\nu),$$

where $c \in \mathbb{R}$ is any constant. For instance, choosing $c = d$, one obtains the standard discrete Laplacian,

$$\sum_{y \in \Lambda} \alpha(x - y)\psi(y) = \frac{1}{2} \sum_{v=1}^{d} (2\psi(x) - \psi(x - e_v) - \psi(x + e_v)) ,$$

where e_v denotes the unit vector in direction v. For vectors ψ which are obtained by taking values of a slowly varying function $\psi : \mathbb{R}^d \to \mathbb{C}$, the right-hand side can be approximated by $-\frac{1}{2}\nabla^2\psi(x)$. Therefore, in this case one may also think of $H_0^{(n)}$ as a discrete approximation of the standard free n-particle Hamiltonian, with particle mass normalised to one.

We construct a pair interaction potential $V(x; L)$ analogously, starting from its Fourier transform $\widehat{V} : \mathbb{T}^d \to \mathbb{C}$ and defining

$$V(x; L) := \int_{\Lambda^*} dk\, e^{i2\pi k \cdot x} \widehat{V}(k) , \qquad x \in \Lambda . \tag{2}$$

To make the potential real-valued and symmetric, we assume that $\widehat{V}$ is real-valued and symmetric. The n-particle pair interaction potential $V^{(n)}$ is then defined via the formula

$$V^{(n)}(x_1, \ldots, x_n; L) := \frac{1}{2} \sum_{i',i=1; i' \neq i}^{n} V(x_{i'} - x_i; L) .$$

The potential function acts as an multiplication operator on wave vectors, and we do not make any distinction in the notation between the function and the operator. Thus, if $\psi : \Lambda^n \to \mathbb{C}$ is an n-particle wave vector, then

$$V^{(n)}\psi(x_1, \ldots, x_n) = V^{(n)}(x_1, \ldots, x_n; L)\psi(x_1, \ldots, x_n) .$$

Naturally, if $n = 1$, we have $V^{(n)} = 0$.

After these preliminaries, we define the full n-particle Hamiltonian by choosing an interaction strength $\lambda \geq 0$, setting $H_\lambda^{(0)} = 0$, and for $n \geq 1$ defining

$$H_\lambda^{(n)} := H_0^{(n)} + \lambda V^{(n)} .$$

The corresponding evolution equation for n-particle wave vectors $\psi(t)$ is

$$\partial_t \psi(t) = -iH_\lambda^{(n)}\psi(t) .$$

The n-particle Hilbert space is here finite-dimensional, $\mathscr{H}_n = (\mathbb{C}^\Lambda)^{\otimes n} = \mathbb{C}^{\Lambda^n}$. By construction, the evolution preserves particle number and each $H_\lambda^{(n)}$ is a bounded self-adjoint operator on $\mathscr{H}_n$. Thus their direct sum $H_\lambda := \bigoplus_{n=0}^{\infty} H_\lambda^{(n)}$ defines a self-adjoint operator on the full Fock space $\mathscr{F} := \bigoplus_{n=0}^{\infty} \mathscr{H}_n$. More precisely, the domain of the operator is

$$D(H_\lambda) := \left\{ \Psi \in \mathscr{F} \,\middle|\, \sum_{n=0}^{\infty} \|H_\lambda^{(n)} \Psi_n\|^2 < \infty \right\},$$

and the action of H_λ on $\Psi = (\Psi_0, \Psi_1, \ldots) \in D(H_\lambda)$ yields the vector $(H_\lambda^{(n)} \Psi_n)_{n=0}^{\infty} \in \mathscr{F}$. (The proof of these properties can be found for instance in [2, Theorem 2.23].) An analogous construction holds for the potential terms $V^{(n)}$ alone, and the corresponding full Fock space operator is denoted by V; clearly, $\lambda V = H_\lambda - H_0$ on the domain of H_λ.

Each $H_0^{(n)}$ and $V^{(n)}$ clearly commutes with permutations of particle labels (i.e. with all of the operators Q_π defined by $(Q_\pi \psi)(x_1, \ldots, x_n) = \psi(x_{\pi(1)}, \ldots, x_{\pi(n)})$, where π is any permutation of $\{1, 2, \ldots, n\}$). Thus H_λ leaves invariant both the fermionic Fock space $\mathscr{F}_-$, containing those $\Psi \in \mathscr{F}$ for which each Ψ_n is antisymmetric under permutations of particle labels, and the bosonic Fock space $\mathscr{F}_+$, containing only symmetric Ψ_n.

From now on, we focus on the corresponding *fermionic* lattice system which is defined by wave vectors $\Psi(t) \in \mathscr{F}_-$ and the semigroup generated by the restriction of H_λ to $\mathscr{F}_-$. Since wave vectors with only finitely many non-zero particle sectors belong to $D(H_\lambda)$ and form a dense set in $\mathscr{F}_-$, we find that for any $\Psi(0) \in \mathscr{F}_-$, the n-particle sector of the time-evolved wave function can be obtained by solving the matrix evolution equation

$$\partial_t \Psi_n(t) = -\mathrm{i} H_\lambda^{(n)} \Psi_n(t),$$

with initial data $\Psi_n(0)$.

2.1 Dynamics in Terms of Creation and Annihilation Operators

Antisymmetry of wave vectors is one of the most important features of fermionic quantum systems, and it can alter the properties of time evolution significantly. Controlling the effect of antisymmetry is difficult in the above formulation of the time evolution. A better alternative is offered by representing the time evolution as an evolution equation of the corresponding fermionic creation and annihilation operators. We summarise their main properties below and refer to [3, Section 5.2] for more mathematical details.

In the present finite lattice case, the Fock space has been constructed using a one-particle space $\mathfrak{h} := \mathbb{C}^\Lambda$ and the corresponding (distinguishable) n-particle sectors $\mathcal{H}_n := \mathfrak{h}^{\otimes n} = \mathbb{C}^{\Lambda^n}$. Let $P_-^{(n)}$ denote the orthogonal projection onto the subspace of antisymmetric functions in $\mathcal{H}_n$; explicitly,

$$(P_-^{(n)}\psi)(x_1, \ldots, x_n) = \frac{1}{n!} \sum_{\pi \in S_n} (-1)^\pi \psi(x_{\pi(1)}, \ldots, x_{\pi(n)}),$$

where S_n denotes the group of permutations of the set $\{1, 2, \ldots, n\}$ and $(-1)^\pi$ is the sign of the permutation $\pi \in S_n$. Since we consider a system of identical fermions, at any time, a wave vector $\Psi \in \mathcal{F}_-$ satisfies $P_-^{(n)}\Psi_n = \Psi_n$ for all n.

Given a one-particle wave vector $g \in \mathfrak{h}$, we define the corresponding *annihilation operator* $a(g)$ as the map which takes a vector $\Psi \in \mathcal{F}_-$ and removes the first particle from each of its sectors, with a weight proportional to the overlap with g. More precisely, for a fixed particle number $n \geq 1$, there is a unique bounded linear map $A_n(g) : \mathcal{H}_n \to \mathcal{H}_{n-1}$ such that for any collection of one-particle wave vectors $f_j \in \mathfrak{h}$,

$$A_n(g)\left(\bigotimes_{j=1}^n f_j\right) = \sqrt{n}\langle g, f_1 \rangle \bigotimes_{j=2}^n f_j,$$

where $\langle g, f \rangle$ is the one-particle scalar product, defined here conjugate linear in the *first* argument, i.e. $\langle g, f \rangle = \sum_{x \in \Lambda} g(x)^* f(x)$. We then define the fermionic annihilation operator $a(g) : \mathcal{F}_- \to \mathcal{F}_-$ by the rule

$$(a(g)\Psi)_n = P_-^{(n)} A_{n+1}(g) P_-^{(n+1)} \Psi_{n+1} = P_-^{(n)} A_{n+1}(g) \Psi_{n+1}, \qquad n \geq 0, \ \Psi \in \mathcal{F}_-.$$

In general, annihilation operators are unbounded on the appropriate Fock space, and one has to worry about the domain of the operator in its definition. However, it is a remarkable consequence of the antisymmetrisation that $a(g)$ is in fact a bounded operator on $\mathcal{F}_-$, and the normalisation $\sqrt{n}$ added above guarantees that its operator norm is the same as the norm of the wave vector g, i.e. we always have $\|a(g)\| = \|g\|_\mathfrak{h}$.

The adjoint of $a(g)$, which we denote here by $a^*(g)$, is called the *creation operator* at the vector $g \in \mathfrak{h}$. The creation operator can indeed be interpreted as creating a particle with wave vector g at the first position (and hence shifting the labels of the existing particles by one). This interpretation is based on a more direct construction analogous to the one for $a(g)$ above. Namely, there is a unique bounded linear map $C_n(g) : \mathcal{H}_n \to \mathcal{H}_{n+1}$ such that for any collection of one-particle wave vectors $f_j \in \mathfrak{h}$,

$$C_n(g)\left(\bigotimes_{j=1}^n f_j\right) = \sqrt{n+1}\, g \otimes f_1 \otimes \cdots \otimes f_n.$$

We also set $C_0(g)1 = g \in \mathscr{H}_1$. The fermionic creation operator is then given by $c(g) : \mathscr{F}_- \to \mathscr{F}_-$, and it satisfies $(c(g)\Psi)_0 = 0$, and

$$(c(g)\Psi)_n = P_-^{(n)} C_{n-1}(g) P_-^{(n-1)} \Psi_{n-1} = P_-^{(n)} C_{n-1}(g) \Psi_{n-1}, \qquad n \geq 1, \ \Psi \in \mathscr{F}_-.$$

One can check that then indeed $c(g) = a^*(g)$ which implies that also $\|c(g)\| = \|g\|_{\mathfrak{h}}$.

One important reason why working with the creation and annihilation operators simplifies the analysis of time evolution is that they satisfy fairly simple algebraic rules for swapping the order of any two such operators. Namely, they satisfy the following *canonical anticommutation relations*: for any one-particle vectors $f, g \in \mathfrak{h}$, we have

$$a(f)a(g) + a(g)a(f) = 0 = a(f)^*a(g)^* + a(g)^*a(f)^*,$$
$$a(f)a(g)^* + a(g)^*a(f) = \langle f, g \rangle 1, \tag{3}$$

where '1' denotes the identity operator on $\mathscr{F}_-$. In particular, $a(f)^2 = 0 = a^*(f)^2$, and if (e_ℓ) is any orthonormal basis of $\mathfrak{h}$, we have

$$a(e_\ell)a(e_{\ell'})^* + a(e_{\ell'})^*a(e_\ell) = \mathbb{1}_{\{\ell=\ell'\}} 1,$$

with $\mathbb{1}_{\{P\}}$ denoting the *generic characteristic function* of the condition P: we define $\mathbb{1}_{\{P\}} = 1$, if P is true, and $\mathbb{1}_{\{P\}} = 0$, if P is false.

Moreover, tensor products in $\mathscr{H}_n$ are conveniently expressed in terms of products of creation operators acting on the vacuum $\Omega = (1, 0, 0, \ldots) \in \mathscr{F}_-$. Namely, if $g_j \in \mathfrak{h}$, $j = 1, 2, \ldots, n$, are given, then $\otimes_j g_j \in \mathscr{H}_n$ after antisymmetrisation defines a vector $\Psi \in \mathscr{F}_-$ by setting all other components to zero, i.e. setting $\Psi_n = P_-^{(n)}(\otimes_j g_j)$ and $\Psi_m = 0$, for $m \neq n$. This vector can also be obtained from

$$\Psi = \frac{1}{\sqrt{n!}} a^*(g_1) \cdots a^*(g_n)\Omega. \tag{4}$$

The collection of creation and annihilation operators corresponding to the standard unit vector orthonormal basis $(e_x)_{x \in \Lambda}$, where $(e_x)_y = \mathbb{1}_{\{x=y\}}$ for all $x, y \in \Lambda$, is of particular interest to us. We employ the following standard shorthand notations:

$$a(x) := a(e_x), \quad a^*(x) := a^*(e_x) = a(x)^*, \qquad x \in \Lambda. \tag{5}$$

These operators can be thought of as annihilating or creating a particle at the site x. By (3), they satisfy the following simple anticommutation relations for any $x, y \in \Lambda$,

$$a(x)a(y) + a(y)a(x) = 0 = a(x)^*a(y)^* + a(y)^*a(x)^*,$$
$$a(x)a(y)^* + a(y)^*a(x) = \mathbb{1}_{\{x=y\}} 1. \tag{6}$$

We can also use the creation operators to generate an *orthonormal basis* for $\mathscr{F}_-$. For this, first define

$$e(x_1, \ldots, x_n) := a^*(x_1) \cdots a^*(x_n)\Omega, \qquad x_i \in \Lambda, \ i = 1, 2, \ldots, n.$$

The orthonormal basis may be constructed by collecting all non-repeating sequences of arbitrary length and then choosing one representative for each collection of sequences which differ by a permutation of particle labels. The actual choice does not play much role: if $(x_i) \in \Lambda^n$ and $\pi \in S_n$ is some permutation, then by the anti-commutation relations

$$e(x_{\pi(1)}, \ldots, x_{\pi(n)}) = (-1)^\pi e(x_1, \ldots, x_n),$$

and hence the choice merely affects signs of the basis vectors.

After these preliminaries, it is straightforward to check that wave vectors and interaction potentials may also be represented using the creation and annihilation operators. Namely, if $\Psi \in \mathscr{F}_-$, $n \in \mathbb{N}$, and $x \in \Lambda^n$, we have

$$\Psi_n(x_1, \ldots, x_n) = \langle \otimes_{i=1}^n e_{x_i}, \Psi_n \rangle_{\mathscr{H}_n} = \langle P_-^{(n)}(\otimes_{i=1}^n e_{x_i}), \Psi_n \rangle_{\mathscr{H}_n},$$

and hence by (4),

$$\Psi_n(x_1, \ldots, x_n) = \frac{1}{\sqrt{n!}} \langle a^*(x_1) \cdots a^*(x_n)\Omega, \Psi \rangle_{\mathscr{F}_-}.$$

Moreover, the anticommutation relations imply that if x, y and $x_i \in \Lambda$, $i = 1, 2, \ldots, n$, then

$$a^*(x)a(y)a^*(x_1) \cdots a^*(x_n)\Omega$$

$$= \sum_{i=1}^n \mathbb{1}_{\{y=x_i\}} a^*(x_1) \cdots a^*(x_{i-1})a^*(x)a^*(x_{i+1}) \cdots a^*(x_n)\Omega. \tag{7}$$

Using these two properties it is now straightforward to check that the earlier defined operators H_0 and V on the fermionic Fock space have the following representations in terms of creation and annihilation operators:

$$H_0 = \sum_{x,y \in \Lambda} \alpha(x - y; L)a(x)^*a(y), \tag{8}$$

$$V = \frac{1}{2} \sum_{x,y \in \Lambda} V(x - y; L)a(x)^*a(y)^*a(y)a(x). \tag{9}$$

The above right-hand sides are finite sums in the Banach space of bounded operators on $\mathscr{F}_-$, and thus H_0, V, and $H_\lambda = H_0 + \lambda V$ are also bounded operators on the fermionic Fock space.

The time evolution of any initial data $\Psi(0) \in \mathscr{F}_-$ under the semigroup $U_t := e^{-itH_\lambda}$ can be solved if we can solve the time evolution of the annihilation operators, i.e. it suffices to study

$$a(x, t) := e^{itH_\lambda} a(x) e^{-itH_\lambda} ,$$

and its adjoint

$$a^*(x, t) := e^{itH_\lambda} a^*(x) e^{-itH_\lambda} .$$

This follows from our definition that the Hamiltonian acts trivially on the vacuum sector, $(H_\lambda)_0 = 0$, and thus

$$a^*(x_1, t) \cdots a^*(x_n, t)\Omega = e^{itH_\lambda} a^*(x_1) \cdots a^*(x_n)\Omega ,$$

implying that

$$\Psi_n(x_1, \ldots, x_n, t) = \frac{1}{\sqrt{n!}} \langle a^*(x_1) \cdots a^*(x_n)\Omega, e^{-itH_\lambda}\Psi(0) \rangle_{\mathscr{F}_-}$$
$$= \frac{1}{\sqrt{n!}} \langle a^*(x_1, t) \cdots a^*(x_n, t)\Omega, \Psi(0) \rangle_{\mathscr{F}_-} .$$

Since the Hamiltonian is a bounded operator, we can directly differentiate the definition and obtain

$$\partial_t a(x, t) = -i e^{itH_\lambda} [a(x), H_\lambda] e^{-itH_\lambda} .$$

The computation of the commutator is straightforward using the anticommutation relations, yielding

$$[a(x), H_\lambda] = \sum_{y \in \Lambda} \alpha(x - y; L)a(y) + \lambda \sum_{y \in \Lambda} V(x - y; L)a(y)^*a(y)a(x) .$$

Therefore, we find that in order to solve the original (linear) evolution equation in the fermionic Fock space, it suffices to solve the following non-linear operator evolution equation on the space of bounded operators on $\mathscr{F}_-$,

$$\partial_t a(x, t) = -i \sum_{y \in \Lambda} \alpha(x - y; L)a(y, t) - i\lambda \sum_{y \in \Lambda} V(x - y; L)a^*(y, t)a(y, t)a(x, t) .$$

$$(10)$$

In Fourier variables, after defining

$$\widehat{a}(k, t) := \sum_{x \in \Lambda} e^{-i2\pi x \cdot k} a(x, t) ,$$

we obtain

$$\partial_t \widehat{a}(k, t) = -i\omega(k)\widehat{a}(k, t)$$

$$- i\lambda \int_{(\Lambda^*)^3} dk_1 dk_2 dk_3 \, \delta_\Lambda(k - k_1 - k_2 - k_3) \widehat{V}(k_1 + k_2) \widehat{a^*}(k_1, t) \widehat{a}(k_2, t) \widehat{a}(k_3, t),$$

$$(11)$$

where $\delta_\Lambda(k) := |\Lambda| \mathbb{1}_{\{k=0 \bmod \Lambda^*\}}$ is a 'discrete Dirac δ-function' and $[\widehat{a}(k, t)]^* = \widehat{a^*}(-k, t)$.

2.2 Fermionic Systems with Spin Interactions and the Hubbard Model

Spin is an integral part of description of quantum mechanical particles. For instance, by the spin–statistics relation, all fermionic particles possess a half-integer spin. In particular, the spin cannot be zero, so the above fermionic description is not yet completely adequate for physical fermions.

Spin is a one-particle property, and hence affects the definition of the one-particle Hilbert space $\mathfrak{h}$ above. It is determined by a half-integer value $S \in \mathbb{N}_0/2$, resulting in $2S + 1$ new 'internal' degrees of freedom which are labelled by values in $\sigma_S :=$ $\{-S, -S + 1, \ldots, S\}$. There are several equivalent ways of defining the wave vector of a particle with a non-zero spin: one can either think that they are multicomponent wave vectors, $\psi(x) \in \mathbb{C}^{\sigma_S}$, or that each lattice site is augmented with D extra degrees of freedom, $\psi(x, \sigma) \in \mathbb{C}$, $\sigma \in \sigma_S$. These descriptions are quantum mechanically equivalent since the identification

$$\psi(x)_\sigma = \phi(x, \sigma) = \langle e_x \otimes e_\sigma, \phi \rangle$$

provides a mapping $\psi \to \phi$ which turns out to be a Hilbert space isomorphism between $\oplus_{\sigma \in \sigma_S} L^2(\Lambda)$ and $L^2(\Lambda \times \sigma_S)$. The second equality above also yields an isomorphism, namely, the standard one between $L^2(\Lambda \times \sigma_S)$ and $L^2(\Lambda) \otimes L^2(\sigma_S)$.

Hence, most of the discussion in the previous sections holds verbatim if we replace $x \in \Lambda$ by $(x, \sigma) \in \Lambda \times \sigma_S$. The main differences come from the physical restrictions for the spin interactions which have no need to be 'translation invariant' in the spin degrees of freedom. Thus, Fourier transforming the spin degrees is not helpful and, instead, one should try to aim at simplifications by finding other unitary transformations which diagonalise at least part of the Hamiltonian.

One case which reduces to the discussion without spin occurs when the total Hamiltonian H can be diagonalised with respect to the spin degrees of freedom, i.e. if there is a unitary transformation U for which $U^* H U = \oplus_{\sigma \in \sigma_S} H_\sigma$. Then after the unitary transformation each spin component evolves independently from the others and thus it satisfies the 'spinless' equations of the previous section.

Spatially translation invariant generalisations of the previous weakly interacting Hamiltonians are determined by the operators

$$H_0 = \sum_{x,y \in \Lambda} \sum_{\sigma,\sigma' \in \sigma_S} \alpha_{\sigma\sigma'}(x - y; L) a(x,\sigma)^* a(y,\sigma'), \tag{12}$$

$$V = \frac{1}{2} \sum_{x,y \in \Lambda} \sum_{\sigma,\sigma' \in \sigma_S} V_{\sigma\sigma'}(x - y; L) a(x,\sigma)^* a(y,\sigma')^* a(y,\sigma') a(x,\sigma). \tag{13}$$

The functions $\alpha_{\sigma\sigma'}(x; L)$ and $V_{\sigma\sigma'}(x; L)$ are constructed as in (1) and (2), using some given $\omega_{\sigma\sigma'} : \mathbb{T}^d \to \mathbb{R}$ and $\widehat{V}_{\sigma\sigma'} : \mathbb{T}^d \to \mathbb{R}$, for each σ, σ'. We require H_0 to be self-adjoint and the interaction symmetric under spatial inversions, and this is guaranteed by assuming that $\omega(-k) = \omega(k) = \omega(k)^*$, as $S \times S$-matrices. Similarly, the self-adjointness of V can be guaranteed by assuming that each $\widehat{V}(k)$ is a Hermitian matrix and that they satisfy an additional symmetry property $\widehat{V}_{\sigma\sigma'}(-k) = \widehat{V}_{\sigma'\sigma}(k)$ related to particle permutation invariance.

One well-studied example of this type is the *Hubbard model* which concerns spin-$\frac{1}{2}$ fermions like electrons. Then $S = \frac{1}{2}$ and usually one simplifies the discussion by labelling the spin degrees of freedom $\{-\frac{1}{2}, \frac{1}{2}\}$ using the sign, i.e. using the set $2\sigma_S = \{-1, 1\}$ for labelling. In the Hubbard model, the free evolution is taken to be fully spin rotation invariant,

$$H_0 = \sum_{x,y \in \Lambda} \sum_{\sigma = \pm 1} \alpha(x - y; L) a(x,\sigma)^* a(y,\sigma), \tag{14}$$

and thus depending only on one dispersion relation function which is typically chosen to be nearest neighbour, $\omega(k) = -\sum_{\nu=1}^{d} \cos(2\pi k_\nu)$. The pair interactions in the Hubbard model are taken to be onsite only,

$$V = \frac{1}{2} \sum_{x \in \Lambda} \sum_{\sigma,\sigma' = \pm 1} V_{\sigma\sigma'} a(x,\sigma)^* a(x,\sigma')^* a(x,\sigma') a(x,\sigma), \tag{15}$$

and, since $a(x,\sigma)^2 = 0$ and $V_{\sigma\sigma'} = V_{\sigma\sigma'}(0; L)$, $\sigma, \sigma' \in \{\pm 1\}$, form a real symmetric 2×2 matrix, without loss of generality, we may set $V_{\sigma\sigma} = 0$ and use $V_{+-} = V_{-+}$ as the sole real parameter. It is usually included in the definition of the coupling λ, and thus the general fermionic spin-$\frac{1}{2}$ onsite interactions are covered by the interaction[1]

$$V_{\text{Hubbard}} = \sum_{x \in \Lambda} a(x,+)^* a(x,-)^* a(x,-) a(x,+). \tag{16}$$

---

[1]The most standard notation for the Hubbard model uses the potential $U \sum_x n(x,+) n(x,-)$ where $n(x,\sigma) := a(x,\sigma)^* a(x,\sigma)$. This is seen to be equivalent to the present case after setting $U = \lambda$ and using the anticommutation relations.

Let us point out that onsite potentials fall into the class of translation invariant potentials studied in the previous subsection. Namely, they correspond to choosing potentials whose Fourier transforms are constant, $\widehat{V}_{\sigma\sigma'}(k) = V_{\sigma\sigma'}$ for all $k \in \mathbb{T}^d$.

The main difficulties compared to deriving the evolution equations in the earlier discussed case are notational. We skip the parts which are similar to the earlier computations, and merely record the outcome in a form which is easy to use in computations involving products of creation and annihilation operators.

We label annihilation operators with an additional label $\tau = -1$ and creation operators with $\tau = +1$, and consider their dynamics after Fourier transform of the spatial degrees of freedom. Explicitly, we define

$$a(k, \sigma, -1, t) := \widehat{a}(k, \sigma, t) = \sum_{x \in \Lambda} e^{-i2\pi x \cdot k} a(x, \sigma, t), \tag{17}$$

$$a(k, \sigma, +1, t) := \widehat{a^*}(k, \sigma, t) = \sum_{x \in \Lambda} e^{-i2\pi x \cdot k} a^*(x, \sigma, t). \tag{18}$$

These operators are connected via operator adjoints, $[a(k, \sigma, \tau, t)]^* = a(-k, \sigma, -\tau, t)$. Since now

$$\partial_t a(x, \sigma, t) = -i \sum_{x' \in \Lambda} \sum_{\sigma' \in \sigma_S} \alpha_{\sigma\sigma'}(x - x'; L) a(x', \sigma', t)$$

$$- i\lambda \sum_{x' \in \Lambda} \sum_{\sigma' \in \sigma_S} V_{\sigma\sigma'}(x - x'; L) a^*(x', \sigma', t) a(x', \sigma', t) a(x, \sigma, t),$$

the above operators satisfy the following closed evolution equations:

$$\partial_t a(k, \sigma, \tau, t) = i\tau \sum_{\sigma' \in \sigma_S} \omega_{\sigma\sigma'}(k; \tau) a(k, \sigma', \tau, t)$$

$$+ i\tau\lambda \sum_{\sigma_1, \sigma_2, \sigma_3 \in \sigma_S} \int_{(\Lambda^*)^3} dk_1 dk_2 dk_3 \, \delta_\Lambda(k - k_1 - k_2 - k_3)$$

$$\times \widehat{V}_{\sigma,\sigma_1,\sigma_2,\sigma_3}(k_1, k_2, k_3; \tau) a(k_1, \sigma_1, 1, t) a(k_2, \sigma_2, \tau, t) a(k_3, \sigma_3, -1, t), \tag{19}$$

where $\omega_{\sigma\sigma'}(k; -1) := \omega_{\sigma\sigma'}(k)$, $\omega_{\sigma\sigma'}(k; +1) := \omega_{\sigma'\sigma}(k)$, and

$$\widehat{V}_{\sigma,\sigma_1,\sigma_2,\sigma_3}(k_1, k_2, k_3; -1) = \mathbb{1}_{\{\sigma_1=\sigma_2, \sigma_3=\sigma\}} \widehat{V}_{\sigma\sigma_2}(k_1 + k_2),$$

$$\widehat{V}_{\sigma,\sigma_1,\sigma_2,\sigma_3}(k_1, k_2, k_3; +1) = \mathbb{1}_{\{\sigma_1=\sigma, \sigma_3=\sigma_2\}} \widehat{V}_{\sigma\sigma_2}(k_2 + k_3).$$

Here we need the above equations only in two special cases. First, if there is no spin, the equation reduces to

$$\partial_t a(k, \tau, t) = i\tau\omega(k)a(k, \tau, t) + i\tau\lambda \int_{(\Lambda^*)^3} dk_1 dk_2 dk_3 \, \delta_\Lambda(k - k_1 - k_2 - k_3)$$
$$\times \widehat{V}(k_1, k_2, k_3; \tau)a(k_1, 1, t)a(k_2, \tau, t)a(k_3, -1, t), \tag{20}$$

with $\widehat{V}(k_1, k_2, k_3; -1) = \widehat{V}(k_1 + k_2)$ and $\widehat{V}(k_1, k_2, k_3; 1) = \widehat{V}(k_2 + k_3)$. Second, for the Hubbard model, the equations can be simplified into

$$\partial_t a(k, \sigma, \tau, t) = i\tau\omega(k)a(k, \sigma, \tau, t) + i\tau\lambda \int_{(\Lambda^*)^3} dk_1 dk_2 dk_3 \, \delta_\Lambda(k - k_1 - k_2 - k_3)$$
$$\times a(k_1, \tau\sigma, 1, t)a(k_2, -\sigma, \tau, t)a(k_3, -\tau\sigma, -1, t). \tag{21}$$

3 States, Reduced Density Matrices and Truncated Correlation Functions

A state in classical mechanics is a probability measure describing the distribution of positions and velocities of the particles at some fixed time. Thus it can be used to compute the statistics of all observables, i.e. measurable functions of the positions and velocities at that time. In Hamiltonian mechanics, an initial state given at time $t = 0$ determines the state at all times $t \in \mathbb{R}$. Often, it is simpler to study the evolution of physical properties of the system by inspecting the evolution starting from some suitably chosen random initial state rather than from a deterministic state with fixed values for the initial positions and velocities of the particles.

A *state* at time t in quantum mechanics is defined as a map ρ_t which associates to each observable A a number $\rho_t[A]$ which gives the limiting value for statistical averages of this observable measured in repeated experiments. This is analogous to the expectation value map under the probability measure which defines the state in the classical case. The more precise mathematical definition of a state takes two ingredients: the collection of observables $\mathscr{A}$, which is assumed to be some subspace of bounded operators, closed under adjoint and containing the identity operator, and a positive linear functional $\rho : \mathscr{A} \to \mathbb{C}$ of norm 1.

For instance, a Borel probability measure μ of wave vectors $\Psi \in \mathscr{H}$, $\|\Psi\| = 1$, generates a state by setting for any bounded operator A on $\mathscr{H}$

$$\rho[A] := \int \mu(d\psi) \, \langle \psi, A\psi \rangle.$$

Most often a state is determined by giving a trace-class operator ρ on $\mathscr{H}$ such that ρ is positive, $\text{Tr } \rho = 1$, and setting $\rho[A] = \text{Tr}[\rho A]$ for all $A \in \mathscr{A}$. Such an operator ρ is called the *density matrix* of the state (note that we do not make a distinction in the notation between the state and its density matrix). If the Hilbert space is separable, such as our Fock spaces are, then for instance all states given by the above Borel probability measures have a density matrix associated with them.

The n-th *reduced density matrix* ρ_n is an analogous quantity which is obtained from the full density matrix by taking a partial trace over the degrees of freedom which concern particle labels higher than n. The general construction is discussed in [3, Section 6.3.3] and in [1, Section 3], but there is a more direct definition available for the present system of lattice fermions: Given a state ρ on the fermionic Fock space, we first define

$$\rho_n(z_1, z_1', \ldots, z_n, z_n') := \rho[a^*(z_1') \cdots a^*(z_n') a(z_n) \cdots a(z_1)] . \tag{22}$$

Here each z_i and z_i' belongs to the one-particle label set, i.e. $z_i \in \Lambda$ in the spinless case and $z_i \in \Lambda \times \sigma_S$ for spin-S particles. The collection of these complex numbers defines the reduced density matrix ρ_n, which is a positive operator on $\mathfrak{h}^{\otimes n}$, via the formula

$$\langle \otimes_i z_i, \rho_n(\otimes_i z_i') \rangle = \rho_n(z_1, z_1', \ldots, z_n, z_n') .$$

In quantum mechanics, given an initial density matrix $\rho(0) = \rho$, the expectation of a time-evolved observable $A(t) = U_t^* A U_t$ satisfies

$$\rho[A(t)] = \mathrm{Tr}[\rho U_t^* A U_t] = \mathrm{Tr}[U_t \rho U_t^* A] ,$$

by cyclicity of trace. Hence, we define the time-evolved density matrix $\rho(t) := U_t \rho U_t^*$ for which $\rho(t)[A] = \rho[A(t)]$. The reduced time-evolved density matrices may thus be obtained as expectations of time-evolved creation and annihilation operators: by replacing each $a(z)$ in (22) by $a(z, t) = U_t^* a(z) U_t$, we obtain the reduced density matrix $\rho(t)_n$.

Considering the earlier observation that time-evolved annihilation operators suffice to determine the time evolution of wave vectors, it is not surprising that reduced density matrices play an important role in the physics of quantum fluids. For instance, the expectation of the hopping Hamiltonian H_0 may be computed from $\rho(t)_1$ by the formula

$$\rho(t)[H_0] = \sum_{x, y \in \Lambda} \sum_{\sigma, \sigma' \in \sigma_S} \alpha_{\sigma\sigma'}(x - y; L) \rho(t)_1((x, \sigma), (y, \sigma')) .$$

Indeed, for kinetic theory, the central goal is to describe the evolution of $\rho(t)_1$, a positive operator on $\mathfrak{h}$, in the limit of weak coupling.

In fact, there is a class of fermionic states, called *quasifree states*, for which ρ_1 uniquely determines all other reduced density matrices: if ρ is quasifree, then for all $n \geq 1$ the corresponding density matrix is given as a determinant of an $n \times n$ matrix,

$$\rho_n(z_1, z_1', \ldots, z_n, z_n') = \det(\rho_1(z_i, z_j'))_{i, j=1, \ldots, n} .$$

To simplify analysis of states which are not quasifree but close to such, one can introduce *truncated correlation functions* ρ^T which are analogous to cumulants of random variables in classical probability theory. The construction below applies to a state ρ on a fermionic system which is *even*: it is assumed that an expectation of any observable remains invariant if we change $a(z)$ to $-a(z)$ for all z. As explained in more detail in [3, pp. 42–43], given an even state ρ to each even length sequence $(a_1, a_2, \ldots, a_m)$ of creation and annihilation operators one may associate a truncated expectation $\rho^T[a_1, a_2, \ldots, a_m]$ such that the expectation of any product of even length can be expressed as a sum over partitions. Explicitly,

$$\rho[a^I] = \sum_{\Pi \in \mathscr{P}_2(I)} \varepsilon(\Pi) \prod_{S \in \Pi} \rho^T[a_S], \tag{23}$$

where $I = (1, 2, \ldots, n)$, $a^I := a_1 \cdots a_n$, $\mathscr{P}_2(I)$ denotes the collection of partitions of I into even length subsequences, $\varepsilon(\Pi)$ is the sign of the permutation which takes I to $\Pi = (S_1, \ldots, S_m)$, and for a subsequence $S = (s_1, \ldots, s_m)$ of I we have used the shorthand notation $a_S = (a_{s_1}, \ldots, a_{s_m})$. Note that odd sequences for even states have always zero expectation, so this is the antisymmetrised analogue of the moments-to-cumulants formula of probability.

The above definition requires careful consideration of the signs of each term. The following identity can also serve as a basis for a recursive definition of the truncated expectations:

$$\rho[a^I] = \sum_{m \in S \subset I} \varepsilon(S, I \setminus S) \rho^T[a_S] \rho[a^{I \setminus S}], \tag{24}$$

where $m \in I$ is any fixed label and $\varepsilon(S, I \setminus S)$ is the sign of the permutation $I \to (S, I \setminus S)$. (Note that all terms where S has an odd length are zero in the sum, since then also $I \setminus S$ is odd, so we could have restricted the sum to even subsequences here.) For instance, $\rho^T[a_1, a_2] = \rho[a_1 a_2]$, and for $n = 4$ we have

$$\rho[a_1 a_2 a_3 a_4] = \rho^T[a_1, a_2, a_3, a_4]$$
$$+ \rho^T[a_1, a_2]\rho[a_3 a_4] - \rho^T[a_1, a_3]\rho[a_2 a_4] + \rho^T[a_1, a_4]\rho[a_2 a_3],$$

and thus

$$\rho^T[a_1, a_2, a_3, a_4] := \rho[a_1 a_2 a_3 a_4]$$
$$- \rho[a_1 a_2]\rho[a_3 a_4] + \rho[a_1 a_3]\rho[a_2 a_4] - \rho[a_1 a_4]\rho[a_2 a_3],$$

and, in accordance with (23), also

$$\rho[a_1 a_2 a_3 a_4] = \rho^T[a_1, a_2, a_3, a_4]$$
$$+ \rho^T[a_1, a_2]\rho^T[a_3, a_4] - \rho^T[a_1, a_3]\rho^T[a_2, a_4] + \rho^T[a_1, a_4]\rho^T[a_2, a_3]. \tag{25}$$

The truncated correlation functions can be used to characterise quasifree states: *an even state ρ is quasifree if and only if $\rho^T[a_1, a_2, \ldots, a_n] = 0$ for all $n > 2$.* This is completely analogous with characterisation of Gaussian measures by vanishing of their higher order cumulants. Even for states which are not quasifree, the truncated correlation functions enjoy properties which are typically not valid for direct expectations:

1. If $n > 2$, then $\rho^T[a_1, a_2, \ldots, a_n]$ is completely antisymmetric with respect to permutation of its arguments: if $\pi \in S_n$, we have $\rho^T[a_{\pi(1)}, a_{\pi(2)}, \ldots, a_{\pi(n)}] = (-1)^\pi \rho^T[a_1, a_2, \ldots, a_n]$. (For a proof, consider a basic odd permutation which swaps two neighbouring labels m and m', and then use (24) and the anticommutation relations.)

2. If ρ is an equilibrium Gibbs state at sufficiently small activity and corresponding to a short-range interaction, all reduced density matrices are typically decaying summably in the separation of their spatial arguments. For a precise statement and assumptions under which this result holds, see [3, Theorem 6.3.21], and further discussion can be found in [4]. In particular, keeping one of the sites fixed, Fourier transforms of the reduced density matrices are typically uniformly bounded in the lattice size L, unlike those of the corresponding expectations.

4 Weak Coupling Limit and Quantum Kinetic Theory

For kinetic theory, we are interested in the evolution of the first truncated reduced density matrix $\rho_1(x', \sigma', x, \sigma; t) = \rho^T[a^*(x', \sigma', t), a(x, \sigma, t)]$. There is no difference between the truncated and direct reduced density matrices for the first reduced density matrix of an even state of fermions but for higher order density matrices there is a difference in their properties. Most notably, for systems which are eventually well approximated by Gibbs states of the type discussed in item 3 at the end of Sect. 3, one would expect the *truncated* correlation functions to decay in the distance. Then, Fourier transforms in these variables are given by 'nice' functions, for instance, uniformly bounded in the lattice size or with a uniformly bounded $L^2(dk)$-norm. In contrast, the Fourier transform of the corresponding moments would be a fairly complicated sum over 'δ_Λ-distributions'.

Here, we consider only initial data which are both *gauge invariant* and *translation invariant*. The first condition means that the initial data does not contain correlations between different particle sectors, and this property is preserved by the present type of evolution. It simplifies the resulting analysis since for gauge invariant states all moments, which do not have the same number of creation and annihilation operators, are zero. For instance, then $\rho[a(y, \sigma'', t)a(x, \sigma, t)] = 0 = \rho[a^*(y, \sigma'', t)a^*(x, \sigma, t)]$.

For translation invariance, we require that all moments are invariant under periodic spatial translations of the lattice Λ. For the present translation invariant H_0 and V, also this property is preserved by the time evolution. As a consequence, any one of the spatial arguments of the correlation functions can be translated to the origin. In particular, there is a function $F : \Lambda \times \mathbb{R} \to \mathbb{C}^{2 \times 2}$ for which

$$\rho_1(x', \sigma', x, \sigma; t) = F_{\sigma'\sigma}(x' - x, t).$$

The *Wigner function* is defined as the discrete Fourier transform of F,

$$W_{\sigma'\sigma}(k, t) := \sum_{y \in \Lambda} e^{-i2\pi y \cdot k} F_{\sigma'\sigma}(y, t) = \int_{\Lambda^*} dk' \, \rho[a(k, \sigma', 1, t)a(k', \sigma, -1, t)].$$

$$(26)$$

Using the properties of adjoints, it is straightforward to check that the so-defined $\sigma_S \times \sigma_S$ matrix $W(k, t)$ is always Hermitian. In addition, translation invariance may be invoked to prove that

$$\rho[a(k, \sigma', 1, t)a(k', \sigma, -1, t)] = W_{\sigma'\sigma}(k, t)\delta_\Lambda(k + k'). \tag{27}$$

We also introduce the related notation $\tilde{W}$ for the corresponding expectation where the order of the operators has been swapped. More precisely, we define as matrices

$$\tilde{W}(k, t) := 1 - W(k, t), \tag{28}$$

where 1 denotes the diagonal unit matrix. By the anticommutation relations, then

$$\rho[a(k', \sigma, -1, t)a(k, \sigma', 1, t)] = \tilde{W}_{\sigma'\sigma}(k, t)\delta_\Lambda(k + k'). \tag{29}$$

The quantum kinetic equation will concern the time evolution of the above Hermitian matrix-valued Wigner functions. There are a number of differences in the computations depending on whether there are spin interactions present or not, and we have split the discussion accordingly below.

4.1 Fermionic Boltzmann–Nordheim Equation

We begin with a case in which the spin degrees of freedom evolve independently. As mentioned above, this case can be handled ignoring the spin degrees of freedom and thus we can use the spinless results and notations. We adapt here the method introduced in [5] for derivation of a phonon Boltzmann equation for the weakly non-linear discrete Schrödinger equation from the evolution hierarchy of *truncated* correlation functions. For comparison, a derivation of the Boltzmann–Nordheim equation using direct perturbation expansions of moments and their graph representations can be found in [1].

It should be stressed that neither method currently produces a mathematically rigorous derivation of fermionic kinetic theory. In particular, it is not yet known which precise assumptions are needed for the kinetic approximation to work nor are there any rigorous bounds for the accuracy of the approximation. From the point of view of the truncated correlation function hierarchy, the key missing ingredient is a control of the evolution of decay properties of correlation functions. Here, we do not go into any detail about the role played by the terms ignored in the derivations below but more details about why their effects are in general expected to be lower order in the weak coupling limit $\lambda \to 0$ can be found in [1, 5].

Let us also point out one case in which rigorous control has been possible: in [6], the kinetic scaling limit of time correlations of equilibrium distributed fields with discrete non-linear Schrödinger evolution are proven to follow the above scenario. In this case, the state itself is stationary and the good decay properties of the truncated correlation functions are provided by the initial data which can be studied with methods from equilibrium statistical mechanics.

Differentiating (26) and recalling the adjoint relations yields the following representation for the time derivative of the Wigner function of translation invariant states:

$$
\begin{aligned}
\partial_t W_{\sigma'\sigma}(k, t) \\
= \int_{\Lambda^*} dk' \left(\rho[\partial_t a(k, \sigma', 1, t)a(k', \sigma, -1, t)] + \rho[\partial_t a(-k', \sigma, 1, t)a(-k, \sigma', -1, t)]^* \right) \\
= \int_{\Lambda^*} dk' \left(\rho[\partial_t a(k, \sigma', 1, t)a(k', \sigma, -1, t)] + \rho[\partial_t a(k, \sigma, 1, t)a(k', \sigma', -1, t)]^* \right) .
\end{aligned} \quad (30)
$$

Thus for a translation invariant states of fermions without spin, we have

$$
\partial_t W(k, t) = 2 \operatorname{Re} \left(\int_{\Lambda^*} dk' \, \rho[\partial_t a(k, 1, t)a(k', -1, t)] \right) . \quad (31)
$$

We use (20) to compute the derivative, yielding

$$
\begin{aligned}
\int_{\Lambda^*} dk' \, \rho[\partial_t a(k, 1, t)a(k', -1, t)] = i\omega(k) \int_{\Lambda^*} dk' \, \rho[a(k, 1, t)a(k', -1, t)] \\
+ i\lambda \int_{(\Lambda^*)^4} dk_1 dk_2 dk_3 dk_4 \, \widehat{V}(k_2 + k_3)\delta_\Lambda(k - k_1 - k_2 - k_3) \\
\times \rho[a(k_1, 1, t)a(k_2, 1, t)a(k_3, -1, t)a(k_4, -1, t)] .
\end{aligned} \quad (32)
$$

The first term on the right is purely imaginary and does not contribute to the real part. In the second term, the expectation is antisymmetric with respect to the swap $k_1 \leftrightarrow k_2$, and thus we can conclude that

$$
\begin{aligned}
\partial_t W(k, t) = \operatorname{Re}\Big[i\lambda \int_{(\Lambda^*)^4} dk_1 dk_2 dk_3 dk_4 \left(\widehat{V}(k_2 + k_3) - \widehat{V}(k_1 + k_3) \right) \\
\times \delta_\Lambda(k - k_1 - k_2 - k_3)\rho[a(k_1, 1, t)a(k_2, 1, t)a(k_3, -1, t)a(k_4, -1, t)] \Big] .
\end{aligned} \quad (33)
$$

We represent the remaining expectation in terms of truncated expectations using (25). Since $\widehat{V}$ is real, all terms involving second-order truncated correlation functions produce terms which are purely imaginary and, hence, they do not contribute to the derivative of the Wigner function. Therefore,

$$
\partial_t W(k, t) = \mathrm{Re}\Big[i\lambda \int_{(\Lambda^*)^4} dk_1 dk_2 dk_3 dk_4 \left(\widehat{V}(k_2 + k_3) - \widehat{V}(k_1 + k_3)\right)
$$

$$
\times \delta_\Lambda(k - k_1 - k_2 - k_3)\rho^T[a(k_1, 1, t), a(k_2, 1, t), a(k_3, -1, t), a(k_4, -1, t)]\Big].
$$
(34)

Computation of derivatives of higher order truncated correlation functions would be simplified by introducing the associated Wick polynomials, as was observed in [5] for commuting fields. However, it is still possible to work out the necessary combinatorics and cancellations by hand for the fourth-order terms which are needed to compute the collision operator of kinetic theory. Namely, after a somewhat lengthy computation employing the symmetry of the function $\widehat{V}$, one finds that

$$
\partial_t \left(e^{-it(\omega_1+\omega_2-\omega_3-\omega_4)}\rho^T[a(k_1, 1, t), a(k_2, 1, t), a(k_3, -1, t), a(k_4, -1, t)]\right)
$$

$$
= i\lambda e^{-it(\omega_1+\omega_2-\omega_3-\omega_4)}\delta_\Lambda(k_1 + k_2 + k_3 + k_4)\left(\widehat{V}(k_2 + k_3) - \widehat{V}(k_1 + k_3)\right)
$$

$$
\times \Big[\tilde{W}(k_2)W(-k_3)W(-k_4) - W(k_1)W(-k_3)W(-k_4)
$$

$$
+ W(k_1)W(k_2)W(-k_4) - W(k_1)W(k_2)\tilde{W}(-k_3)\Big]
$$

$$
+ \text{(higher order truncated functions)},
$$
(35)

where we have introduced the shorthand notations $\omega_i := \omega(k_i)$, $\tilde{W} = 1 - W$, and each W and $\tilde{W}$ factor is evaluated at t.

We then integrate the above time derivatives from 0 to t. The terms involving higher order truncated functions (4:th and 6:th in (35)), as well as the substitution term involving the fourth-order truncated correlation at time 0, are expected to contribute only terms which are subleading in λ at the kinetic timescales $t \propto \lambda^{-2}$, due to the 'integrals' over the oscillatory phase factors. The remaining terms yield the approximation

$$
W(k, t) - W(k, 0) \approx \int_0^t dt' \int_0^{t'} ds \, \mathrm{Re}\Big\{-\lambda^2 \int_{(\Lambda^*)^4} dk_1 dk_2 dk_3 dk_4 \, e^{i(t'-s)(\omega_1+\omega_2-\omega_3-\omega_4)}
$$

$$
\times \left(\widehat{V}(k_2 + k_3) - \widehat{V}(k_1 + k_3)\right)^2 \delta_\Lambda(k - k_1 - k_2 - k_3)\delta_\Lambda(k_1 + k_2 + k_3 + k_4)
$$

$$
\times \Big[\tilde{W}(k_2)W(-k_3)W(-k_4) - W(k_1)W(-k_3)W(-k_4)
$$

$$
+ W(k_1)W(k_2)W(-k_4) - W(k_1)W(k_2)\tilde{W}(-k_3)\Big]\Big\},
$$
(36)

where each W and $\tilde{W}$ factor is evaluated at s. Inside the integrand $-k_4 = k$. Hence, integration over k_4 is straightforward and swapping the sign of k_3, the order of time integrals, and denoting $W_i := W(k_i, s)$ and $\tilde{W}_i := 1 - W_i$, we arrive at the approximation

$$
\begin{aligned}
W(k_0, t) - W(k_0, 0) &\approx \lambda^2 \int_0^t ds \int_{(\Lambda^*)^3} dk_1 dk_2 dk_3 \, \mathrm{Re} \int_0^{t-s} dr \, e^{ir(\omega_1 + \omega_2 - \omega_3 - \omega_0)} \\
&\times \left(\widehat{V}(k_2 - k_3) - \widehat{V}(k_1 - k_3) \right)^2 \delta_\Lambda(k_0 - k_1 - k_2 + k_3) \\
&\times \left[-\tilde{W}_2 W_3 W_0 + W_1 W_3 W_0 - W_1 W_2 W_0 + W_1 W_2 \tilde{W}_3 \right],
\end{aligned} \tag{37}
$$

The real part of the remaining oscillatory time integral formally convergences to $\pi \delta(\omega_0 - \omega_3 - \omega_1 - \omega_2)$ as $t \to \infty$. In fact, the δ-function approximation should only be used after the thermodynamic limit $L \to \infty$ has been taken; for a finite lattice, also values for which $\omega_1 + \omega_2 - \omega_3 - \omega_0$ is not exactly zero but close enough to zero (e.g. $o(L^{-2})$) will contribute to the collision term. Assuming that the thermodynamic limit of the function W exists and using the same notation for the limit, we obtain

$$
W(k_0, t) - W(k_0, 0) \approx \int_0^t ds \, \mathscr{C}_{\mathrm{fBN}}[W(\cdot, s)](k_0), \tag{38}
$$

where a relabelling $k_1 \leftrightarrow k_3$ yields the following more standard form of a fermionic Boltzmann–Nordheim collision operator:

$$
\begin{aligned}
\mathscr{C}_{\mathrm{fBN}}[W](k_0) &:= \pi \lambda^2 \int_{(\mathbb{T}^d)^3} dk_1 dk_2 dk_3 \, \delta(\omega_0 + \omega_1 - \omega_2 - \omega_3) \\
&\times \left(\widehat{V}(k_1 - k_2) - \widehat{V}(k_1 - k_3) \right)^2 \delta_{\mathbb{T}^d}(k_0 + k_1 - k_2 - k_3) \\
&\times \left[\tilde{W}_1 W_2 W_3 - W_0 W_2 W_3 - W_0 W_1 \tilde{W}_2 + W_0 W_1 W_3 \right].
\end{aligned} \tag{39}
$$

The kinetic equation obtained by replacing the approximation sign in (38) by an equals sign is called the (spatially homogeneous) fermionic Boltzmann–Nordheim equation. The term in square brackets in (39) is then usually written in a more symmetric form as
$$
\tilde{W}_0 \tilde{W}_1 W_2 W_3 - W_0 W_1 \tilde{W}_2 \tilde{W}_3.
$$

However, it should be noted that, since the highest order terms indeed cancel, the collision operator has a non-linearity of third order, not of fourth order.

The above lattice kinetic theories have two conserved quantities, $\int dk \, \omega(k) W(k, t)$ related to energy and $\int dk \, W(k, t)$ related to particle density. The mathematical properties of their solutions have mainly been studied in the continuum case for which instead of the lattice wave number $k \in \mathbb{T}^d$ one uses the particle velocity $v \in \mathbb{R}^d$ and the dispersion relation is $\omega(v) = v^2$ in the nonrelativistic case. For the existence and uniqueness of solutions in the continuum case, we refer to [7, 8], while the corresponding issues for a lattice model will be discussed in the next section, based on [9].

4.2 Kinetic Theory of the Spatially Homogeneous Hubbard Model

We next repeat the above computations for the Hubbard model which has a simple onsite potential but includes spin interactions. By (30),

$$\partial_t W_{\sigma'\sigma}(k, t) = \int_{\Lambda^*} dk' \, \rho[\partial_t a(k, \sigma', 1, t) a(k', \sigma, -1, t)] + (\text{h.c.}) , \qquad (40)$$

where 'h.c.' denotes a Hermitian conjugate with respect to the spin degrees of freedom. Employing (21) we find

$$\int_{\Lambda^*} dk' \, \rho[\partial_t a(k, \sigma', 1, t) a(k', \sigma, -1, t)] = i\omega(k) \int_{\Lambda^*} dk' \, \rho[a(k, \sigma', 1, t) a(k', \sigma, -1, t)]$$
$$+ i\lambda \int_{(\Lambda^*)^4} dk_1 dk_2 dk_3 dk_4 \, \delta_\Lambda(k - k_1 - k_2 - k_3)$$
$$\times \rho[a(k_1, \sigma', 1, t) a(k_2, -\sigma', 1, t) a(k_3, -\sigma', -1, t) a(k_4, \sigma, -1, t)] . \qquad (41)$$

The first term on the right is antisymmetric with respect to the Hermitian conjugate, and hence does not contribute to the time derivative of W. We represent the remaining expectation in terms of truncated expectations using (25). In contrast to the spinless case, the second-order terms need no longer cancel: explicitly, they contribute to (41) the term

$$i\lambda \int_{\Lambda^*} dk' \left(W_{\sigma'\sigma}(k) W_{-\sigma',-\sigma'}(k') - W_{-\sigma',\sigma}(k) W_{\sigma',-\sigma'}(k') \right) . \qquad (42)$$

It depends on the expectation

$$\Sigma_{\sigma'\sigma} := \int_{\Lambda^*} dk' \, W_{\sigma'\sigma}(k') = \rho[a^*(0, \sigma') a(0, \sigma)] = \frac{1}{|\Lambda|} \sum_{x \in \Lambda} \rho[a^*(x, \sigma') a(x, \sigma)] , \qquad (43)$$

i.e. on the spin correlation matrix. These expectations are conserved by the time evolution of the Hubbard model, and hence the matrix $\Sigma_{\sigma'\sigma}$ is time-independent. Therefore, the dominant term in the time derivative (40) is given by

$$i\lambda \left(W_{\sigma'\sigma}(k) \Sigma_{-\sigma',-\sigma'} - W_{-\sigma',\sigma}(k) \Sigma_{\sigma',-\sigma'} - W_{\sigma'\sigma}(k) \Sigma_{-\sigma,-\sigma} + W_{\sigma',-\sigma}(k) \Sigma_{-\sigma,\sigma} \right) , \qquad (44)$$

which is most conveniently written as the (σ', σ) -component of the commutator

$$-\mathrm{i}\lambda[\Sigma, W(k, t)].$$

New terms arise also in the computation of the second-order term in λ. The computations are in principle completely analogous to those in the previous subsection but one has to carefully consider the propagation of the spin variable. After taking the thermodynamic limit $L \to \infty$ and neglecting terms which are expected to be higher order in λ, new features compared to the spinless case arise. Most importantly, since one takes a Hermitian, not complex, conjugate of (41), the imaginary part of the oscillatory time integral also contributes in the evolution equation. In other words, one needs to use here the formal identification

$$\int_0^\infty \mathrm{d}r \, \mathrm{e}^{\mathrm{i}r\omega} = \pi\delta(\omega) + \mathrm{i}\,\mathrm{P.V.}\frac{1}{\omega},$$

where 'P.V.' denotes a Cauchy principal value when integrating over the real variable ω. The terms arising from the imaginary part do not resemble usual collision integrals. Instead, they combine into conservative Vlasov-type terms, similarly to what occurred above for the lowest order contribution.

The final evolution equation is most conveniently written as an evolution equation for the Hermitian 2×2 -matrix $W(k, t)$, $k \in \mathbb{T}^d$. It reads

$$\partial_t W(k, t) = \mathscr{C}_{\mathrm{Hubb}}[W(\cdot, t)](k) - \mathrm{i}\left[H^{\mathrm{eff}}[W(\cdot, t)](k), W(k, t)\right], \qquad (45)$$

where the collision operator may be written as

$$\mathscr{C}_{\mathrm{Hubb}}[W](k_0) := \lambda^2\pi \int_{(\mathbb{T}^d)^3} \mathrm{d}k_1\mathrm{d}k_2\mathrm{d}k_3 \,\delta(k_0 + k_1 - k_2 - k_3)\delta(\omega_0 + \omega_1 - \omega_2 - \omega_3)$$
$$\times \left(\tilde{W}_0 W_2 J[\tilde{W}_1 W_3] + J[W_3 \tilde{W}_1] W_2 \tilde{W}_0 - W_0 \tilde{W}_2 J[W_1 \tilde{W}_3] - J[\tilde{W}_3 W_1] \tilde{W}_2 W_0\right)$$
$$(46)$$

using the matrix operation $J[A] := 1\, \mathrm{Tr}\, A - A \in \mathbb{C}^{2\times 2}$. The 'effective Hamiltonian' in the matrix commutator term is given by

$$H^{\mathrm{eff}}[W](k_0) := \lambda\Sigma + \lambda^2\mathrm{P.V.}\int_{(\mathbb{T}^d)^3} \mathrm{d}k_1\mathrm{d}k_2\mathrm{d}k_3\delta(k_0 + k_1 - k_2 - k_3)$$
$$\times \frac{1}{\omega_0 + \omega_1 - \omega_2 - \omega_3}\left(\tilde{W}_2 J[W_1 \tilde{W}_3] + W_2 J[\tilde{W}_1 W_3]\right). \qquad (47)$$

Also the Hubbard–Boltzmann equation (45) can be derived using direct perturbation expansions and their graph representations, as has been done in [10] for more general spin interaction potentials and with a slightly different splitting between the terms in H_0 and V operators. Neither of these derivations provides rigorous estimates of how accurately the solutions to the Hubbard–Boltzmann equation describe the original fermionic reduced density matrices. The principal value integral, in particular, is somewhat troublesome from a mathematical point of view.

The precise mathematical meaning of the terms appearing in the Hubbard–Boltzmann equation (45), as well as the existence and uniqueness of its solutions for physically relevant initial data, has been studied in [9]. It is shown there that for the nearest neighbour Hubbard model with a sufficiently high dimension, $d \geq 3$, any Lebesgue measurable initial data $W_0(k)$ satisfying the matrix constraint $0 \leq W_0(k) \leq 1$ allows a global solution to (45) which is also unique among solutions satisfying the constraint $0 \leq W(k, t) \leq 1$. (The constraint is physically related to the Pauli exclusion principle and it can be checked to follow from the earlier mentioned properties of the fermionic creation and annihilation operators.) This solution is also proven to conserve energy and total spin. More precisely, the real observable $\int dk\, \omega(k)\, \mathrm{Tr}\, W(k, t)$ and the matrix observable $\int dk\, W(k, t)$ are constants along the solutions. Together these properties show that the approximations leading to the Hubbard–Boltzmann equation are consistent, and the resulting kinetic equation should have range of validity similar to the more standard kinetic theories such as the Boltzmann–Nordheim equation derived earlier.

5 Thermalization in Spatially Homogeneous Kinetic Theory

For ergodic systems, time averages of observables will converge to ensemble averages when the averaging period is taken to infinity. In fact, the ensembles covered by such limits could be identified with thermal equilibrium states of the system. However, for system with local conservation laws the approach to global equilibrium typically takes a very long time, often diverging when the system size is increased: for instance, for systems with normal heat conductivity, heat relaxation occurs diffusively and thus involves timescales of order L^2 for systems of spatial diameter L.

For physical transport phenomena, one is interested in the state of the system at *mesoscopic* timescales, i.e. times which are long in microscopic units but short on the macroscopic scale. If the system has only short-range interactions, even though its state could not yet be well approximated by the global equilibrium state, often time averages of observables local to a point in space can be ever better approximated by one of the equilibrium states. This allows describing the evolution of the state of the system by first parametrizing its equilibrium states and then inspecting the evolution of these parameters. A common example would be introduction of space–time-dependent temperature function related to the temperature parameter of the canonical Gibbs state for those systems where total energy is conserved by the evolution.

Systems, which have the above local approximation property, are said to be in *local thermal equilibrium*, and *thermalization* refers to the approach to one of the local thermal equilibrium states from the given initial state. The *thermalization time*, i.e. the time it takes for local thermal equilibrium states to become good approximations, is typically mesoscopic, not macroscopic.

In fact, kinetic theory provides a method of estimating the thermalization process and times. We focus here on thermalization of spatially homogeneous states. This simplifies the analysis since the slow processes associated with spatial relaxation of the equilibrium parameters are then absent. As explained below, kinetic theory indicates that the Wigner function relaxes to stationary states labelled by a few parameters and hence one would expect local equilibrium or quasi-equilibrium to be reached already at kinetic timescales proportional to λ^{-2}. The key to these properties is finding an entropy functional satisfying an H-theorem for the appropriate kinetic evolution. The vanishing of entropy production restricts the functional form of stationary solutions and allows their explicit parametrisation.

5.1 Thermalization Without Spin Interactions

The entropy functional associated with the spatially homogeneous fermionic Boltzmann–Nordheim equation,

$$\partial_t W(k, t) = \mathscr{C}_{\mathrm{fBN}}[W(\cdot, t)](k),$$

where the collision operator is defined in (39), is given by

$$S[W] := -\int_{\mathbb{T}^d} dk \left(W(k) \log W(k) + \widetilde{W}(k) \log \widetilde{W}(k) \right). \qquad (48)$$

Computing the time derivative, one obtains

$$\frac{d}{dt} S[W(t)] = \sigma[W(t)],$$

where the *entropy production functional* is

$$\sigma[W] = \pi \int_{(\mathbb{T}^d)^4} dk_1 dk_2 dk_3 dk_4 \delta(k_1 + k_2 - k_3 - k_4) \delta(\omega_1 + \omega_2 - \omega_3 - \omega_4)$$
$$\times \left(\widehat{V}(k_2 - k_3) - \widehat{V}(k_2 - k_4) \right)^2 G(\widetilde{W}_1 \widetilde{W}_2 W_3 W_4, W_1 W_2 \widetilde{W}_3 \widetilde{W}_4), \qquad (49)$$

with $G(x, y) = (x - y) \ln(x/y)$. Since $\sigma[W] \geq 0$ for physical Wigner functions with $W, \widetilde{W} \geq 0$, this proves that S satisfies an analogue of the *H-theorem* of classical rarefied gas Boltzmann equation.

In particular, any stationary solution to the kinetic equation needs to satisfy $\sigma[W^{(\mathrm{eq})}] = 0$. For sufficiently non-degenerate $\widehat{V}$ and ω, the only regular solutions to this equation are given by the two-parameter family

$$W_{\beta,\mu}^{(eq)}(k) = \left(e^{\beta(\omega(k)-\mu)} + 1\right)^{-1}, \tag{50}$$

where the values of the parameters $\beta, \mu \in \mathbb{R}$ could also be fixed by giving the values for the conserved energy and particle density observables. These Wigner functions can also be obtained by considering the one-particle reduced density matrix of the standard grand canonical Fermi–Dirac states after setting $\lambda = 0$, cf. [3, Proposition 5.2.23]. These states are gauge invariant and quasifree, and thus the Wigner function determines all other reduced density matrices.

It is clear that $\widehat{V}(k)$ cannot be a constant since then $\mathscr{C}_{fBN}[W] = 0$, but otherwise the function $\widehat{V}$ can be fairly arbitrary for this result to hold; one merely needs that the difference $\widehat{V}(k_2 - k_3) - \widehat{V}(k_2 - k_4)$ is non-zero almost everywhere on the manifold defined by the two δ-constraints. The conditions on the dispersion relation ω are more intricate but in two and higher dimensions quite generally the above solutions should be the only stationary ones, see [1, Appendix B.1] and [11] for detailed conditions and more discussion on the topic.

In case $\widehat{V}$ and ω are such that the only stationary solutions are given by (50), one expects that for any regular initial data the solution of the fermionic Boltzmann–Nordheim equation converges as $t \to \infty$ to the unique function $W_{\beta,\mu}^{(eq)}$ where $\beta, \mu \in \mathbb{R}$ are determined by the initial energy and particle number. Unlike for the corresponding bosonic equation, the solutions cannot diverge since they satisfy $0 \leq W \leq 1$ at all times. Thus the space of regular stationary solutions should suffice to cover all asymptotic limits of the solutions. The convergence to a regular stationary solution has been proven for certain continuum models and initial data in [12].

The above results suggest that thermalization timescale for weakly interacting spinless lattice fermions is in great generality given by the kinetic timescale, $t \propto \lambda^{-2}$. It is also consistent with the hypotheses that, apart from special degenerate interactions, the only equilibrium parameters are related to the conservation of energy and particle number. More precisely, one can use β and μ of the standard grand canonical Fermi–Dirac states on the fermionic Fock space as parameters.

5.2   Thermalization in the Hubbard Model

The spin structure of the Hubbard–Boltzmann equation (45) leads to some new phenomena compared to the above spinless Boltzmann–Nordheim case. The entropy functional needs to be generalised to

$$S[W] := -\int dk \left(\text{Tr } W \ln W + \text{Tr } \tilde{W} \ln \tilde{W} \right), \tag{51}$$

where W is a 2×2 Hermitian matrix. Computing its derivative requires some effort, yielding

$$\frac{d}{dt} S[W(t)] = \sigma[W(t)],$$

where the entropy production functional is again positive, $\sigma[W] \geq 0$. To write down the entropy production, let us first diagonalise the matrices $W(k)$, yielding an eigensystem $(\lambda_a(k), \psi_a(k))$, $a = 1, 2$, for each $k \in \mathbb{T}^d$. Then

$$
\sigma[W](k_1) := \frac{\pi}{4} \int d^4k \, \delta(k_1 + k_2 - k_3 - k_4) \delta(\omega_1 + \omega_2 - \omega_3 - \omega_4) \sum_{a \in \{1,2\}^4}
$$

$$
\times \left(\tilde\lambda_1 \tilde\lambda_2 \lambda_3 \lambda_4 - \lambda_1 \lambda_2 \tilde\lambda_3 \tilde\lambda_4 \right) \ln \frac{\tilde\lambda_1 \tilde\lambda_2 \lambda_3 \lambda_4}{\lambda_1 \lambda_2 \tilde\lambda_3 \tilde\lambda_4} \, |\langle \psi_1, \psi_3 \rangle \langle \psi_2, \psi_4 \rangle - \langle \psi_1, \psi_4 \rangle \langle \psi_2, \psi_3 \rangle|^2 ,
$$

where $\psi_i := \psi_{a_i}(k_i)$, $\lambda_i := \lambda_{a_i}(k_i)$ and $\tilde\lambda := 1 - \lambda$.

The solution of the condition $\sigma[W] = 0$ is no longer quite as straightforward as before, and one has to consider a few degenerate cases separately. However, if $d \geq 2$, the non-degeneracy conditions mentioned earlier are satisfied for the nearest neighbour interaction of the Hubbard model, and thus the analysis of the two δ-constraints is simplified. As derived in [13], then one of the following possibilities needs to be realised by physical stationary solutions $W^{(eq)}(k)$ which are Hermitian matrices satisfying $0 \leq W(k) \leq 1$ for every $k \in \mathbb{T}^d$. First, choose a spin basis such that the total spin correlation matrix Σ is diagonal. Then one of the following cases holds:

1. There are grand canonical parameters β, μ_+, μ_-, fixed by the diagonal matrix Σ and the energy, such that

$$
W^{(eq)}(k) = \begin{pmatrix} g_+(k) & 0 \\ 0 & g_-(k) \end{pmatrix} , \tag{52}
$$

   where $g_\pm(k) := (1 + e^{\beta(\omega(k) - \mu_\pm)})^{-1}$ are standard Fermi–Dirac distributions.
2. One of the bands is empty and the other is arbitrary: there is a function $f(k)$ with $0 \leq f(k) \leq 1$ and $\sigma \in \{\pm 1\}$ such that $W_{\sigma\sigma}(k) = f(k)$ and all other elements of $W(k)$ are zero.
3. One of the bands is full and the other is arbitrary: there is a function $f(k)$ with $0 \leq f(k) \leq 1$ and $\sigma \in \{\pm 1\}$ such that $W_{\sigma\sigma}(k) = f(k)$, $W_{-\sigma,-\sigma}(k) = 1$, and all off-diagonal elements of $W(k)$ are zero.

These solutions are expected to behave differently when occurring as asymptotic stationary states in the Hubbard model. If the initial data is such that both bands are partially filled, i.e. if one can find β, μ_+, μ_- and a unitary matrix U such that the function $W^{(eq)}$ in (52) satisfies $\int dk \, U^* W(k, t) U = \int dk \, W^{(eq)}(k)$ and $\int dk \, \omega(k) \operatorname{Tr} W(k, t) = \int dk \, \omega(k) \operatorname{Tr} W^{(eq)}(k)$ initially, and hence for all t, then one expects $W(k, t) \to U W^{(eq)}(k) U^*$ as $t \to \infty$.

However, if one of the bands is either empty or full initially, then no thermalization can be expected. In fact, this property is not only an artefact of the kinetic theory but it can also be realised in the original Hubbard model. Consider an initial wave vector for which there are no particles with '$-$'-spin. Then the pair interaction V acting on the vector produces zero and, since the free Hamiltonian does not mix the two bands, one can check that Hubbard model evolution equations are satisfied by the solution of the free evolution generated by H_0. The free semigroup leaves for instance all quasifree states invariant and one can choose the Wigner function of the '+' -component arbitrarily.

The above situation is radically changed if $d = 1$. This case is known to be integrable, see [14] for a review of the one-dimensional Hubbard model, and the large number of conserved quantities is reflected also in the kinetic evolution. As shown in [13], in this case one may take in the stationary solutions in item 1 above instead of the standard Fermi–Dirac distributions $g_\pm$ any functions which are of the form $(1 + e^{\beta(f(k)-\mu_\pm)})^{-1}$ for some real periodic function f which satisfies the antisymmetry condition $f(\frac{1}{2} - k) = -f(k)$. Hence, one needs infinitely many parameters to describe the stationary solutions. The various scenarios for the convergence towards a steady state are explored numerically in [13]. There it is also observed that adding a next-to-nearest neighbour term to the free evolution appears to lift the degeneracy, leaving only the standard Fermi–Dirac distributions as possible limits, similarly to what was stated above for the cases with $d \geq 2$.

6 Concluding Remarks

Reliable study of large-scale evolution of a system of weakly interacting fermions is a challenge both to numerical simulations and to theoretical analysis. We advocate here using kinetic equations not only to reproduce standard folklore results, such as convergence towards Fermi–Dirac distribution, but as a tool for *systematic* study of the approach to equilibrium and thermalization in these systems. Even lacking complete mathematical control over the accuracy and applicability of the kinetic approximation, analysis of kinetic equations can provide testable predictions and reveal possible sources of 'anomalies' and other degeneracies. For instance, the role of the dispersion relation and dimensionality in the Hubbard model revealed in the above references encourages such studies in other models.

The almost unreasonable usefulness of kinetic theory begs for better understanding of its underpinnings, in particular, of what is the most accurate connection between the microsopic evolution and the kinetic theory and what are the most appropriate kinetic equations for this purpose. These questions lie in the realm of mathematically rigorous study of scaling limits producing observables which exactly follow some kinetic equation. However, ultimately the goal should be in also extracting practical information about the error in such approximations and how well the approximations extend beyond their apparent regions of applicability, as dictated by the convergence of the scaling limits.

For instance, finding answers to the following open questions could benefit from mathematically rigorous approaches:

1. For which initial data does the corresponding solution to the kinetic equation converge towards the stationary solution determined by the values of the conserved quantities? Could one estimate the rate of convergence?
2. How would the kinetic equations and their solutions change for general spin interactions, including also interactions with external magnetic fields?
3. If the initial state of the system is not spatially homogeneous, when does its evolution follow an inhomogeneous Boltzmann equation? Are there ways of improving the accuracy of the model, for instance, by including a Vlasov–Poisson-type correction?
4. Could one improve the accuracy of the kinetic equation by 'renormalizing' the microscopic observables? How much?

Acknowledgements I am most grateful to Herbert Spohn for our collaboration and many discussions about validity and properties of kinetic theory. Most of the results here are based on his works and on our joint collaborations. The related research has been made possible by support from the Academy of Finland and also partially supported by the French Ministry of Education through the grant ANR (EDNHS).

References

1. Lukkarinen, J., Spohn, H.: Not to normal order–Notes on the kinetic limit for weakly interacting quantum fluids. J. Stat. Phys. **134**(5), 1133–1172 (2009)
2. Teschl, G.: Mathematical Methods in Quantum Mechanics: With Applications to Schrödinger Operators. Graduate studies in mathematics, 1st edn., vol. 99. American Mathematical Society (2009)
3. Bratteli, O., Robinson, D.W.: Operator Algebras and Quantum Statistical Mechanics II. Springer, New York (1981)
4. Salmhofer, M.: Clustering of fermionic truncated expectation values via functional integration. J. Stat. Phys. **134**(5), 941–952 (2009)
5. Lukkarinen, J., Marcozzi, M.: Wick polynomials and time-evolution of cumulants. J. Math. Phys. **57**(8), 083301:1–27 (2016)
6. Lukkarinen, J., Spohn, H.: Weakly nonlinear Schrödinger equation with random initial data. Invent. Math. **183**(1), 79–188 (2011)
7. Dolbeault, J.: Kinetic models and quantum effects: a modified Boltzmann equation for Fermi-Dirac particles. Arch. Ration. Mech. Anal. **127**, 101–131 (1994)
8. Escobedo, M., Mischler, S., Valle, M.A.: Homogeneous Boltzmann equation in quantum relativistic kinetic theory. Electron. J. Differ. Equ. Monogr. **04**, 1–85 (2003)
9. Lukkarinen, J., Mei, P., Spohn, H.: Global well-posedness of the spatially homogeneous Hubbard-Boltzmann equation. Commun. Pure Appl. Math. **68**(5), 758–807 (2015)
10. Fürst, M.L.R., Lukkarinen, J., Mei, P., Spohn, H.: Derivation of a matrix-valued Boltzmann equation for the Hubbard model. J. Phys. A: Math. Theor. **46**(48), 485002 (2013)
11. Spohn, H.: Collisional invariants for the phonon Boltzmann equation. J. Stat. Phys. **124**, 1131–1135 (2006)
12. Lu, X., Wennberg, B.: On stability and strong convergence for the spatially homogeneous Boltzmann equation for Fermi-Dirac particles. Arch. Ration. Mech. Anal. **168**(1), 1–34 (2003)
13. Fürst, M.L.R., Mendl, C.B., Spohn, H.: Matrix-valued Boltzmann equation for the Hubbard chain. Phys. Rev. E **86**, 031122 (2012)
14. Essler, F.H.L., Frahm, H., Göhmann, F., Klümper, A., Korepin, V.E.: The One-Dimensional Hubbard Model. Cambridge University Press, Cambridge (2005)

The BCS Critical Temperature in a Weak External Electric Field via a Linear Two-Body Operator

Rupert L. Frank and Christian Hainzl

Dedicated to Herbert Spohn on the occasion of his seventieth birthday

Abstract We study the critical temperature of a superconductive material in a weak external electric potential via a linear approximation of the BCS functional. We reproduce a similar result as in Frank et al. (Commun Math Phys 342(1):189–216, 2016, [5]) using the strategy introduced in Frank et al. (The BCS critical temperature in a weak homogeneous magnetic field, [2]), where we considered the case of an external constant magnetic field.

1 Introduction and Main Result

1.1 Objective and Background

In this paper, we want to consider a linear two-body operator which determines the critical temperature of a superconductive or superfluid system. This linear opera-

R. L. Frank
Mathematisches Institut der Universität München, Theresienstr. 39,
80333 Munich, Germany
e-mail: rlfrank@caltech.edu

R. L. Frank
Mathematics 253-37, Caltech, Pasadena, CA 91125, USA

C. Hainzl (✉)
Mathematisches Institut, Universität Tübingen, Auf der Morgenstelle 10,
72076 Tübingen, Germany
e-mail: christian.hainzl@uni-tuebingen.de

© Springer Nature Switzerland AG 2018   29
D. Cadamuro et al. (eds.), *Macroscopic Limits of Quantum Systems*,
Springer Proceedings in Mathematics & Statistics 270,
https://doi.org/10.1007/978-3-030-01602-9_2

tor was studied recently in connection with the influence of a constant magnetic field on the critical temperature [2]. The analysis of this operator was significantly complicated by the unboundedness of the magnetic vector potential as well as the noncommutativity of the components of the magnetic momentum. For this reason, we want to present here the method of [2] in the simplified situation where the external field consists of an electric potential.

We have the following situation in mind. Two particles interact via a two-body potential $-2V(x - y)$ and both particles are placed in an external electric potential $h^2 W(hx)$, where $h > 0$ is a small parameter. Thus, the external field is weak of order h^2 and varies on the scale of order $1/h$, whereas both the strength and the scale of the interaction are of order one determined by V. The energy is given by the linearized BCS (Bardeen–Cooper–Schrieffer) functional at positive temperature $T = 1/\beta$.

Therefore, we are interested in the infimum of the spectrum of the two-body operator

$$\frac{p_x^2 + h^2 W(hx) + p_y^2 + h^2 W(hy) - 2\mu}{\tanh\left(\frac{\beta}{2}\left(p_x^2 + h^2 W(hx) - \mu\right)\right) + \tanh\left(\frac{\beta}{2}\left(p_y^2 + h^2 W(hy) - \mu\right)\right)} - V(x - y) \tag{1}$$

acting in

$$L^2_{\mathrm{symm}}(\mathbb{R}^3 \times \mathbb{R}^3) = \left\{\alpha \in L^2(\mathbb{R}^3 \times \mathbb{R}^3) : \ \alpha(x, y) = \alpha(y, x) \text{ for all } x, y \in \mathbb{R}^3\right\}.$$

Here, $p_x = -i\nabla_x$ and $p_y = -i\nabla_y$. The interaction potential $-2V(x - y)$ between the two particles is assumed to be spherically symmetric, i.e., to depend only on the distance $|x - y|$. (We will also assume that the interaction potential is nonpositive and the minus sign, as opposed to the more usual plus sign, will simplify some formulas.) Moreover, $\mu \in \mathbb{R}$ is the chemical potential. We are interested in the dependence of the operator on two parameters, namely, the inverse temperature $\beta > 0$ and the scale ratio $h > 0$. More precisely, we are interested in identifying regimes of temperatures $T = \beta^{-1}$ such that the infimum of the spectrum of the above operator is positive or negative for all sufficiently small $h > 0$.

As we explained in detail in [2] and will repeat below, the motivation for this question comes from the BCS theory of superconductivity and the operator (1) arises through the linearization of the Bogoliubov–de Gennes equation around the normal state. Therefore, the question whether the infimum of the spectrum of the operator (1) is positive or negative corresponds to the local stability or instability of the normal state. In that sense, it is not hard to imagine that the BCS critical temperature corresponds to the value of T for which the infimum of the spectrum of this operator is exactly zero.

To describe our main result, we introduce the effective one-body operator

$$\frac{(-i\nabla_r)^2 - \mu}{\tanh\left(\frac{\beta}{2}\left((-i\nabla_r)^2 - \mu\right)\right)} - V(r) \tag{2}$$

acting in

$$L^2_{\text{symm}}(\mathbb{R}^3) = \{\alpha \in L^2(\mathbb{R}^3) : \alpha(-r) = \alpha(r) \text{ for all } r \in \mathbb{R}^3\}.$$

Later on, we will see that the variable $r \in \mathbb{R}^3$ arises as the relative coordinate $r = x - y$ of the two particles at x and y. We will *assume* that the operator $|(-i\nabla_r)^2 - \mu| - V(r)$ has a negative eigenvalue. Then, it is easy to see (see, e.g., [8]) that there is a unique $\beta_c \in (0, +\infty)$ such that the operator (2) is nonnegative for $\beta \le \beta_c$ and has a negative eigenvalue for $\beta > \beta_c$. Let $T_c = \beta_c^{-1}$. Then, our main result is, roughly speaking, that the infimum of the spectrum of the two-particle operator (1) is negative for $T \le T_c + c_0 h^2 + o(h^2)$ and positive for $T \ge T_c + c_0 h^2 - o(h^2)$. Here, c_0 is a positive constant which we compute explicitly in terms of the zero-energy ground state of (2) at $\beta = \beta_c$. (In fact, $c_0 = -T_c D_c$ with D_c from (9).) Thus, the external electric field $h^2 W(hx)$ changes the critical temperature by an amount $c_0 h^2 + o(h^2)$. Informally (that is, ignoring issues like the possible nonuniqueness of a critical temperature), this says that

$$T_c(h) = T_c + c_0 h^2 + o(h^2).$$

The mathematical challenge of this problem is that low-energy states of the two-particle operator (1) exhibit a two-scale structure. As function of the relative coordinate $r = x - y$ and the center-of-mass coordinate $X = (x + y)/2$ they vary on a scale of order one with respect to r and on a (much larger) scale of order $1/h$ with respect to X. The variation on the former scale is responsible for the leading order term T_c for the critical temperature, whereas the variation on the latter scale is responsible for the subleading correction $c_0 h^2$. This subleading correction is determined by an effective linear Ginzburg–Landau functional which emerges on the macroscopic scale $1/h$ determined by the external potential. We hereby recover a similar result for the critical temperature as in the full nonlinear BCS theory in [5]. This is of course not unexpected since we deal with the second derivative around the normal state of the BCS functional.

The work [5] relied on [4] where the Ginzburg–Landau functional was derived from the BCS functional close to the critical temperature by means of a rather intricate proof. In view of this, the goal of the present paper is twofold. First, we explain the strategy from [2] in a simpler setting, and second, we derive the linearized Ginzburg–Landau equation in a simpler way as in the full nonlinear case [4]. One difference compared to the work [4, 5] is the fact that we do not restrict ourselves to a finite box, and therefore omit the periodicity assumptions. Further, we work in relative and center-of-mass coordinates which is natural in terms of the before-mentioned two-scale structure.

As in [2] we will not work directly with the two-particle operator (1), but rather with its Birman–Schwinger version.

Before we describe the precise setup of our analysis, we would like to stress that in this paper we work with the BCS functional and its linearization around the normal state. This should not be confused with what is often called the BCS Hamiltonian or the BCS model and which was investigated, for instance, by Haag, Thirring and Wehrl from the point of view of algebraic quantum field theory. The BCS Hamiltonian is a many-body Hamiltonian which corresponds to a regularization of a δ interaction. The BCS functional arises as an effective nonlinear functional by restricting the BCS Hamiltonian to quasi-free states and dropping the direct and exchange terms. We do allow, however, for more general interaction potentials. It remains an open problem to understand from a mathematically rigorous point of view the relation between the BCS functional and many-body quantum mechanics. Nevertheless, our analysis leads to quantitative estimates which agree with physics.

1.2   Model and Main Result

Our model has the following ingredients:

Assumption 1. (1) External electric potential $h^2 W(hx)$ such that $W \in W^{1,\infty}(\mathbb{R}^3)$.
(2) Inverse temperature $\beta = T^{-1} > 0$.
(3) Chemical potential $\mu \in \mathbb{R}$.
(4) Nonnegative, spherically symmetric interaction potential V such that $V \in L^\infty(\mathbb{R}^3)$ and $|r|V \in L^\infty(\mathbb{R}^3)$.

We recall that the Sobolev space $W^{1,\infty}(\mathbb{R}^3)$ consists of all bounded, Lipschitz continuous functions with a finite global Lipschitz constant.

The nonnegativity assumption on V is for technical convenience. To simplify notation and since the precise meaning is always clear from the context, we use the same symbol V also for the corresponding multiplication operators on $L^2_{\text{symm}}(\mathbb{R}^3)$ (i.e., $(V\alpha)(r) = V(r)\alpha(r)$) and on $L^2_{\text{symm}}(\mathbb{R}^3 \times \mathbb{R}^3)$ (i.e., $(V\alpha)(x, y) = V(x - y)\alpha(x, y)$).

The corresponding single-particle Hamiltonian, acting in $L^2(\mathbb{R}^3)$, is defined by

$$\mathfrak{h}_W = p^2 + h^2 W(hx) - \mu. \tag{3}$$

with the notation $p = -i\nabla$. The locations of the two particles are represented by coordinates $x, y \in \mathbb{R}^3$. If we want to emphasize the variables on which the operators act, we write

$$\mathfrak{h}_{W,x} = p_x^2 + h^2 W(hx) - \mu, \qquad \mathfrak{h}_{W,y} = p_y^2 + h^2 W(hy) - \mu.$$

As in [2] we introduce a function $\Xi_\beta : \mathbb{R}^2 \to \mathbb{R}$ by

$$\Xi_\beta(E, E') := \frac{\tanh \frac{\beta E}{2} + \tanh \frac{\beta E'}{2}}{E + E'}$$

if $E + E' \neq 0$ and $\Xi_\beta(E, -E) = (\beta/2)/\cosh^2(\beta E/2)$. Since the operators $\mathfrak{h}_{W,x}$ and $\mathfrak{h}_{W,y}$ commute, we can define the operator

$$L_{T,W} = \Xi_\beta(\mathfrak{h}_{W,x}, \mathfrak{h}_{W,y}).$$

We will always consider this operator in the Hilbert space $L^2_{\text{symm}}(\mathbb{R}^3 \times \mathbb{R}^3)$. Note that, with this notation, the operator in (1) can be written as $L^{-1}_{T,W} - V$.

Next, in order to formulate our assumption on the critical temperature, we introduce the function $\chi_\beta : \mathbb{R} \to \mathbb{R}$ by

$$\chi_\beta(E) := \frac{\tanh \frac{\beta E}{2}}{E}$$

and set $\chi_\infty(E) := |E|^{-1}$. We consider the compact operator

$$V^{1/2}\chi_\beta(p_r^2 - \mu)V^{1/2}$$

in $L^2_{\text{symm}}(\mathbb{R}^3)$, where

$$p_r = -i\nabla_r$$

denotes the momentum operator. (The operator $\chi_\beta(p_r^2 - \mu)$ is denoted by K_T^{-1} in [8] and several works thereafter.)

Assumption 2. sup spec $V^{1/2}\chi_\infty(p_r^2 - \mu)V^{1/2} > 1$.

Since $\beta \mapsto \chi_\beta(E)$ is strictly increasing for each fixed $E \in \mathbb{R}$, Assumption 2 implies that there is a unique $\beta_c \in (0, \infty)$ such that

$$\begin{aligned}
\text{sup spec } V^{1/2}\chi_\beta(p_r^2 - \mu)V^{1/2} &\leq 1 \quad \text{if } \beta \leq \beta_c, \\
\text{sup spec } V^{1/2}\chi_\beta(p_r^2 - \mu)V^{1/2} &> 1 \quad \text{if } \beta > \beta_c.
\end{aligned}$$

We set $T_c = \beta_c^{-1}$. Note that the operator $V^{1/2}\chi_{\beta_c}(p_r^2 - \mu)V^{1/2}$ has eigenvalue 1.

Assumption 3. The eigenvalue 1 of the operator $V^{1/2}\chi_{\beta_c}(p_r^2 - \mu)V^{1/2}$ is simple.

We denote by φ_*, a normalized eigenfunction of $V^{1/2}\chi_\beta(p_r^2 - \mu)V^{1/2}$ corresponding to the eigenvalue 1 which, by assumption, is unique up to a phase. Since p_r^2 and V are real operators, so is $V^{1/2}\chi_\beta(p_r^2 - \mu)V^{1/2}$ and we can assume that φ_* is real valued.

The spherical symmetry of V from Assumption 1 and the non-degeneracy from Assumption 3 imply that φ_* is spherically symmetric.

From a physics point of view, Assumption 3 restricts us to potentials giving rise to s-wave superconductivity. It is known that this assumption is fulfilled for a large class of potentials, including those which have a nonnegative Fourier transform [6]. For partial results in the case where Assumption 3 is violated, we refer to [1].

As the final preliminary before stating our main result, we will introduce some constants. They are defined in terms of the auxiliary functions

$$g_0(z) = \frac{\tanh(z/2)}{z},$$

$$g_1(z) = \frac{e^{2z} - 2ze^z - 1}{z^2(e^z + 1)^2} = \frac{1}{2z^2} \frac{\sinh z - z}{\cosh^2(z/2)},$$

$$g_2(z) = \frac{2e^z(e^z - 1)}{z(e^z + 1)^3} = \frac{1}{2z} \frac{\tanh(z/2)}{\cosh^2(z/2)}, \tag{4}$$

as well as the function

$$t(p) := \|\chi_{\beta_c}((-i\nabla_r)^2 - \mu)V^{1/2}\varphi_*\|^{-1} 2(2\pi)^{-3/2} \int_{\mathbb{R}^3} dx\, V(x)^{1/2}\varphi_*(x)e^{-ip\cdot x}. \tag{5}$$

(The prefactor in front of the integral is irrelevant for us and only introduced for consistency with the definition in [5].) We now set

$$\Lambda_0 := \frac{\beta_c^2}{16} \int_{\mathbb{R}^3} \frac{dp}{(2\pi)^3} |t(p)|^2 \left(g_1(\beta_c(p^2 - \mu)) + \frac{2}{3}\beta_c p^2 g_2(\beta_c(p^2 - \mu)) \right), \tag{6}$$

$$\Lambda_1 := \frac{\beta_c^2}{4} \int_{\mathbb{R}^3} \frac{dp}{(2\pi)^3} |t(p)|^2 g_1(\beta_c(p^2 - \mu)), \tag{7}$$

$$\Lambda_2 := \frac{\beta_c}{8} \int_{\mathbb{R}^3} \frac{dp}{(2\pi)^3} |t(p)|^2 \cosh^{-2}(\beta_c(p^2 - \mu)/2). \tag{8}$$

The constants Λ_0 and Λ_2 are positive (for a proof for Λ_0 see [4]). Note that the quotient Λ_0/Λ_2, which will appear in our main result, has the dimension of an inverse temperature.

We set

$$D_c := \frac{\Lambda_0}{\Lambda_2} \inf \operatorname{spec} \left(p_X^2 + \frac{\Lambda_1}{\Lambda_0} W(X) \right), \tag{9}$$

where the operator on the right side is considered as an operator in $L^2(\mathbb{R}^3)$ and where $p_X = -i\nabla_X$.

The following is our main theorem.

Theorem 4. *Under Assumptions 1–3, the following holds.*

(1) Let $0 < T_1 < T_c$. Then, there are constants $h_0 > 0$ and $C > 0$ such that for all $0 < h \le h_0$ and all $T_1 \le T < T_c(1 - h^2 D_c) - Ch^3$ one has

$$\inf_{\Phi} \langle \Phi, (1 - V^{1/2}L_{T,w}V^{1/2})\Phi \rangle < 0.$$

(2) There are constants $h_0 > 0$ and $C > 0$ such that for all $0 < h \le h_0$ and all $T > T_c(1 - h^2 D_c) + Ch^{5/2}$ one has

$$\langle \Phi, (1 - V^{1/2} L_{T,W} V^{1/2}) \Phi \rangle > 0 \,,$$

unless $\Phi = 0$.

Remark 5. Let us restate this theorem in a heuristic form. Informally, we think of the critical temperature $T_c(h)$ as the value of the parameter T such that

$$\sup \mathrm{spec} V^{1/2} L_{T,W} V^{1/2} = 1 \,.$$

This is not a precise definition because in contrast to the one-body operator $V^{1/2} \chi_\beta(p_r^2 - \mu) V^{1/2}$ it is not clear whether the two-body operator $V^{1/2} L_{T,W} V^{1/2}$, or at least the infimum of its spectrum, is monotone in T, and therefore the uniqueness of the value of T such that $\sup \mathrm{spec} V^{1/2} L_{T,W} V^{1/2} = 1$ is not guaranteed. Ignoring this issue, as well as some technicalities connected with T_1 in part (1) which we discuss below, we see that our main theorem says that

$$T_c(h) = T_c(1 - D_c h^2) + o(h^2) \,.$$

Note that concerning the potential nonuniqueness of the critical temperature the theorem implies that, if it occurs at all, it occurs only in a temperature interval of size $o(h^2)$.

Remark 6. Observe that D_c can have either sign, depending on W. Thus, an external electric field $h^2 W(hx)$ can both raise and lower the critical temperature by an amount of order h^2. This is in contrary to the influence of magnetic fields where the critical temperature always goes down.

Remark 7. Let us compare our results here with those in [5] where we also computed the shift of the critical temperature. The results of [5] concern a definition of the critical temperature in the nonlinear BCS functional, whereas here we base our definition of critical temperature on a quadratic approximation to the BCS functional around the normal state. Both notions lead to the same result to order h^2. A minor difference is that the setting in [5] is a finite sample, whereas here we work on the whole space. Technically, the methods of proof in the two approaches are quite different.

Remark 8. The assumption in part (1) that the temperature is bounded away from zero is probably only technical. Note, however, that our result is valid for arbitrarily small $T_1 > 0$, as long as it is uniform in h. The reason for this restriction is that our expansions diverge as the temperature goes to zero. Remarkably, there is no such restriction in part (2) of the theorem.

Remark 9. Let us emphasize that our definition of the critical temperature T_c coincides with that in [8] (and therefore with that in [4, 5]) and that our Assumptions 2 and 3 coincide with [4, Assumption 2]. This is a consequence of the Birman–Schwinger principle, which also implies that, if α_* denotes a normalized, real-valued eigenfunction of the operator (2), then

$$V^{1/2}\alpha_* = \pm \|\chi_{\beta_c}((i\nabla_r)^2 - \mu)V^{1/2}\varphi_*\|^{-1}\varphi_* .$$

(To get the normalization constant, we apply $\chi_{\beta_c}((-i\nabla_r)^2 - \mu)V^{1/2}$ to both sides and use the equation for α_* and its normalization.)

Remark 10. In the physics literature, the two-body interaction V is usually replaced by a local contact interaction. With this modification, the linear two-body operator (1) was studied earlier in the literature, in particular, in the school by Gorkov and coauthors. In the presence of a constant magnetic field, this operator was used by Werthamer et al. [9, 14] in their study of the upper critical field. This approach was later extended in different directions, see e.g., [11–13]. In particular, [10] relaxed the local approximation and was an initial motivation for our work [2].

1.3 Connection to BCS Theory

In this subsection, we repeat our argument from [2] and describe how the two-body operators (1) and $L_{T,W}$ arise in a problem in superconductivity. Our purpose here is to give a motivation and our presentation in this subsection will be informal. For background and references on the mathematical study of BCS theory we refer to our earlier works [1–6, 8] and, in particular, to the review [7].

We consider a superconducting sample occupying all of $\mathbb{R}^3$ at inverse temperature $\beta > 0$ and chemical potential $\mu \in \mathbb{R}$. The particles interact through a two-body potential $-2V(x - y)$ and are placed in an external electric field with potential $h^2 W(hx)$. In BCS theory, the state of a system is described by two operators γ and α in $L^2(\mathbb{R}^3)$, representing the one-body density matrix and the Cooper pair wave function, respectively. The operator γ is assumed to be Hermitian and the operator α is assumed to satisfy $\alpha^* = \overline{\alpha}$, where for a general operator A we write $\overline{A} = \mathcal{C}A\mathcal{C}$ with $\mathcal{C}$ denoting complex conjugation. Moreover, it is assumed that

$$0 \le \begin{pmatrix} \gamma & \alpha \\ \overline{\alpha} & 1 - \overline{\gamma} \end{pmatrix} \le 1.$$

In an equilibrium state, the operators γ and α satisfy the (nonlinear) Bogoliubov–de Gennes equation

$$\begin{pmatrix} \gamma & \alpha \\ \overline{\alpha} & 1 - \overline{\gamma} \end{pmatrix} = \left(1 + \exp\left(\beta H_{\Delta_{V,\alpha}}\right)\right)^{-1},$$

$$\text{where} \quad \Delta_{V,\alpha}(x, y) = -2V(x - y)\alpha(x, y) \quad \text{and} \quad H_\Delta = \begin{pmatrix} \mathfrak{h}_W & \Delta \\ \overline{\Delta} & -\overline{\mathfrak{h}_W} \end{pmatrix}.$$

Here, Δ is considered as an integral operator in $L^2(\mathbb{R}^3)$ with integral kernel $\Delta(x, y)$. Moreover, $\mathfrak{h}_W$ is the one-particle operator introduced in (3).

Note that one solution of the equation is $\gamma = (1 + \exp(\beta\mathfrak{h}_W))^{-1}$ and $\alpha = 0$. This is the *normal state*. We are interested in the local stability of this solution, and therefore will linearize the equation around it.

It is somewhat more convenient to write the equation in the equivalent form

$$\begin{pmatrix} \gamma & \alpha \\ \alpha & 1 - \gamma \end{pmatrix} = \frac{1}{2} - \frac{1}{2}\tanh\left(\frac{\beta}{2}H_{\Delta_{V,\alpha}}\right) .$$

Then, in view of the partial fraction expansion (also known as Mittag-Leffler series)

$$\tanh z = \sum_{n\in\mathbb{Z}} \frac{1}{z - i(n + 1/2)\pi}$$

(where we write $\sum_{n\in\mathbb{Z}}$ short for $\lim_{N\to\infty}\sum_{n=-N}^{N}$ for conditionally convergent sums like this one; convergence becomes manifest by combining the $+n$ and $-n$ terms),

$$\tanh\left(\frac{\beta}{2}H_\Delta\right) = -\frac{2}{\beta}\sum_{n\in\mathbb{Z}} \frac{1}{i\omega_n - H_\Delta}$$

with the *Matsubara frequencies*

$$\omega_n = \pi(2n + 1)T , \qquad n \in \mathbb{Z}. \tag{10}$$

Using this formula, we can expand the operator $\tanh(\beta H_\Delta/2)$ in powers of Δ. Since

$$\frac{1}{i\omega_n - H_\Delta} = \frac{1}{i\omega_n - H_0} + \frac{1}{i\omega_n - H_0}\begin{pmatrix} 0 & \Delta \\ \Delta & 0 \end{pmatrix}\frac{1}{i\omega_n - H_0} + \cdots$$

$$= \begin{pmatrix} (i\omega_n - \mathfrak{h}_W)^{-1} & 0 \\ 0 & (i\omega_n + \bar{\mathfrak{h}}_W)^{-1} \end{pmatrix}$$

$$+ \begin{pmatrix} 0 & (i\omega_n - \mathfrak{h}_W)^{-1}\Delta(i\omega_n + \bar{\mathfrak{h}}_W)^{-1} \\ (i\omega_n + \bar{\mathfrak{h}}_W)^{-1}\bar{\Delta}(i\omega_n - \mathfrak{h}_W)^{-1} & 0 \end{pmatrix} + \cdots ,$$

the Bogoliubov–de Gennes equation for the Cooper pair wave function becomes

$$\alpha = \frac{1}{\beta}\sum_{n\in\mathbb{Z}}(i\omega_n - \mathfrak{h}_W)^{-1}\Delta_{V,\alpha}(i\omega_n + \bar{\mathfrak{h}}_W)^{-1} + \cdots ,$$

where $\ldots$ stands for terms that are higher order in $\Delta_{V,\alpha}$. The key observation now is that

$$\frac{1}{\beta}\sum_{n\in\mathbb{Z}}(i\omega_n - \mathfrak{h}_W)^{-1}\Delta_{V,\alpha}(i\omega_n + \bar{\mathfrak{h}}_W)^{-1} = L_{T,W}V\alpha. \tag{11}$$

(Here, $V\alpha$ on the right side is considered as a two-particle wave function, defined by $(V\alpha)(x, y) = V(x - y)\alpha(x, y)$.) This identity follows by writing

$$-\frac{2}{\beta}\sum_{n\in\mathbb{Z}}(i\omega_n - E)^{-1}(i\omega_n + E')^{-1} = -\frac{2}{\beta}\sum_{n\in\mathbb{Z}}\frac{1}{E + E'}\left(\frac{1}{i\omega_n - E} - \frac{1}{i\omega_n + E'}\right)$$

(12)

and using the partial fraction expansion of tanh to recognize the right side as $\Xi_\beta(E, E')$.

Thus, the linearized Bogoliubov–de Gennes equation becomes

$$\alpha = L_{T,W}V\alpha.$$

There are two ways to make the operator appearing in this equation self-adjoint. The first one is to apply the operator $L_{T,W}^{-1}$ to both sides and to subtract $V\alpha$. In this way, we obtain the operator (1). The other way is to multiply both sides of the equation by $V^{1/2}$, to subtract $V^{1/2}L_{T,W}V\alpha$ and to call $\Phi = V^{1/2}\alpha$. In this way, we arrive at the operator $1 - V^{1/2}L_{T,W}V^{1/2}$ which appears in our main result, Theorem 4.

The upshot of this discussion is that positivity of the operator (1) (or, equivalently, of $1 - V^{1/2}L_{T,W}V^{1/2}$) corresponds to local stability of the normal state and the existence of negative spectrum of (1) corresponds to local instability. If we define two critical local temperatures $\overline{T_c^{loc}}(h)$ as the smallest temperature above which the normal state is always stable and $\underline{T_c^{loc}}(h)$ as the largest temperature below which the normal state is never stable, then our theorems say that both $\overline{T_c^{loc}}(h)$ and $\underline{T_c^{loc}}(h)$ are equal to $T_c(1 - D_ch^2) + O(h)$ as $h \to 0$.

2 A Representation Formula for the Operator $L_{T,W}$

In this section, we derive a useful representation formula for the operator $L_{T,W}$ as a sum over contributions from the individual Matsubara frequencies ω_n from (10). Moreover, we express the formula in terms of center-of-mass and relative coordinates,

$$r = x - y, \qquad X = (x + y)/2.$$

We recall that the corresponding momenta are denoted by $p_r = -i\nabla_r$ and $p_X = -i\nabla_X$.

Our starting point is (11), which can be written in the form

$$\left(L_{T,W}\Delta\right)(x, y) = -\frac{2}{\beta}\sum_{n\in\mathbb{Z}}\left(\frac{1}{i\omega_n - \mathfrak{h}_W}\Delta\frac{1}{i\omega_n + \mathfrak{h}_W}\right)(x, y).$$

(13)

(Here, we used the fact that $\mathfrak{h}_W = \overline{\mathfrak{h}_W}$.) This formula means that as an operator on $L^2(\mathbb{R}^3 \times \mathbb{R}^3)$, we have

$$L_{T,W} = -\frac{2}{\beta} \sum_{n \in \mathbb{Z}} \frac{1}{i\omega_n - \mathfrak{h}_{W,x}} \frac{1}{i\omega_n + \mathfrak{h}_{W,y}}.$$

The strategy now will be to expand the operators $1/(i\omega_n \mp \mathfrak{h}_W)$ with respect to W. Clearly the leading term is

$$L_{T,0} = -\frac{2}{\beta} \sum_{n \in \mathbb{Z}} \frac{1}{i\omega_n - \mathfrak{h}_{0,x}} \frac{1}{i\omega_n + \mathfrak{h}_{0,y}}$$

and the subleading correction is h^2 times

$$N_{T,W} := -\frac{2}{\beta} \sum_{n \in \mathbb{Z}} \left(-\frac{1}{i\omega_n - \mathfrak{h}_{0,x}} \frac{1}{i\omega_n + \mathfrak{h}_{0,y}} W(hy) \frac{1}{i\omega_n + \mathfrak{h}_{0,y}} \right.$$
$$\left. + \frac{1}{i\omega_n + \mathfrak{h}_{0,x}} W(hx) \frac{1}{i\omega_n + \mathfrak{h}_{0,x}} \frac{1}{i\omega_n - \mathfrak{h}_{0,y}} \right).$$

The following lemma justifies this formal expansion.

Lemma 11. *As an operator on $L^2(\mathbb{R}^3 \times \mathbb{R}^3)$, we have*

$$\|L_{T,W} - L_{T,0}\| \lesssim \beta^3 h^2$$

and

$$\|L_{T,W} - L_{T,0} - h^2 N_{T,W}\| \lesssim \beta^5 h^4$$

Proof. Using the resolvent identity, we write

$$\frac{1}{i\omega_n - \mathfrak{h}_{W,x}} \frac{1}{i\omega_n + \mathfrak{h}_{W,y}} = \frac{1}{i\omega_n - \mathfrak{h}_{0,x}} \frac{1}{i\omega_n + \mathfrak{h}_{0,y}}$$
$$- \frac{1}{i\omega_n - \mathfrak{h}_{0,x}} \frac{1}{i\omega_n + \mathfrak{h}_{W,y}} h^2 W(hy) \frac{1}{i\omega_n + \mathfrak{h}_{0,y}}$$
$$+ \frac{1}{i\omega_n + \mathfrak{h}_{W,x}} h^2 W(hx) \frac{1}{i\omega_n + \mathfrak{h}_{0,x}} \frac{1}{i\omega_n - \mathfrak{h}_{W,y}}.$$

The first term on the right side, when summed with respect to n, corresponds to the operator $L_{T,0}$. In the remaining terms, we use $W \in L^\infty(\mathbb{R}^3)$ and bound each resolvent in norm by $|\omega_n|^{-1}$. The resulting bound is summable with respect to n. This proves the first bound. For the proof of the second bound we expand the resolvents once more. □

In the remainder of this section, we will do two things, namely, bring the operator $L_{T,0}$ in a more explicit form and extract the leading term from the operator $N_{T,W}$. While in Lemma 11, we considered $L_{T,W}$ as an operator on $L^2(\mathbb{R}^3 \times \mathbb{R}^3)$, we will from now on restrict it to the subspace $L^2_{\text{symm}}(\mathbb{R}^3 \times \mathbb{R}^3)$.

In order to investigate the operator $L_{T,0}$, we denote by g^z the integral kernel of $1/(z - \mathfrak{h}_0)$, that is,

$$\frac{1}{z - \mathfrak{h}_0}(x, x') = g^z(x - x').$$

Using center-of-mass and relative coordinates we can rewrite (13) as

$$
\begin{aligned}
(L_{T,0}\Delta)\,(X + \frac{r}{2}, X - \frac{r}{2}) &= -\frac{2}{\beta}\sum_{n\in\mathbb{Z}}\iint_{\mathbb{R}^3\times\mathbb{R}^3} dY ds\,\Delta(Y + \frac{s}{2}, Y - \frac{s}{2}) \\
&\quad\times g^{i\omega_n}(X - Y + \frac{r-s}{2})g^{-i\omega_n}(X - Y - \frac{r-s}{2}) \\
&= \iint_{\mathbb{R}^3\times\mathbb{R}^3} dZ ds\, k_T(Z, r - s)\Delta(X - Z + \frac{s}{2}, X - Z - \frac{s}{2})
\end{aligned}
$$

with

$$k_T(Z, \rho) := -\frac{2}{\beta}\sum_{n\in\mathbb{Z}} g^{i\omega_n}(Z + \frac{\rho}{2})g^{-i\omega_n}(Z - \frac{\rho}{2}).$$

Next, we use the fact that $\psi(X - Z) = (e^{-iZ\cdot p_X}\psi)(X)$ to write

$$(L_{T,0}\Delta)\,(X + \frac{r}{2}, X - \frac{r}{2}) = \iint_{\mathbb{R}^3\times\mathbb{R}^3} dZ ds\, k_T(Z, r - s)\left(e^{-iZ\cdot p_X}\Delta\right)(X + \frac{s}{2}, X - \frac{s}{2}). \quad (14)$$

We claim that in this formula we can replace $e^{-iZ\cdot p_X}$ by $\cos(Z \cdot p_X)$. To do so, we change variables $Z \mapsto -Z, r \mapsto -r$, and $s \mapsto -s$ and use $\Delta(x, y) = \Delta(y, x)$ and $k_T(-Z, -r + s) = k_T(Z, r - s)$ in order to obtain the same formula as in (14), but with $e^{-iZ\cdot p_X}$ replaced by $e^{+iZ\cdot p_X}$. Adding the two formulas we finally find

$$(L_{T,0}\Delta)\,(X + \frac{r}{2}, X - \frac{r}{2}) = \iint_{\mathbb{R}^3\times\mathbb{R}^3} dZ ds\, k_T(Z, r - s)\,(\cos(Z \cdot p_X)\Delta)\,(X + \frac{s}{2}, X - \frac{s}{2}). \quad (15)$$

Next, we derive a convenient representation of $k_T(Z, \rho)$. Setting $\ell = p + q$ and $k = (p - q)/2$ and recalling (11) and (12), we calculate

$$
\begin{aligned}
k_T(Z, \rho) &= -\frac{2}{\beta}\sum_{n\in\mathbb{Z}}\iint_{\mathbb{R}^3\times\mathbb{R}^3} \frac{dp}{(2\pi)^3}\frac{dq}{(2\pi)^3}\frac{e^{ip\cdot(Z+\frac{\rho}{2})}}{i\omega_n - p^2 + \mu}\frac{e^{iq\cdot(Z-\frac{\rho}{2})}}{i\omega_n + q^2 - \mu} \\
&= \iint_{\mathbb{R}^3\times\mathbb{R}^3} \frac{dp}{(2\pi)^3}\frac{dq}{(2\pi)^3} L(p, q)e^{ip\cdot(Z+\frac{\rho}{2})+iq\cdot(Z-\frac{\rho}{2})} \\
&= \iint_{\mathbb{R}^3\times\mathbb{R}^3} \frac{d\ell}{(2\pi)^3}\frac{dk}{(2\pi)^3} L(k + \frac{\ell}{2}, k - \frac{\ell}{2})e^{i\ell\cdot Z+ik\cdot\rho} \quad (16)
\end{aligned}
$$

with

$$L(p, q) := \frac{\tanh\frac{\beta(p^2-\mu)}{2} + \tanh\frac{\beta(q^2-\mu)}{2}}{p^2 - \mu + q^2 - \mu}. \quad (17)$$

Let us explain the intuition for the following. Since the external field is varying on the scale $1/h$, which is much larger than the typical distance between the particles, each momentum p_X will pick up an additional factor of h. Therefore, we expect the leading term in (15) to be given by the corresponding operator with $\cos(Z \cdot p_X)$ replaced by 1. We will justify this approximation in the following lemma. The next order, namely $-(1/2)(Z \cdot p_X)^2$, which will ultimately give rise to the Laplacian in Ginzburg–Landau theory, will be discussed in the following section.

In order to compute the right side of (15) with $\cos(Z \cdot p_X)$ replaced by 1, we first compute, using (16),

$$\int_{\mathbb{R}^3} dZ\, k_T(Z, \rho) = \int_{\mathbb{R}^3} \frac{dk}{(2\pi)^3} L(k, k) e^{ik \cdot \rho} . \tag{18}$$

This implies that

$$\iint_{\mathbb{R}^3 \times \mathbb{R}^3} dZ\, ds\, k_T(Z, r - s) \Delta(X + \frac{s}{2}, X - \frac{s}{2}) = \left(\chi_\beta(p_r^2 - \mu) \Delta \right)(X + r/2, X - r/2),$$

that is,

$$L_{T,0} = \chi_\beta(p_r^2 - \mu) - \int_{\mathbb{R}^3} dZ\, k_T(Z)\,(1 - \cos(Z \cdot p_X)) , \tag{19}$$

where $k_T(Z)$ denotes the operator in $L^2_{\text{symm}}(\mathbb{R}^3)$ with integral kernel $k_T(Z, r - s)$.

We now quantify the replacement of $\cos(Z \cdot p_X)$ by 1.

Lemma 12.

$$\left\| \left(L_{T,0} - \chi_\beta(p_r^2 - \mu) \right) \Delta \right\| \lesssim \beta^3 \left\| p_X^2 \Delta \right\|$$

Proof. We have to bound the integral on the right side of (19). For this we consider a single term in the definition of $k_T(Z, \rho)$. For fixed $r \in \mathbb{R}^3$, we estimate using Minkowski's inequality

$$\left(\int_{\mathbb{R}^3} dX \left| \int_{\mathbb{R}^6} dZ\, ds\, g^{i\omega_n}(Z + (r - s)/2) g^{-i\omega_n}(Z - (r - s)/2) \right. \right.$$

$$\left. \left. \times \left((1 - \cos(Z \cdot p_X)) \Delta \right)(X + s/2, X - s/2) \right|^2 \right)^{1/2}$$

$$\leq \int_{\mathbb{R}^6} dZ\, ds\, \left| g^{i\omega_n}(Z + (r - s)/2) g^{-i\omega_n}(Z - (r - s)/2) \right|$$

$$\times \left(\int_{\mathbb{R}^3} dX \left| ((1 - \cos(Z \cdot p_X)) \Delta)(X + s/2, X - s/2) \right|^2 \right)^{1/2} .$$

Now, we bound for fixed $Z, s \in \mathbb{R}^3$

$$\left(\int_{\mathbb{R}^3} dX \left|((1 - \cos(Z \cdot p_X))\Delta)(X + s/2, X - s/2)|^2\right)^{1/2}\right.$$

$$\leq \left\|\frac{1 - \cos(Z \cdot p_X)}{(Z \cdot p_X)^2}\right\| \left(\int_{\mathbb{R}^3} dX \left|((Z \cdot p_X)^2 \Delta)(X + s/2, X - s/2)|^2\right)^{1/2}\right.$$

$$\lesssim |Z|^2 t(s)$$

where

$$t(s) := \left(\int_{\mathbb{R}^3} dX \left|(p_X^2 \Delta)(X + s/2, X - s/2)|^2\right)^{1/2}.\right.$$

Thus, the quantity we are interested in is bounded by a constant times

$$\int_{\mathbb{R}^6} dZ\, ds \left|g^{i\omega_n}(Z + (r - s)/2)g^{-i\omega_n}(Z - (r - s)/2)\right| |Z|^2 t(s).$$

Using

$$|Z|^2 \leq \frac{1}{2}\left(\left|Z + \frac{r - s}{2}\right|^2 + \left|Z - \frac{r - s}{2}\right|^2\right)$$

we can bound the above quantity by

$$\frac{1}{2}\left(\left((|\cdot|^2 g^{i\omega_n}) * g^{-i\omega_n} * t\right)(r) + \left(g^{i\omega_n} * (|\cdot|^2 g^{-i\omega_n}) * t\right)(r)\right).$$

The L^2 norm of this term with respect to r is bounded according to Young's convolution inequality by

$$\frac{1}{2}\left(\left\||\cdot|^2 g^{i\omega_n}\right\|_1 \left\|g^{-i\omega_n}\right\|_1 + \left\|g^{i\omega_n}\right\|_1 \left\||\cdot|^2 g^{-i\omega_n}\right\|_1\right) \|t\|_2.$$

By [2, Lemma 9], this expression is summable with respect to n, and therefore the left side in the lemma is bounded by a constant times $\|t\|_2 = \|p_X^2 \Delta\|$, as claimed. $\qquad \square$

This concludes our discussion of the leading term $L_{T,0}$. We now aim at extracting the leading term from the operator $N_{T,W}$ and we concentrate on a term of the form

$$\iint_{\mathbb{R}^3 \times \mathbb{R}^3} dx'dy' \left(\frac{1}{i\omega_n - \mathfrak{h}_0} W(h \cdot)\frac{1}{i\omega_n - \mathfrak{h}_0}\right)(x, x')\frac{1}{-i\omega_n - \mathfrak{h}_0}(y, y')\Delta(x', y').$$

We introduce again center-of-mass and relative coordinates $X = (x + y)/2$, $r = x - y$, $Y = (x' + y')/2$, and $s = x' - y'$. In order to obtain concise expressions, we introduce the abbreviation

$$\zeta_X^r = X + r/2, \quad \zeta_Y^{-s} = Y - s/2,$$

where the second term should just show the consistency of the symbol. With these definitions, we obtain

$$
\int_{\mathbb{R}^6} dx'dy' \left(\frac{1}{i\omega_n - \mathfrak{h}_0} W(h\cdot) \frac{1}{i\omega_n - \mathfrak{h}_0} \right)(x, x') \frac{1}{-i\omega_n - \mathfrak{h}_0}(y, y')\Delta(x', y')
$$
$$
= \int_{\mathbb{R}^9} dY ds dz' \, g^{i\omega_n}(\zeta_X^r - z')W(hz')g^{i\omega_n}(z' - \zeta_Y^s)g^{-i\omega_n}(\zeta_{X-Y}^{s-r})\Delta(\zeta_Y^s, \zeta_Y^{-s})
$$
$$
= \int_{\mathbb{R}^9} dY ds dz \, g^{i\omega_n}(\tfrac{r}{2} - z)W(hX + hz)g^{i\omega_n}(z + \zeta_{X-Y}^{-s})g^{-i\omega_n}(\zeta_{X-Y}^{s-r})\Delta(\zeta_Y^s, \zeta_Y^{-s})
$$
$$
= \int_{\mathbb{R}^9} dZ ds dz \, g^{i\omega_n}(\tfrac{r}{2} - z)W(hX + hz)g^{i\omega_n}(z + \zeta_Z^{-s})g^{-i\omega_n}(\zeta_Z^{s-r})\Delta(\zeta_{X-Z}^s, \zeta_{X-Z}^{-s})
$$
$$
= \int_{\mathbb{R}^9} dZ ds dz \, g^{i\omega_n}(\tfrac{r}{2} - z)W(hX + hz)g^{i\omega_n}(z + \zeta_Z^{-s})g^{-i\omega_n}(\zeta_Z^{s-r}) \left(e^{-iZ\cdot p_X}\Delta \right)(\zeta_X^s, \zeta_X^{-s}),
$$
$$(20)$$

where in the last step we used again

$$
\alpha(\zeta_{X-Z}^s, \zeta_{X-Z}^{-s}) = \alpha(X - Z + s/2, X - Z - s/2)
$$
$$
= \left(e^{-iZ\cdot p_X}\alpha \right)(X + s/2, X - s/2)
$$
$$
= \left(e^{-iZ\cdot p_X}\alpha \right)(\zeta_X^s, \zeta_X^{-s}).
$$

We claim that to leading order, we can replace $W(hX + hz)$ in this integral by $W(hX)$. Therefore, we define

$$
\left(\tilde{N}_{T,W}\Delta \right)(\zeta_X^r, \zeta_X^{-r}) := W(hX) \iint_{\mathbb{R}^3 \times \mathbb{R}^3} dZ ds \, \ell_T(Z, r - s) \left(e^{-iZ\cdot p_X}\Delta \right)(\zeta_X^s, \zeta_X^{-s})
$$
$$(21)$$

and

$$
\ell_T(Z, \rho) := \frac{2}{\beta} \sum_{n \in \mathbb{Z}} \left(\left(g^{i\omega_n} * g^{i\omega_n} \right)(\zeta_Z^\rho)g^{-i\omega_n}(\zeta_Z^{-\rho}) + g^{i\omega_n}(\zeta_Z^\rho)\left(g^{-i\omega_n} * g^{-i\omega_n} \right)(\zeta_Z^{-\rho}) \right). \quad (22)
$$

Lemma 13.
$$
\left\| \left(N_{T,W} - \tilde{N}_{T,W} \right)\Delta \right\| \lesssim h \left(\|\Delta\| + \||r|\Delta\| \right).
$$

Proof. In (20), we write

$$
W(hX + hz) = W(hX) + h \int_0^1 z \cdot \nabla W(hX + thz)\, dt
$$

and then we have to estimate the norm of the error term coming from the t-integral. In order to calculate the $L^2(\mathbb{R}^3 \times \mathbb{R}^3)$-norm of the corresponding expression in the (X, r)-variables, we first fix $r \in \mathbb{R}^3$ and consider the following term, which has a prefactor of h in front,

$$\left(\int_{\mathbb{R}^3} dX \left| \int_{\mathbb{R}^9} dZ ds dz \, g^{i\omega_n}(\tfrac{r}{2} - z) \int_0^1 z \cdot \nabla W(hX + thz) dt \times \right. \right.$$
$$\left. \left. \times g^{i\omega_n}(z + \zeta_Z^{-s}) g^{-i\omega_n}(\zeta_Z^{s-r}) \left(e^{-iZ \cdot p_X} \Delta \right) (\zeta_X^s, \zeta_X^{-s}) \right|^2 \right)^{1/2}. \tag{23}$$

Using Minkowski's inequality, we can bound this by

$$\int_{\mathbb{R}^9} dZ ds dz \, |g^{i\omega_n}(\tfrac{r}{2} - z)| |g^{i\omega_n}(z + \zeta_Z^{-s})| |g^{-i\omega_n}(\zeta_Z^{s-r})|$$
$$\times \left(\int_{\mathbb{R}^3} dX \left| \int_0^1 z \cdot \nabla W(hX + thz) dt \left(e^{-iZ \cdot p_X} \Delta \right) (\zeta_X^s, \zeta_X^{-s}) \right|^2 \right)^{1/2}$$
$$\leq \int_{\mathbb{R}^9} dZ ds dz \, |g^{i\omega_n}(\tfrac{r}{2} - z)| |g^{i\omega_n}(z + \zeta_Z^{-s})| |g^{-i\omega_n}(\zeta_Z^{s-r})|$$
$$\times |z| \|\nabla W\|_\infty \left(\int_{\mathbb{R}^3} dX \left| \left(e^{-iZ \cdot p_X} \Delta \right) (X + s/2, X - s/2) \right|^2 \right)^{1/2}$$
$$= \int_{\mathbb{R}^9} dZ ds dz \, |g^{i\omega_n}(\tfrac{r}{2} - z)| |g^{i\omega_n}(z + \zeta_Z^{-s})| |g^{-i\omega_n}(\zeta_Z^{s-r})| |z| \|\nabla W\|_\infty m(s)$$

where

$$m(s) := \left(\int_{\mathbb{R}^3} dX \, |\Delta(X + s/2, X - s/2)|^2 \right)^{1/2}$$

and where we used the unitarity of $e^{-iZ \cdot p_X}$ in the last equality.

The inequality

$$|z| \leq \frac{1}{2}|z - r/2| + \frac{1}{2}|z + Z - s/2| + \frac{1}{2}|Z - (r - s)/2| + \frac{1}{2}|s|$$

leads to four terms, which we bound separately. The term with $|z - r/2|$ can be bounded by

$$\|\nabla W\|_\infty \int_{\mathbb{R}^9} dZ ds dz \, |g^{i\omega_n}(r/2 - z)| |r/2 - z| |g^{i\omega_n}(z + Z - s/2)| |g^{-i\omega_n}(Z + (s - r)/2)| m(s)$$
$$= \|\nabla W\|_\infty \left(\left| |\cdot| g^{i\omega_n} \right| * |g^{i\omega_n}| * |g^{-i\omega_n}| * m \right)(r).$$

According to Young's inequality, the L^2 norm of this term is bounded by $\|\nabla W\|_\infty$ times

$$\| |\cdot| g^{i\omega_n}\|_1 \|g^{i\omega_n}\|_1 \|g^{-i\omega_n}\|_1 \|m\|_2 = \| |\cdot| g^{i\omega_n}\|_1 \|g^{i\omega_n}\|_1 \|g^{-i\omega_n}\|_1 \|\Delta\|_2.$$

According to [2, Lemma 9], this expression is summable with respect to n, and therefore the contribution of this term to $\left(N_{T,W} - \tilde{N}_{T,W}\right)\Delta$ is bounded by a constant times $h\|\Delta\|_2$.

The argument for the terms with $|z + Z - s/2|$ and $|Z - (r-s)/2|$ is similar. The term with $|s|$ can be bounded by

$$\|\nabla W\|_\infty \int_{\mathbb{R}^9} dZ ds dz \, |g^{i\omega_n}(r/2 - z)||s||g^{i\omega_n}(z + Z - s/2)||g^{-i\omega_n}(Z + (s - r)/2)|m(s)$$

$$= \|\nabla W\|_\infty \left(|g^{i\omega_n}| * |g^{i\omega_n}| * |g^{-i\omega_n}| * (|\cdot|m)\right)(r),$$

According to Young's inequality, the L^2 norm of this term is bounded by $\|\nabla W\|_\infty$ times

$$\|g^{i\omega_n}\|_1 \|g^{i\omega_n}\|_1 \|g^{-i\omega_n}\|_1 \||\cdot| m\|_2 = \|g^{i\omega_n}\|_1 \|g^{i\omega_n}\|_1 \|g^{-i\omega_n}\|_1 \||\cdot| \Delta\|_2 .$$

Again by [2, Lemma 9], this expression is summable with respect to n, and therefore the contribution of this term to $\left(N_{T,W} - \tilde{N}_{T,W}\right)\Delta$ is bounded by a constant times $h \||\cdot| \Delta\|_2$. This proves the lemma.   □

3   Representation of $L_{T,W}$ on the States $\Delta = \Psi(X)\tau(r)$

We will argue below that we are able to restrict to a specific class of states, which are of the form $\Delta(X + r/2, X - r/2) = \psi(X)\tau(r)$. Due to the symmetry of Δ, τ has to be an even function, but in fact we will later see that τ can be assumed as radial, and for the proof of our main theorem τ will be proportional to $V^{1/2}(r)\varphi_*(r)$, where $\varphi_*(r)$ is the zero eigenstate of $1 - V^{1/2}\chi_{\beta_c}(p_r^2 - \mu)V^{1/2}$.

The following corollary is an immediate consequence of the bounds in the previous section.

Corollary 14. *If $\Delta(X + r/2, X - r/2) = \psi(X)\tau(r)$ with τ even, then*

$$\langle \Delta, L_{T,W}\Delta \rangle = \langle \psi, \psi \rangle \langle \tau, \chi_\beta(p_r^2 - \mu)\tau \rangle$$

$$- \int_{\mathbb{R}^3} dZ \, \langle \psi, (1 - \cos(Z \cdot p_X))\psi \rangle \iint_{\mathbb{R}^3 \times \mathbb{R}^3} drds \, \overline{\tau(r)} k_T(Z, r - s)\tau(s)$$

$$+ h^2 \int_{\mathbb{R}^3} dZ \, \langle \psi, W(hX)e^{-iZ \cdot p_X}\psi \rangle \iint_{\mathbb{R}^3 \times \mathbb{R}^3} drds \, \overline{\tau(r)} \ell_T(Z, r - s)\tau(s)$$

$$+ O(h^3)\|\psi\|^2\|\tau\|\| \cdot |\tau\|. \tag{24}$$

We remark that with slightly more work, we could replace the error term $\|\tau\|\|| \cdot |\tau\|$ by $\||\cdot|^{1/2}\tau\|^2$.

The second term on the right side of (24) is given to leading order by the same expression with $1 - \cos(Z \cdot p_X)$ replaced by $(Z \cdot p_X)^2/2$. Under the assumption that

τ is a radial function, we therefore obtain $\langle \psi, p_X^2 \psi \rangle$ times a constant depending on τ.

The third term on the right side of (24) is given to leading order by the same expression with $e^{-iZ \cdot px}$ replaced by 1. We therefore obtain $h^2 \langle \psi, W(h \cdot) \psi \rangle$ times a constant depending on τ.

This tells us that the center-of-mass fluctuations are governed by a one-body operator of the form $c_1 p_X^2 + c_2 h^2 W(hX)$, which is unitarily equivalent to the operator

$$h^2 \left(c_1 p_X^2 + c_2 W(X) \right).$$

The precise value of the constants c_1, c_2 depends on the specific choice of τ.

As we will show below, the errors made in these two approximations can be controlled by $\| p_X^2 \psi \|^2$ and $h^2 \| p_X \psi \| \| \psi \|$. In order to get an intuition why the error terms are indeed of higher order in h we recall the heuristic picture of our chosen scaling. The external field W varies on the scale $1/h$. Therefore, we expect the optimal function ψ to match this behavior and vary as well on the macroscopic scale. More precisely, we expect that ψ will be of the form $\psi(X) = h^{3/2} \tilde{\psi}(hX)$ with a function $\tilde{\psi}$ which is bounded in H^2 uniformly for small h. Therefore the error bounds $\| p_X^2 \psi \|^2$ and $h^2 \| p_X \psi \| \| \psi \|$ are $o(h^2)$.

Next, we formulate this intuitive picture as a precise mathematical statement.

Theorem 15. *There is a constant C such that for Δ of the form*

$$\Delta(X + r/2, X - r/2) = \psi(X) \tau(r)$$

with τ radial, one has

$$\left| \langle \Delta, L_{T,W} \Delta \rangle - A_T^{(0)}[\tau] \| \psi \|^2 - A_T^{(1)}[\tau] \langle \psi, p_X^2 \psi \rangle - h^2 A_T^{(2)}[\tau] \langle \psi, W(h \cdot) \psi \rangle \right|$$
$$\leq C \left(\| \tau \|^2 \| p_X^2 \psi \|^2 + h^2 \| \tau \|^2 \| p_X \psi \| \| \psi \| + h^3 \| \psi \|^2 \| | \cdot | \tau \| \| \tau \| \right) \quad (25)$$

with

$$A_T^{(0)}[\tau] = \beta \int_{\mathbb{R}^3} dp \, |\hat{\tau}(p)|^2 \, g_0(\beta(p^2 - \mu)),$$

$$A_T^{(1)}[\tau] = -\frac{\beta^2}{4} \int_{\mathbb{R}^3} dp \, |\hat{\tau}(p)|^2 \left(g_1(\beta(p^2 - \mu)) + \frac{2}{3} \beta p^2 g_2(\beta(p^2 - \mu)) \right),$$

$$A_T^{(2)}[\tau] = \frac{\beta^2}{4} \int_{\mathbb{R}^3} dp \, |\hat{\tau}(p)|^2 \, g_1(\beta(p^2 - \mu))$$

in terms of the functions g_0, g_1, and g_2 from (4).

Proof. This theorem is essentially a consequence of (24). We first notice that

$$\langle \tau, \chi_\beta(p_r^2 - \mu)\tau \rangle = A_T^{(0)}[\tau].$$

Moreover, using arguments as in the previous subsection one can verify that

$$\left| \int_{\mathbb{R}^3} dZ \, \langle \psi, (1 - \cos(Z \cdot p_X) - (Z \cdot p_X)^2/2)\psi \rangle F_\tau(Z) \right| \lesssim \|\tau\|^2 \|p_X^2 \psi\|^2$$

where we have introduced

$$F_\tau(Z) := \iint_{\mathbb{R}^3 \times \mathbb{R}^3} dr \, ds \, \overline{\tau(r)} k_T(Z, r - s)\tau(s).$$

Since τ is radial, so is F_τ, and therefore

$$\frac{1}{2} \int_{\mathbb{R}^3} dZ \, \langle \psi, (Z \cdot p_X)^2 \psi \rangle F_\tau(Z) = \frac{1}{6} \int_{\mathbb{R}^3} dZ \, Z^2 F_\tau(Z) \langle \psi, p_X^2 \psi \rangle.$$

Now using (16),

$$\int_{\mathbb{R}^3} dZ \, Z^2 F_\tau(Z) = - \int_{\mathbb{R}^3} dk \, \nabla_\ell^2|_{\ell=0} L(k + \frac{\ell}{2}, k - \frac{\ell}{2}) |\hat{\tau}(k)|^2,$$

and a tedious, but straightforward computation yields

$$\nabla_\ell^2|_{\ell=0} L(k + \frac{\ell}{2}, k - \frac{\ell}{2}) = -\frac{3\beta^2}{2} \left(g_1(\beta(k^2 - \mu)) + \frac{2}{3}\beta k^2 g_2(\beta(k^2 - \mu)) \right),$$

which shows that

$$-\frac{1}{2} \int_{\mathbb{R}^3} dZ \, \langle \psi, (Z \cdot p_X)^2 \psi \rangle F_\tau(Z) = A_T^{(1)}[\tau] \langle \psi, p_X^2 \psi \rangle.$$

Finally, by estimating $1 - e^{iZ \cdot p_X}$, we obtain

$$\left| \int_{\mathbb{R}^3} dZ \, \langle \psi, W(hX) \left(e^{-iZ \cdot p_X} - 1 \right) \psi \rangle G_\tau(Z) \right| \lesssim \|\tau\|^2 \|p_X \psi\| \|\psi\|$$

with

$$G_\tau(Z) := \iint_{\mathbb{R}^3 \times \mathbb{R}^3} dr ds \, \overline{\tau(r)} \ell_T(Z, r - s)\tau(s).$$

Rewriting (22) in Fourier space and summing over the Matsubara frequencies gives

$$\int_{\mathbb{R}^3} dZ\, \ell_T(Z, \rho) = \int_{\mathbb{R}^3} \frac{dk}{(2\pi)^3} \frac{\beta^2}{4} g_1(\beta(k^2 - \mu)) e^{ik\cdot\rho}$$

and therefore

$$\int_{\mathbb{R}^3} dZ\, G_\tau(Z) = A_T^{(2)}[\tau].$$

This concludes the proof of the theorem.   □

4   Lower Bound on the Critical Temperature

We now provide the *Proof of part (1) of Theorem 4*, which will be a rather straight-forward consequence of Theorem 15. We will work under Assumptions 1 and 2. Assumption 3 is not needed in this part of Theorem 4.

We fix a parameter T_1 with $0 < T_1 < T_c$ and restrict ourselves to temperatures $T \geq T_1$. We consider functions Φ in $L^2_{\mathrm{symm}}(\mathbb{R}^3 \times \mathbb{R}^3)$ of the form

$$\Phi(x, y) = \varphi(x - y) h^{3/2} \psi(h(x + y)/2),$$

where the functions $\varphi \in L^2_{\mathrm{symm}}(\mathbb{R}^3)$ and $\psi \in L^2(\mathbb{R}^3)$ are still to be determined. At the moment we only require that $\|\psi\| = 1$ and $\|p_X^2 \psi\| < \infty$.

We first assume, in addition, that $T \geq T_c - Mh^2$ for some constant M independent of h. In this case we choose φ radial and then, applying the expansion from Theorem 15 with $\tau(r) = V(r)^{1/2}\varphi(r)$, we find that

$$\begin{aligned}
\langle \Phi, (1 - V^{1/2} L_{T,W} V^{1/2})\Phi \rangle &= \|\varphi\|^2 - \langle \tau(r)\psi(X), L_{T,W}\tau(r)\psi(X) \rangle \\
&\leq \|\varphi\|^2 - A_T^{(0)}[\tau] - h^2 A_T^{(1)}[\tau]\langle \psi, p_X^2 \psi \rangle \\
&\quad - h^2 A_T^{(2)}[\tau]\langle \psi, W\psi \rangle + Ch^3.
\end{aligned} \tag{26}$$

The constant C here depends only on upper bounds on $\|p_X^2 \psi\|$, $\|\tau\|$ and $\||\cdot|\tau\|$ (as well as on M). The leading order term on the right side is

$$\|\varphi\|^2 - A_T^{(0)}[\tau] = \langle \varphi, \left(1 - V^{1/2}\chi_\beta(p_r^2 - \mu)V^{1/2}\right)\varphi \rangle. \tag{27}$$

We choose

$$\varphi := (2\pi)^{-3/2} \|\chi_{\beta_c}(p^2 - \mu)V^{1/2}\varphi_*\|\, \varphi_*,$$

which makes (27) equal to zero at $T = T_c$. With this choice of φ we therefore obtain

$$\begin{aligned}
\langle \Phi, (1 - V^{1/2} L_{T,W} V^{1/2})\Phi \rangle &\leq A_{T_c}^{(0)}[\tau] - A_T^{(0)}[\tau] \\
&\quad - h^2 \left(A_T^{(1)}[\tau]\langle \psi, p_X^2 \psi \rangle + A_T^{(2)}[\tau]\langle \psi, W\psi \rangle\right) + Ch^3.
\end{aligned} \tag{28}$$

In order to proceed, we note the fact that $\tau = V^{1/2}\varphi = (2\pi)^{-3/2}V\alpha_*$, and therefore, in terms of the function t from (5),

$$\hat{\tau} = (1/2)(2\pi)^{-3/2}t. \tag{29}$$

It follows from this identity that

$$\frac{d}{dT}|_{T=T_c}A_T^{(0)}[\tau] = -T_c^{-1}\Lambda_2,$$

and some simple analysis of the function g_0 shows that

$$A_{T_c}^{(0)}[\tau] - A_T^{(0)}[\tau] \leq -\Lambda_2\frac{T_c - T}{T_c} + C(T_c - T)^2$$

for all $T_1 \leq T \leq T_c$. Using (29) once again we also find that

$$A_{T_c}^{(1)}[\tau] = -\Lambda_0 \quad \text{and} \quad A_{T_c}^{(2)}[\tau] = -\Lambda_1,$$

which in turn can be used to prove that

$$A_T^{(1)}[\tau] \geq -\Lambda_0 - C(T_c - T) \quad \text{and} \quad \left|A_T^{(2)}[\tau] + \Lambda_1\right| \leq C(T_c - T)$$

for all $T_1 \leq T \leq T_c$.

Inserting these expansions into (28) we obtain

$$\langle\Phi, (1 - V^{1/2}L_{T,W}V^{1/2})\Phi\rangle \leq -\Lambda_2\frac{T_c - T}{T_c} + h^2\langle\psi, (\Lambda_0 p_X^2 + \Lambda_1 W)\psi\rangle + Ch^3$$

for all $T_1 \leq T \leq T_c$. Note that here we used the assumption $T \geq T_c - Mh^2$, so that the error terms are independent of $T - T_c$.

In order to conclude the proof we assume first, for the sake of simplicity, that inf spec $(\Lambda_0 p_X^2 + \Lambda_1 W(X))$ is an eigenvalue. In this case we simply choose ψ to be a corresponding normalized eigenfunction. With this choice we obtain, recalling the definition of D_c from (9),

$$\langle\Phi, (1 - V^{1/2}L_{T,W}V^{1/2})\Phi\rangle \leq -\Lambda_2\frac{T_c - T}{T_c} + h^2\Lambda_2 D_c + Ch^3.$$

The right side is negative if $T < T_c(1 - D_ch^2 + (C/\Lambda_2)h^3)$, as claimed.

In case inf spec $(\Lambda_0 p_X^2 + \Lambda_1 W(X))$ is not an eigenvalue, we choose a sequence of functions ψ_h with $\|\psi_h\| = 1$,

$$\langle\psi_h, (\Lambda_0 p_X^2 + \Lambda_1 W(X))\psi_h\rangle \leq \Lambda_2(D_c + h) \quad \text{and} \quad \|p_X^2\psi_h\| \leq C$$

for some C independent of h. Such a sequence is obtained by choosing elements in the spectral subspace of $\Lambda_0 p_X^2 + \Lambda_1 W(X)$ corresponding to the intervals $[\Lambda_2 D_c, \Lambda_2 (D_c + h)]$. Since $\Lambda_0 p_X^2 + \Lambda_1 W(X)$ has the same operator domain as p_X^2 we conclude that

$$\|p_X^2 \psi_h\| \lesssim \left\| \left(\Lambda_0 p_X^2 + \Lambda_1 W(X) + C' \right) \psi_h \right\| \leq \Lambda_2 (D_c + h) + C',$$

which proves the last requirement.

We can now repeat the proof with ψ replaced by ψ_h. Since all constants were uniform in ψ as long as $\|\psi\| = 1$ and $\|p_X^2 \psi\| \leq C$, we arrive at the same conclusion as before. This proves the assertion in case $T \geq T_c - Mh^2$ for some fixed M independent of h.

Thus, in order to complete the proof of part (1) in the theorem, we show that there is an $M > 0$ such that if $T < T_c - Mh^2$, then there are φ and ψ such that the Φ defined as above satisfies $\langle \Phi, (1 - V^{1/2} L_{T,W} V^{1/2}) \Phi \rangle < 0$.

We proceed similarly as before, but use Corollary 14 instead of Theorem 15. By similar, but simpler estimates as in the proof of Theorem 15 we obtain

$$\langle \Phi, (1 - V^{1/2} L_{T,W} V^{1/2}) \Phi \rangle \leq \|\varphi\|^2 - A_T^{(0)}[\tau] + Ch^2. \tag{30}$$

The constant C here depends only on upper bounds on $\|p_X \psi\|$, $\|\tau\|$ and $\||\cdot|\tau\|$ (as well as on T_1). Thus the leading term on the right side is again (27).

To bound this term, we denote by λ_T the largest eigenvalue of $V^{1/2} \chi_\beta (p_r^2 - \mu) V^{1/2}$ in $L^2_{\text{symm}}(\mathbb{R}^3)$. By definition of T_c we have $\lambda_{T_c} = 1$. Since $\beta \mapsto \chi_\beta(E)$ is monotone for any E with positive derivative, we infer by analytic perturbation theory that there is a $c > 0$ such that

$$\lambda_T \geq \lambda_{T_c} + c(T_c - T) = 1 + c(T_c - T) \qquad \text{for all } 0 \leq T \leq T_c.$$

Let φ_T be a normalized eigenfunction of $V^{1/2} \chi_\beta (p_r^2 - \mu) V^{1/2}$ corresponding to λ_T. With $\varphi = \varphi_T$ and an arbitrary normalized function ψ with $\|p_X^2 \psi\| < \infty$ we obtain, by inserting (27) into (30) and using the above bound,

$$\langle \Phi, (1 - V^{1/2} L_{T,W} V^{1/2}) \Phi \rangle \leq 1 - \lambda_T + Ch^2 \leq -c(T_c - T) + Ch^2.$$

The right side is negative for $T < T_c - (C/c)h^2$, as claimed. This completes the proof of part (1) of Theorem 4. $\qquad\square$

5 The Approximate Form of Almost Minimizers

In this and the following section, we work under Assumptions 1–3.

5.1 The Decomposition Lemma

The remainder of this paper is devoted to proving an upper bound on the critical temperature. As a preliminary step, we prove in this section a decomposition lemma, which says that, if $|T_c - T| \leq C_1 h^2$ and if Φ satisfies $\langle \Phi, (1 - V^{1/2} L_{T,W} V^{1/2}) \Phi \rangle \leq C_2 h^2$ for some fixed constants C_1 and C_2 independent of h, then Φ has, up to a controllable error, the same form as the trial function that we used in the proof of the lower bound on the critical temperature.

Theorem 16. *For given constants C_1, $C_2 > 0$ there are constants $h_0 > 0$ and $C > 0$ such that the following holds. If $T > 0$ satisfies $|T - T_c| \leq C_1 h^2$, if $\Phi \in L^2_{\text{symm}}(\mathbb{R}^3 \times \mathbb{R}^3)$ satisfies $\|\Phi\| = 1$ and*

$$\langle \Phi, (1 - V^{1/2} L_{T,W} V^{1/2}) \Phi \rangle \leq C_2 h^2 ,$$

and if ε satisfies $\varepsilon \in [h^2, h_0^2]$, then there are $\psi_\leq \in L^2(\mathbb{R}^3)$ and $\sigma \in L^2_{\text{symm}}(\mathbb{R}^3 \times \mathbb{R}^3)$ such that

$$\Phi(X + r/2, X - r/2) = \varphi_*(r) \psi_\leq(X) + \sigma ,$$

where

$$\|(p_X^2)^{k/2} \psi_\leq\|^2 \leq C \varepsilon^{k-1} h^2 \quad \text{if } k \geq 1, \tag{31}$$

$$\|\sigma\|^2 \leq C \varepsilon^{-1} h^2 \tag{32}$$

and

$$1 \geq \|\psi_\leq\|^2 \geq 1 - C \varepsilon^{-1} h^2 . \tag{33}$$

Moreover, $\psi_\leq \in \operatorname{ran} \mathbb{1}(p_X^2 \leq \varepsilon)$ and there is a $\psi_> \in L^2(\mathbb{R}^3) \cap \operatorname{ran} \mathbb{1}(p_X^2 > \varepsilon)$ such that

$$\sigma_0(X + r/2, X - r/2) := \frac{\varphi_*(r) \cos(p_X \cdot r/2)}{\sqrt{\int_{\mathbb{R}^3} |\varphi_*(r')|^2 \cos^2(p_X \cdot r'/2) \, dr'}} \psi_>(X)$$

satisfies

$$\|\sigma - \sigma_0\|^2 \leq C h^2 . \tag{34}$$

Thus, Φ is of the form $\psi_\leq(X) \varphi_*(r)$ up to a small error. The parameter ε provides a momentum cutoff similarly as in [4, 5] and ensures that we have control on the expectation of $(p_X^2)^2$ in $\psi_\leq$.

5.2   Upper Bound on $L_{T,W}$

Our goal in this subsection is to obtain an operator lower bound on $1 - V^{1/2} L_{T,W} V^{1/2}$. In [4, 5] such a bound was proved by means of a relative entropy inequality [4, Lemma 3], which controlled a two-particle operator by the sum of two one-particle operators, and by [4, Lemma 5] which showed that the energy of the system is dominated by the kinetic energy of the center-of-mass motion. This was sufficient to recover the corresponding a priori estimates. In [2] this operator bound was performed in the presence of a constant magnetic field. Following the spirit of [4, 5], we had to come up with new ideas in order to overcome the problems of noncommutativity of the components of the magnetic momentum operator. In the present much simpler situation, we can choose a mixture of the two methods [2, 4, 5].

We define the unitary operator

$$U := e^{-i p_X \cdot r/2} \tag{35}$$

in $L^2(\mathbb{R}^3 \times \mathbb{R}^3)$, where, as usual, $r = x - y$ and $X = (x + y)/2$.

Proposition 17. *There is a constant $C > 0$ such that for all $T > 0$,*

$$V^{1/2} L_{T,W} V^{1/2} \leq \frac{1}{2} \left(U V^{1/2} \chi_\beta(p_r^2 - \mu) V^{1/2} U^* + U^* V^{1/2} \chi_\beta(p_r^2 - \mu) V^{1/2} U \right)$$
$$+ C \beta^3 h^2.$$

Proof. Since for any real numbers E and E' one has

$$\Xi_\beta(E, E') \leq \frac{1}{2} \left(\frac{\tanh \frac{\beta E}{2}}{E} + \frac{\tanh \frac{\beta E'}{2}}{E'} \right) = \frac{1}{2} \left(\chi_\beta(E) + \chi_\beta(E') \right),$$

we have

$$L_{T,0} = \Xi_\beta(\mathfrak{h}_{0,x}, \mathfrak{h}_{0,y}) \leq \frac{1}{2} \left(\chi_\beta(\mathfrak{h}_{0,x}) + \chi_\beta(\mathfrak{h}_{0,y}) \right).$$

In the variables $r = x - y$, $X = (x + y)/2$, we have $p_x = p_r + p_X/2$ and $p_y = p_r - p_X/2$, and therefore

$$\mathfrak{h}_{0,x} = (p_r + p_X/2)^2 - \mu = U \left(p_r^2 - \mu \right) U^*, \quad \mathfrak{h}_{0,y} = (p_r - p_X/2)^2 - \mu = U^* \left(p_r^2 - \mu \right) U,$$

so the previous bound can be written as

$$L_{T,0} \leq \frac{1}{2} \left(U \chi_\beta(p_r^2 - \mu) U^* + U^* \chi_\beta(p_r^2 - \mu) U \right).$$

On the other hand, by Lemma 11, we have

$$L_{T,W} \leq L_{T,0} + C\beta^3 h^2 \,.$$

Since V commutes with U we obtain the claimed bound.   □

5.3 A priori Bound on the Critical Temperature and an Operator Inequality

As a first consequence of Proposition 17, we obtain a rough a priori upper bound on the critical temperature.

Corollary 18. *There are constants $h_0 > 0$ and $C > 0$ such that for all $0 < h \leq h_0$ and $T > T_c + Ch^2$ one has*

$$\langle \Phi, (1 - V^{1/2} L_{T,W} V^{1/2}) \Phi \rangle > 0 \,,$$

unless $\Phi = 0$.

Proof. According to Proposition 17 for all $T \geq T_c$,

$$1 - V^{1/2} L_{T,W} V^{1/2} \geq 1 - \frac{1}{2} \left(U V^{1/2} \chi_\beta(p_r^2 - \mu) V^{1/2} U^* + U^* V^{1/2} \chi_\beta(p_r^2 - \mu) V^{1/2} U \right)$$
$$- Ch^2 \,. \tag{36}$$

We next recall that the family of operators $V^{1/2} \chi_\beta(p_r^2 - \mu) V^{1/2}$ is nondecreasing with respect to β and has an eigenvalue 1 at $\beta = \beta_c$. Moreover, since the function $\chi_\beta(E)$ is strictly increasing with respect to β for every $E \in \mathbb{R}$, we learn from analytic perturbation theory that there are $c > 0$ and $T_2 > T_c$ such that for all $T_c \leq T \leq T_2$,

$$V^{1/2} \chi_\beta(p_r^2 - \mu) V^{1/2} \leq 1 - c(T - T_c) \,.$$

Again by monotonicity this implies that for all $T \geq T_c$

$$V^{1/2} \chi_\beta(p_r^2 - \mu) V^{1/2} \leq 1 - c \min\{T - T_c, T_2 - T_c\} \,.$$

Inserting this into the lower bound above, we conclude that

$$1 - V^{1/2} L_{T,W} V^{1/2} \geq c \min\{T - T_c, T_2 - T_c\} - Ch^2 \,.$$

The right side is positive if $T > T_c + (C/c)h^2$ and $h^2 \leq (c/C)(T_2 - T_c)$, which proves the corollary.   □

As a consequence of this corollary and the lower bound on the critical temperature, from now on we may and will restrict ourselves to temperatures T such that $|T - T_c|$ is bounded by a constant times h^2.

Our next goal is to deduce from Proposition 17 a lower bound on the operator $1 - V^{1/2} L_{T,W} V^{1/2}$. We recall that by definition of β_c, the largest eigenvalue of the operator $V^{1/2} \chi_{\beta_c}(p_r^2 - \mu) V^{1/2}$ equals one. Moreover, by Assumption 3, this eigenvalue is simple and φ_* denotes a corresponding real-valued, normalized eigenfunction. We denote by

$$P := |\varphi_*\rangle\langle\varphi_*|$$

the corresponding projection and write $P^\perp = 1 - P$. Since $V^{1/2} \chi_{\beta_c}(p_r^2 - \mu) V^{1/2}$ is a compact operator, there is a $\kappa > 0$ such that

$$V^{1/2} \chi_{\beta_c}(p_r^2 - \mu) V^{1/2} \leq 1 - \kappa P^\perp. \tag{37}$$

Finally, we introduce the operator

$$Q := \frac{1}{2} \left(UPU^* + U^*PU \right). \tag{38}$$

We can now state our operator inequality for $1 - V^{1/2} L_{T,W} V^{1/2}$.

Proposition 19. *Given $C_1 > 0$ and $h_0 > 0$ with $C_1 h_0^2 < T_c$, there is a constant $C > 0$ such that for all $|T - T_c| \leq C_1 h^2$ and $0 < h \leq h_0$ one has*

$$1 - V^{1/2} L_{T,W} V^{1/2} \geq \kappa (1 - Q) - Ch^2. \tag{39}$$

Proof. Our starting point is again inequality (36), which is valid for all $|T - T_c| \leq C_1 h_0^2$. Since the derivative of $\chi_\beta(E)$ with respect to T is bounded uniformly in E for T away from 0, we infer that there is a $C' > 0$ such that for all $|T - T_c| \leq C_1 h_0^2$ and all $E \in \mathbb{R}$,

$$\left| \chi_\beta(E) - \chi_{\beta_c}(E) \right| \leq C' |T - T_c|. \tag{40}$$

This, together with the gap inequality (37), implies that for $|T - T_c| \leq C_1 h^2 \leq C_1 h_0^2$,

$$
\begin{aligned}
1 - V^{1/2} L_{T,W} V^{1/2} &\geq 1 - \frac{1}{2} \left(U V^{1/2} \chi_{\beta_c}(p_r^2 - \mu) V^{1/2} U^* + U^* V^{1/2} \chi_{\beta_c}(p_r^2 - \mu) V^{1/2} U \right) \\
&\quad - C'' |T - T_c| - Ch^2 \\
&\geq \frac{\kappa}{2} \left(U P^\perp U^* + U^* P^\perp U \right) - (C_1 C'' + C)h^2 \\
&= \kappa (1 - Q) - (C_1 C'' + C)h^2,
\end{aligned}
$$

as claimed.   □

Next, we observe that for functions $\Phi \in L^2_{\text{symm}}(\mathbb{R}^3 \times \mathbb{R}^3)$, one can write

$$(Q\Phi)(X + r/2, X - r/2)$$
$$= \varphi_*(r) \cos(p_X \cdot r/2) \int_{\mathbb{R}^3} ds \, \overline{\varphi_*(s)} \cos(p_X \cdot s/2) \Phi(X + s/2, X - s/2)$$
$$=: |A_{p_X}\rangle\langle A_{p_X}|\Phi\rangle$$

with

$$A_p(r) := \varphi_*(r) \cos(p \cdot r/2).$$

(More precisely, the expression $|A_{p_X}\rangle\langle A_{p_X}|$ can be written as a direct integral over the center-of-mass momenta p_X. In the case of magnetic fields [2], this did not work because the components of the magnetic momentum did not commute.)

Now, we use the fact that in each fiber Q can be estimated from above by its largest eigenvalue, and hence we immediately conclude that

$$1 - Q \geq 1 - \langle A_{p_X}|A_{p_X}\rangle = 1 - R \tag{41}$$

with

$$R := \int_{\mathbb{R}^3} dr \, |\varphi_*(r)|^2 \cos^2(r \cdot p_X/2) \tag{42}$$

acting in $L^2(\mathbb{R}^3)$. Since $\cos(r \cdot p_X/2)^2 \leq 1$ and since φ_* is normalized, we have $R \leq 1$ and therefore $1 - R \geq 0$. We now prove a more precise lower bound.

Lemma 20. *There are constants $E_0 > 0$ and $c > 0$ such that*

$$1 - R \geq c \, \frac{p_X^2}{E_0 + p_X^2}.$$

Proof. All operators involved are diagonal in Fourier space, so for the proof we can consider p_X to be a vector in $\mathbb{R}^3$. Using the normalization of φ_* we are thus lead to considering the function

$$1 - R(p_X) = \int_{\mathbb{R}^3} dr \, |\varphi_*(r)|^2 \left(1 - \cos^2(p_X \cdot r/2)\right) = \int_{\mathbb{R}^3} dr \, |\varphi_*(r)|^2 \sin^2(p_X \cdot r/2).$$

First, we have

$$\lim_{p_X \to 0} \frac{1 - R(p_X)}{p_X^2} = \frac{1}{12} \int_{\mathbb{R}^3} dr \, |\varphi_*(r)|^2 r^2 =: c.$$

(The right side is finite, as shown in [4].) Therefore, there is a $\delta > 0$ such that $1 - R(p_X) \geq (c/2)p_X^2$ for $|p_X| \leq \delta$.

Second, by the Riemann–Lebesgue lemma, we have

$$\lim_{|p_X| \to \infty} (1 - R(p_X)) = \frac{1}{2},$$

and therefore there is an $M > 0$ such that $1 - R(p_X) \geq 1/4$ for $|p_X| \geq M$.

Since for any $p_X \neq 0$ the function $r \mapsto \sin^2(p_X \cdot r/2)$ vanishes only on a set of measure zero, we have $1 - R(p_X) > 0$ for all $p_X \neq 0$. Since $p_X \mapsto R(p_X)$ is continuous, there is a $c' > 0$ such that $1 - R(p_X) \geq c'$ for all $\delta \leq |p_X| \leq M$. This proves that

$$1 - R(p_X) \geq \min\{(c/2)p_X^2, c', 1/4\},$$

which immediately implies the lemma. $\square$

5.4   Proof of the Decomposition Lemma

As a consequence of Proposition 19, we now deduce a first decomposition result for almost maximizers Φ of $1 - V^{1/2}L_{T,W}V^{1/2}$.

Let us now define the projection

$$P_Q := \frac{|A_{p_X}\rangle\langle A_{p_X}|}{\langle A_{p_X}|A_{p_X}\rangle},$$

where the last expression is again a direct integral over the momenta p_X. To see how this operator acts define for a given $\Phi \in L^2_{\text{symm}}(\mathbb{R}^3 \times \mathbb{R}^3)$,

$$\psi(X) := \frac{\langle A_{p_X}|}{\|A_{p_X}\|}\Phi = \int_{\mathbb{R}^3} ds \frac{\varphi_*(s)\cos(p_X \cdot s/2)}{\sqrt{\int_{\mathbb{R}^3} |\varphi_*(s')|^2 \cos^2(p_X \cdot s'/2)ds'}}\Phi(X + s/2, X - s/2). \quad (43)$$

Then

$$P_Q\Phi(X + r/2, X - r/2) = \frac{\varphi_*(r)\cos(p_X \cdot r/2)}{\sqrt{\int_{\mathbb{R}^3} |\varphi_*(r')|^2 \cos^2(p_X \cdot r'/2)dr'}}\psi(X),$$

and we define $\xi \in L^2_{\text{symm}}(\mathbb{R}^3 \times \mathbb{R}^3)$ by

$$\Phi = P_Q\Phi + \xi. \quad (44)$$

With these definitions, we can formulate a first version of the decomposition lemma.

Lemma 21. *Given $C_1, C_2 > 0$ there are $h_0 > 0$, $E_0 > 0$, and $C > 0$ with the following properties. If $|T - T_c| \leq C_1h^2 \leq C_1h_0^2$ and if $\Phi \in L^2_{\text{symm}}(\mathbb{R}^3 \times \mathbb{R}^3)$ with $\|\Phi\| = 1$ satisfies*

$$\langle \Phi, (1 - V^{1/2}L_{T,W}V^{1/2})\Phi \rangle \leq C_2h^2, \quad (45)$$

then, with ψ and ξ defined in (43) and (44),

$$\left\langle \psi, \frac{p_X^2}{E_0 + p_X^2} \psi \right\rangle + \|\xi\|^2 \leq Ch^2 .$$

and

$$1 \geq \|\psi\|^2 \geq 1 - Ch^2 .$$

Proof. By Proposition 19 and Assumption (45), we obtain

$$\langle \Phi, (1 - Q)\Phi \rangle \leq \kappa^{-1}(C + C_2)h^2 . \tag{46}$$

By construction, for every fixed value p_X of the Fourier transform with respect to X, $P_Q \Phi$ and ξ are orthogonal as functions of r. Therefore

$$\langle \Phi, (1 - Q)\Phi \rangle = \langle P_Q \Phi, (1 - Q) P_Q \Phi \rangle + \|\xi\|^2 .$$

On the other hand, it is easy to see that

$$\langle P_Q \Phi, (1 - Q) P_Q \Phi \rangle = \langle \psi, (1 - R)\psi \rangle .$$

Therefore, the lower bound on $1 - R$ from Lemma 20 implies the first assertion in the lemma.

In order to prove the second assertion, we note that

$$\|\psi\|^2 = \|P_Q \Phi\|^2 = \|\Phi\|^2 - \|\xi\|^2 = 1 - \|\xi\|^2$$

and use the bound on $\|\xi\|^2$ from the first assertion. □

Proof of Theorem 16. Let ψ be as in Lemma 21. For $\varepsilon \in [h^2, h_0^2]$, we set

$$\psi_\leq := \mathbb{1}(p_X^2 \leq \varepsilon)\psi , \qquad \psi_> := \mathbb{1}(p_X^2 > \varepsilon)\psi .$$

Recall from Lemma 21 that $\langle \psi, p_X^2(E_0 + p_X^2)^{-1}\psi \rangle \leq Ch^2$. This implies that for $k \geq 1$,

$$\left\| (p_X^2)^{k/2} \psi_\leq \right\|^2 \leq \varepsilon^{k-1} \|p_X \psi_\leq\|^2 \leq (E_0 + \varepsilon)\varepsilon^{k-1} \left\langle \psi, \frac{p_X^2}{E_0 + p_X^2} \psi \right\rangle$$

$$\leq C(E_0 + \varepsilon)\varepsilon^{k-1}h^2$$

and

$$\|\psi_>\|^2 \leq \frac{E_0 + \varepsilon}{\varepsilon} \left\langle \psi, \frac{p_X^2}{E_0 + p_X^2} \psi \right\rangle \leq C \frac{E_0 + \varepsilon}{\varepsilon} h^2 . \tag{47}$$

We now define

$$\sigma_0(X + r/2, X - r/2) := \frac{\varphi_*(r)\cos(p_X \cdot r/2)}{\sqrt{\int_{\mathbb{R}^3} |\varphi_*(r')|^2 \cos^2(p_X \cdot r'/2)\, dr'}} \psi_>(X),$$

$$\sigma_1(X + r/2, X - r/2) := -\varphi_*(r)\left(1 - \frac{\cos(p_X \cdot r/2)}{\sqrt{\int_{\mathbb{R}^3} |\varphi_*(r')|^2 \cos^2(p_X \cdot r'/2)\, dr'}}\right)\psi_\leq(X)$$

and

$$\sigma := \sigma_0 + \sigma_1 + \xi,$$

so that, by Lemma 21,

$$\Phi = \varphi_*(r)\psi_\leq(X) + \sigma.$$

According to Lemma 21 and (47), we have

$$\|\psi_\leq\|^2 = \|\psi\|^2 - \|\psi_>\|^2 \geq 1 - Ch^2 - C\varepsilon^{-1}h^2 \geq 1 - C'\varepsilon^{-1}h^2.$$

and, again according to (47), we have

$$\|\sigma_0\|^2 = \|\psi_>\|^2 \leq C\frac{E_0 + \varepsilon}{\varepsilon}h^2.$$

Moreover,

$$\|\sigma_1\|^2 = \langle \psi_\leq, S\psi_\leq \rangle$$

with the operator

$$S := \int_{\mathbb{R}^3} dr\, |\varphi_*(r)|^2 \left(1 - \frac{\cos(p_X \cdot r/2)}{\sqrt{\int_{\mathbb{R}^3} |\varphi_*(r')|^2 \cos^2(p_X \cdot r'/2)\, dr'}}\right)^2$$

$$= 2\int_{\mathbb{R}^3} dr\, |\varphi_*(r)|^2 \left(1 - \frac{\cos(p_X \cdot r/2)}{\sqrt{\int_{\mathbb{R}^3} |\varphi_*(r')|^2 \cos^2(p_X \cdot r'/2)\, dr'}}\right)$$

acting in $L^2(\mathbb{R}^3)$. In the same way as in the proof of Lemma 20, we can show that

$$S \leq C\frac{p_X^2}{E_0 + p_X^2},$$

and therefore

$$\|\sigma_1\|^2 \lesssim \langle \psi_\le, \frac{p_X^2}{E_0 + p_X^2}\psi_\le\rangle \le \langle \psi, \frac{p_X^2}{E_0 + p_X^2}\psi\rangle \lesssim h^2 \,.$$

We conclude that

$$\|\sigma - \sigma_0\| = \|\sigma_1 + \xi\| \le \|\sigma_1\| + \|\xi\| \lesssim h \,.$$

This concludes the proof of the theorem. $\qquad\qquad\square$

6 Upper Bound on the Critical Temperature

In this section, we prove part (2) of Theorem 4. In view of Corollary 18 and the lower bound on the critical temperature it suffices to consider T satisfying $|T - T_c| \le C_1 h^2$. Moreover, it clearly suffices to consider functions Φ with $\|\Phi\| = 1$ satisfying

$$\langle \Phi, (1 - V^{1/2}L_{T,W}V^{1/2})\Phi\rangle \le C_2 h^2$$

(for if there are no such Φ, then the theorem is trivially true). According to Theorem 16, for any parameter $\varepsilon \in [h^2, h_0^2]$, Φ can be decomposed as

$$\Phi = \varphi_*(r)\psi_\le(X) + \sigma \,.$$

Thus,

$$\langle \Phi, (1 - V^{1/2}L_{T,W}V^{1/2})\Phi\rangle = I_1 + I_2 + I_3$$

with

$$\begin{aligned}
I_1 &:= \langle\varphi_*(r)\psi_\le(X), (1 - V^{1/2}L_{T,W}V^{1/2})\varphi_*(r)\psi_\le(X)\rangle, \\
I_2 &:= \langle\sigma, (1 - V^{1/2}L_{T,W}V^{1/2})\sigma\rangle, \\
I_3 &:= 2\operatorname{Re}\langle\sigma, (1 - V^{1/2}L_{T,W}V^{1/2})\varphi_*(r)\psi_\le(X)\rangle.
\end{aligned}$$

The term I_1 is the main term and can be treated exactly as in the proof of the lower bound on the critical temperature. We obtain

$$I_1 \ge -\Lambda_2\frac{T_c - T}{T_c}\|\psi_\le\|^2 + \langle\psi_\le, (\Lambda_0 p_X^2 + \Lambda_1 h^2 W(hX))\psi_\le\rangle - C\varepsilon h^2 \,.$$

The fact that the error h^3 is replaced by εh^2 comes from the bound $\|p_X^2\psi_\le\|^2 \lesssim \varepsilon h^2$ from (31).

Let us therefore bound the error terms I_2 and I_3. Using the operator inequality from Proposition 19, dropping the nonnegative term $\kappa(1 - Q)$ and using the bound (32) on σ, we obtain

$$I_2 \gtrsim -h^2 \|\sigma\|^2 \gtrsim -\varepsilon^{-1} h^4 \, .$$

In order to bound I_3, we use the first bound in Lemma 11 and the bounds (32) and (33) on σ and $\psi_\le$ to obtain

$$
\begin{aligned}
I_3 &\ge 2\,\mathrm{Re}\left\langle \sigma, \left(1 - V^{1/2} L_{T,0} V^{1/2}\right) \varphi_*(r)\psi_\le(X)\right\rangle - Ch^2 \|\sigma\| \|\varphi_*(r)\psi_\le(X)\| \\
&\ge 2\,\mathrm{Re}\left\langle \sigma, \left(1 - V^{1/2} L_{T,0} V^{1/2}\right) \varphi_*(r)\psi_\le(X)\right\rangle - C'\varepsilon^{-1/2} h^3 \, .
\end{aligned}
$$

To bound the first term on the right side we decompose $\sigma = \sigma_0 + (\sigma - \sigma_0)$. We claim that

$$\left\langle \sigma_0, \left(1 - V^{1/2} L_{T,0} V^{1/2}\right) \varphi_*(r)\psi_\le(X)\right\rangle = 0 \, .$$

Indeed, to see this, we note that for fixed r, the Fourier transforms of $\sigma_0(X + r/2, X - r/2)$ and $V(r)^{1/2}\sigma_0(X + r/2, X - r/2)$ with respect to the variable X are supported in $\{p_X^2 > \varepsilon\}$ and likewise the Fourier transforms of $\varphi_*(r)\psi_\le(X)$ and $V(r)^{1/2}\varphi_*(r)\psi_\le(X)$ with respect to the variable X are supported in $\{p_X^2 \le \varepsilon\}$. Thus $\left\langle \sigma_0, \varphi_*(r)\psi_\le(X)\right\rangle = 0$, and the full claim follows by observing that the operator $L_{T,0}$ acts diagonally in Fourier space with respect to the X variables, see (15).

Thus, it remains to bound the term with $\sigma - \sigma_0$. We decompose $L_{T,0} = \chi_\beta(p_r^2 - \mu) + (L_{T,0} - \chi_\beta(p_r^2 - \mu))$ and, using the fact that $(1 - V^{1/2}\chi_{\beta_c}(p_r^2 - \mu)V^{1/2})\varphi_* = 0$, we find

$$
\begin{aligned}
&\left\langle \sigma - \sigma_0, \left(1 - V^{1/2} L_{T,0} V^{1/2}\right) \varphi_*(r)\psi_\le(X)\right\rangle \\
&= \left\langle \sigma - \sigma_0, V^{1/2} \left(\chi_{\beta_c}(p_r^2 - \mu) - \chi_\beta(p_r^2 - \mu)\right) V^{1/2}\varphi_*(r)\psi_\le(X)\right\rangle \\
&\quad - \left\langle \sigma - \sigma_0, V^{1/2} \left(L_{T,0} - \chi_\beta(p_r^2 - \mu)\right) V^{1/2}\varphi_*(r)\psi_\le(X)\right\rangle \, .
\end{aligned}
$$

Using inequality (40), as well as the bounds (34) and (33) on $\sigma - \sigma_0$ and $\psi_\le$, we find

$$
\begin{aligned}
\left\langle \sigma - \sigma_0, V^{1/2} \left(\chi_{\beta_c}(p_r^2 - \mu) - \chi_\beta(p_r^2 - \mu)\right) V^{1/2}\varphi_*(r)\psi_\le(X)\right\rangle &\gtrsim -h^2 \|\sigma - \sigma_0\| \|\psi_\le\| \\
&\gtrsim -h^3 \, .
\end{aligned}
$$

The remaining term we bound similarly using Lemma 12, as well as the bounds (34) and (33) on $\sigma - \sigma_0$ and $\psi_\le$,

$$
\begin{aligned}
-\left\langle \sigma - \sigma_0, V^{1/2} \left(L_{T,0} - \chi_\beta(p_r^2 - \mu)\right) V^{1/2}\varphi_*(r)\psi_\le(X)\right\rangle &\gtrsim -\|\sigma - \sigma_0\| \|p_X^2 \psi_\le\| \\
&\gtrsim -\varepsilon^{1/2} h^2 \, .
\end{aligned}
$$

To summarize, we have shown that

$$\langle \Phi, (1 - V^{1/2} L_{T,W} V^{1/2}) \Phi \rangle \geq -\Lambda_2 \frac{T_c - T}{T_c} \|\psi_\leq\|^2 + \left\langle \psi_\leq, \left(\Lambda_0 p_X^2 + \Lambda_1 h^2 W(hX) \right) \psi_\leq \right\rangle$$
$$- Ch^2 \left(\varepsilon + \varepsilon^{-1} h^2 + \varepsilon^{-1/2} h + h + \varepsilon^{1/2} \right).$$

In order to minimize the error we choose $\varepsilon = h$. With this choice we obtain, recalling also the lower bound on $\|\psi_\leq\|$ from (33),

$$\langle \Phi, (1 - V^{1/2} L_{T,W} V^{1/2}) \Phi \rangle \geq \left\langle \psi_\leq, \left(\Lambda_0 p_X^2 + \Lambda_1 h^2 W(hX) - \Lambda_2 \frac{T_c - T}{T_c} - C'h^{5/2} \right) \psi_\leq \right\rangle$$

By definition of D_c plus a rescaling we can bound the right side from below by

$$\left(h^2 \Lambda_2 D_c - \Lambda_2 \frac{T_c - T}{T_c} - C'h^{5/2} \right) \|\psi_\leq\|^2 .$$

Recalling that $\|\psi_\geq\| \neq 0$, we conclude that this is > 0 provided $T > T_c(1 - D_c h^2 + (C'/\Lambda_2)h^{5/2})$. This concludes the proof of the upper bound on the critical temperature.

Acknowledgements We thank Edwin Langmann who initiated and coauthored our previous work [2] which forms the basis of the present paper. We further thank Robert Seiringer and Jan Philip Solovej for our long-lasting collaboration on BCS theory. Further, partial support by the U.S. National Science Foundation through grant DMS-1363432 (R.L.F.) is acknowledged.

References

1. Frank, R.L., Lemm, M.: Ginzburg-Landau theory with multiple order parameters: microscopic derivation and examples. Ann. H. Poincaré **17**(9), 2285–2340 (2016)
2. Frank, R.L., Hainzl, C., Langmann, E.: The BCS critical temperature in a weak homogeneous magnetic field, to appear in J. Spect. Theory
3. Frank, R.L., Hainzl, C., Naboko, S., Seiringer, R.: The critical temperature for the BCS equation at weak coupling. J. Geom. Anal. **17**, 559–568 (2007)
4. Frank, R.L., Hainzl, C., Seiringer, R., Solovej, J.P.: Microscopic derivation of Ginzburg-Landau theory. J. Am. Math. Soc. **25**(3), 667–713 (2012)
5. Frank, R.L., Hainzl, C., Seiringer, R., Solovej, J.P.: The external field dependence of the BCS critical temperature. Commun. Math. Phys. **342**(1), 189–216 (2016)
6. Hainzl, C., Seiringer, R.: Critical temperature and energy gap in the BCS equation. Phys. Rev. B **77**, 184517 (2008)
7. Hainzl, C., Seiringer, R.: The Bardeen-Cooper-Schrieffer functional of superconductivity and its mathematical properties. J. Math. Phys. **57**, 021101 (2016)
8. Hainzl, C., Hamza, E., Seiringer, R., Solovej, J.P.: The BCS functional for general pair interactions. Commun. Math. Phys. **281**(2), 349–367 (2008)
9. Helfand, E., Werthamer, R.: Temperature and purity dependence of the superconducting critical field, H_{c2}. II. Phys. Rev. **147**(1), 288–294 (1966)
10. Langmann, E.: Theory of the upper critical magnetic field without local approximation. Phys. C **159**, 561 (1989)

11. Langmann, E.: $B_{c2}(T)$ of anisotropic systems: some explicit results. Phys. B **165–166**, 1061 (1990)
12. Langmann, E.: On the upper critical field of anisotropic superconductors. Phys. C **173**, 347 (1991)
13. Schossmann, M., Schachinger, E.: Strong-coupling theory of the upper critical magnetic field H_{c2}. Phys. Rev. B **33**, 6123 (1986)
14. Werthamer, N.R., Helfand, E., Hohenberg, P.C.: Temperature and purity dependence of the superconducting critical field, H_{c2}. III. Phys. Rev. **147**, 295 (1966)

Effective Dynamics of Two Tracer Particles Coupled to a Fermi Gas in the High-Density Limit

Maximilian Jeblick, David Mitrouskas and Peter Pickl

Abstract In a recent paper (Jeblick, Mitrouskas, Petrat and Pickl, Commun. Math. Phys. 356(1), 143–187 (2017) [1]), we studied the dynamics of a particle coupled to a dense and homogeneous ideal Fermi gas in two spatial dimensions. In that paper, closeness of the time evolution to an effective *free* dynamics for *large densities* of the gas was proven. The main point of interest of this result was to consider coupling constants that do not scale with the density of the gas. In the present article, we generalize this result to a system of *two* tracer particles coupled to the ideal Fermi gas. Naively, one might expect this to be a trivial extension of the single particle result. But this is not the case since there is an interesting additional effect: While the dynamics of both tracer particles decouple again from the Fermi gas, there are collision processes which are not present for a single tracer particle and which lead to an effective interaction *between* the two tracer particles. This effective interaction is mediated through the creation and immediate annihilation of electron–hole pairs, and can thus be interpreted as a Lamb shift-type effect.

1 Introduction and Main Result

In this article, we consider the dynamics of two interacting tracer particles coupled to a dense and homogeneous two-dimensional fermionic gas. In order to keep the analysis simple, we neglect the interaction between the gas particles and focus only

M. Jeblick · P. Pickl (✉)
Ludwig-Maximilians-Universität, Mathematisches Institut, Theresienstr. 39,
80333 München, Germany
e-mail: pickl@math.lmu.de

M. Jeblick
e-mail: jeblick@math.lmu.de

D. Mitrouskas
Universität Stuttgart, Fachbereich Mathematik, Pfaffenwaldring 57,
70569 Stuttgart, Germany
e-mail: mitrouskas@mathematik.uni-stuttgart.de

© Springer Nature Switzerland AG 2018   63
D. Cadamuro et al. (eds.), *Macroscopic Limits of Quantum Systems*,
Springer Proceedings in Mathematics & Statistics 270,
https://doi.org/10.1007/978-3-030-01602-9_3

on the interaction between the tracer particles y_1 and y_2 among each other and with the gas particles $x_1, \ldots, x_N$. We consider the Hamiltonian

$$H = -\frac{1}{2m_y}\Delta_{y_1} - \frac{1}{2m_y}\Delta_{y_1} + W(y_1 - y_2) - \sum_{i=1}^{N}\frac{1}{2m_x}\Delta_{x_i} + g\sum_{i=1}^{N}(v(x_i - y_1) + v(x_i - y_2)),$$
(1)

where $v \in C_0^\infty$ (the space of smooth functions with compact support), and g is a coupling constant. The function W has to satisfy weak regularity conditions which we specify below. The time evolution of the (N + 2)-body wave function $\Psi_t \in \mathcal{H}_{y_1,y_2} \otimes \mathcal{H}_N = L^2(\mathbb{T}^4) \otimes L^2(\mathbb{T}^{2N})$, where $\mathbb{T}$ is a one-dimensional torus of length $L \in \mathbb{R}$, and L^2 denotes the space of complex square integrable functions (for simplicity, we neglect spin), is given by the Schrödinger equation

$$i\partial_t \Psi_t = H\Psi_t.$$
(2)

As initial condition we choose a factorized state $\Psi_0 = \varphi_0 \otimes \Omega_0$, where $\varphi_0 \in \mathcal{H}_{y_1,y_2}$ is the initial wave function describing the two tracer particles and $\Omega_0 \in \mathcal{H}_N$ is the free fermionic ground state with periodic boundary conditions in the two-dimensional box of side length L. Note that we do not impose any particular symmetry on φ_0. For analyzing Ψ_t we will take the limit $N, L \to \infty$ with $\varrho = N/L^2 = const.$ in order to remove finite size effects and then consider large gas densities $\varrho \gg 1$. In this situation, the average potential energy of the tracer particles is proportional to $g\varrho$. We choose $g = 1$, such that our analysis is beyond any weak-coupling limit.

In a previous work [1], we analyzed the analogous model with only *one* tracer particle coupled to the Fermi gas. There we could show that the effective description is, up to a global phase, given by the free time evolution of the tracer particle. The global phase was determined by the mean field energy of the system together with a subleading energy correction (while the mean field energy is of order ϱ, the additional energy correction was shown to be of order one). The subleading energy correction had to be included explicitly into the effective Hamiltonian in order to cancel a nonvanishing collision process appearing in the perturbative expansion of the full dynamics. This collision process appears at first time at second order of the expansion and is due to the creation and immediate annihilation of electron–hole pairs.

Now, looking at two tracer particles, we get similar non-negligible terms of second order: Creation of an electron–hole pair and immediate annihilation of that pair while

(a) Either both interactions occur with tracer particle y_1 or both interactions occur with tracer particle y_2.
(b) One of the two interactions occurs with tracer particle y_1, the other interaction occurs with tracer particle y_2.

While, as in the single-tracer-particle case, processes described as in (a) give only rise to a constant in the Hamiltonian and are thus irrelevant for the time evolution, the effect of the (b)-diagrams does depend on the relative coordinates $y_1 - y_2$ (processes described as in (b) are obviously not present if only a single tracer particle is considered). It is not surprising that the respective term in the effective description, which we denote by $W_{\text{eff},\varrho}$, will thus depend on $y_1 - y_2$. The explicit form of $W_{\text{eff},\varrho}$ is given (in the thermodynamic limit) by

$$
W_{\text{eff},\varrho}(y_1 - y_2)
$$
$$
= \frac{2}{(2\pi)^4} \int_{B_{\sqrt{4\pi\varrho}}(0)} d^2k \int_{\mathbb{R}^2 \setminus B_{\sqrt{4\pi\varrho}}(0)} d^2l \, \frac{|\hat{v}(k-l)|^2 \cos((k-l)(y_1 - y_2))}{l^2 - k^2} \theta\left(|l| - |k| - \varrho^{-\frac{1}{2}}\right),
\tag{3}
$$

where $B_r(0) \subset \mathbb{R}^2$ denotes a ball of radius r around zero, $\hat{v}$ is the Fourier transform of the potential $v \in C_0^\infty$, and $\theta(x)$ denotes the usual Heaviside step function, i.e., $\theta(x) = 1$ for $x \geq 0$ and zero otherwise. We expect that $W_{\text{eff},\varrho}$ has a nontrivial limit for $\varrho \to \infty$, leading to an effective interaction of order one between the two tracer particles. However, we do not verify this expectation in the present article but only show that

$$
|W_{\text{eff},\varrho}(y)| \leq C\varrho^{2\varepsilon} + C_\varepsilon \varrho^{-1/\varepsilon},
\tag{4}
$$

for any $\varepsilon > 0$. This bound is sufficient to generalize Theorem 1.1. in [1] (the single-tracer-particle case) in the following way:

Theorem 1.1 *Let the masses $m_x = m_y = 1/2$ and the coupling constant $g = 1$. Let further $\Psi_0 = \varphi_0 \otimes \Omega_0$ where $\varphi_0 \in \mathcal{H}_{y_1,y_2}$ with $\|\varphi_0\| + \|(\nabla_1 + \nabla_2)^4 \varphi_0\| \leq C$ uniformly in ϱ, and Ω_0 is the free fermionic ground state in $\mathbb{T}^2$ with $\|\Omega_0\| = 1$. Let $W \in C^2(\mathbb{T}^2)$. Then, for any small enough $\varepsilon > 0$, there exists a positive constant C_ε such that*

$$
\lim_{\substack{N,L \to \infty \\ \varrho = N/L^2 = const.}} \left\| e^{-iHt}\Psi_0 - e^{-iH^{mf}t}\Psi_0 \right\|_{\mathcal{H}_{y_1,y_2} \otimes \mathcal{H}_N} \leq C_\varepsilon (1+t)^{\frac{3}{2}} \varrho^{-\frac{1}{8}+\varepsilon}
\tag{5}
$$

for all $t > 0$, where

$$
H^{mf} = -\Delta_{y_1} - \Delta_{y_2} + W(y_1 - y_2) - W_{\text{eff},\varrho}(y_1 - y_2) - \sum_{i=1}^N \Delta_{x_i} + 2\varrho\hat{v}(0) - W_{\text{eff},\varrho}(0)
\tag{6}
$$

is the effective Hamiltonian with constant mean field $\varrho\hat{v}(0) = \langle \Omega_0, \sum_{i=1}^N v(x_i - y)\Omega_0 \rangle_{\mathcal{H}_N}$ and $W_{\text{eff},\varrho}(y_1 - y_2)$ as defined in (3).

The proof of the theorem is to a large extent very similar to the proof of [1, Theorem 1.1], and we thus omit many of the details here. In Sect. 1.2, we give a description of the idea of the proof, and then, in Sects. 2.1–2.4, we state the key ingredients for proving Theorem 1.1. In Sect. 2.5, eventually, we carry out in some more detail the part of the proof which is different compared to the proof of [1, Theorem 1.1].

Remark 1.2 We want to emphasize that in Theorem 1.1, all scales are fixed except for the density ϱ and the time t. The bound in (5) becomes thus meaningful for any pair of t and ϱ such that the error on the r.h.s. is small compared to one (hence in particular for $\varrho \gg 1$ and $t \ll \varrho^{1/12}$).

Remark 1.3 The contribution $W_{\text{eff},\varrho}(0)$ was denoted $E_{re}(\varrho)$ in [1]. As we already explained, this contribution arises from immediate recollisions of the same tracer particle with one of the gas particles. It was shown that for any $\varepsilon > 0$ there are positive constants C_1, C_2, and C_ε such that

$$C_1 \leq E_{re}(\varrho) \leq C_2 \varrho^{2\varepsilon} + C_\varepsilon \varrho^{-1/\varepsilon}. \tag{7}$$

Since $|W_{\text{eff},\varrho}(y)| \leq E_{re}(\varrho)$, we immediately obtain the upper bound stated in (4).

Remark 1.4 Since $2\varrho\hat{v}(0) - W_{\text{eff},\varrho}(0)$ is constant as a function of the coordinates $y_1, y_2, x_1, ..., x_N$, the time evolution $e^{-iH^{\text{mf}}t}$ is physically equivalent to the dynamics generated by $H^{\text{eff}}_{y_1,y_2} - \sum_{i=1}^N \Delta_{x_i}$, where

$$H^{\text{eff}}_{y_1,y_2} = -\Delta_{y_1} - \Delta_{y_2} + W(y_1 - y_2) - W_{\text{eff},\varrho}(y_1 - y_2)$$

describes the motion of the two tracer particles decoupled from the Fermi gas, but with modified pair interaction. If one regards Ω_0 as a reference vacuum, as it is for example done in Bogoliubov theory, one might interpret $E_{re}(\varrho)$ and $W_{\text{eff},\varrho}(y_1 - y_2)$ to be caused by quantum fluctuations (creation and annihilation of virtual particles). From this perspective, one might see our result as a (rigorous) many-body model explanation for a Lamb shift-type effect. The Lamb shift, originally discussed by Bethe in [2], describes the modification of the interaction potential between two particles due to polarization of the surrounding vacuum.

Remark 1.5 A similar statement as Theorem 1.1 can be proven also for spatial dimension $d = 1$. In the one-dimensional case, however, it is not difficult to show that the correspondingly defined potential $W_{\text{eff},\varrho}(y)$ vanishes, uniformly in $y \in \mathbb{R}$, when ϱ tends to ∞. This can be understood in terms of energy and momentum conservation which in one dimension suppresses the creation of electron-hole pairs for $\varrho \gg 1$. Whether a similar result is true also for $d = 3$ is not clear. The method we use to prove Theorem 1.1 seems to be not applicable in this case. For a detailed comparison of the analysis in different dimensions, we refer to [1, Section 1.2.2, Appendices A and B].

1.1 The Model in More Detail

We model the potential between the tracer particles and each of the gas particles by an infinitely differentiable function with compact support (uniformly in L), i.e., $v \in C_0^\infty(\mathbb{T}^2) \cap C_0^\infty(\mathbb{R}^2)$. We abbreviate the total interaction term in H by $V = \sum_{i=1}^N (v(x_i - y_1) + v(x_i - y_2))$. Since V is bounded, H defines a self-adjoint operator on the second Sobolev space $H^2(\mathbb{T}^{2(N+2)}) \subset \mathcal{H}_{y_1,y_2} \otimes \mathcal{H}_N$. For the corresponding time evolution, we write $U(t) = e^{-iHt}$. The initial wave function φ_0 of the tracer particles is restricted to be an element of $H^4(\mathbb{T}^4) \subset \mathcal{H}_{y_1,y_2}$ with $\|\varphi_0\|_{H^4} < C$ for all values of ϱ. The initial state of the gas is assumed to be given by the ground state of the ideal Fermi gas which is described by the antisymmetric product of N one-particle plane waves,

$$\Omega_0(x_1, ..., x_N) = \frac{1}{\sqrt{N!}} \sum_{\tau \in S_N} (-1)^\tau \prod_{i=1}^N \varphi_{p_{\tau(i)}}(x_i), \tag{8}$$

with $\varphi_p(x) = L^{-1} e^{ip \cdot x} \in L^2(\mathbb{T}^2)$, and $(p_j)_{j=1}^N$ the N pairwise different elements of $(2\pi/L)\mathbb{Z}^2$ with smallest absolute value. S_N denotes the group of permutations of integers $\{1, ..., N\}$, and $(-1)^\tau$ is the sign of the permutation τ. Since the system is defined on a torus of side length L (with periodic boundary conditions), the set of possible momenta in the gas is given by the lattice $(2\pi/L)\mathbb{Z}^2$. We label the momenta such that for $j_1, j_2 \geq 1$ we have $j_1 < j_2 \Leftrightarrow |p_{j_1}| \leq |p_{j_2}|$. The wave function Ω_0 corresponds thus to the lowest possible kinetic energy given by $\sum_{k=1}^N p_k^2$.

An important quantity that characterizes the state Ω_0 is the Fermi momentum k_F. It is defined as $k_F = |p_N|$ where p_N belongs to the set of momenta $\{p_k \in (2\pi/L)\mathbb{Z}^2 : k = 1, ..., N\}$ which minimizes the kinetic energy $\sum_{i=1}^N p_{k_i}^2$. The value k_F defines the so-called Fermi sphere and is related in two space dimensions to the average density ϱ via

$$\varrho = \frac{1}{L^2} \sum_{k=1}^N = \int_{|p| \leq k_F} \frac{d^2 p}{(2\pi)^2} = \frac{k_F^2}{4\pi} \qquad \Leftrightarrow \qquad k_F = \sqrt{4\pi\varrho}. \tag{9}$$

We study the model in the thermodynamic limit, i.e., for $N, L \to \infty$, and $\varrho = N/L^2 = const$. This simplifies the analysis because it allows us to ignore additional effects which are due to the chosen boundary conditions. For very large systems, i.e., in particular for $L/\text{supp}(v) \gg 1$, such boundary effects are not expected to be physically relevant which justifies the analysis in the thermodynamic limit. We emphasize that for the result we are interested in this work, it is really the parameter $\varrho \gg 1$ which is the physically interesting one. We expect a very similar result to hold if one repeats all estimates for fixed but large values of N and L, and then considers the regime in which $N \gg L$.

Let us next discuss the effective model. The effective dynamics is described by the Schrödinger equation with mean field Hamiltonian H^{mf}. Note that H^{mf} is also self-adjoint on $H^2(\mathbb{T}^{2(N+2)})$ and the corresponding mean field time evolution is denoted as $U^{\text{mf}}(t)$. For each of the two tracer particles, the average potential w.r.t. Ω_0 that acts at position $y \in \mathbb{T}^2$ is given by

$$E(y) = \langle \Omega_0, \sum_{i=1}^N v(x_i - y)\Omega_0 \rangle_{\mathcal{H}_N}(y) = \varrho \hat{v}(0), \tag{10}$$

which is spatially constant. Since there are two tracer particles in the gas, this explains the $2\varrho\hat{v}(0)$ in the effective Hamiltonian. We will denote $E = E(y_1) + E(y_2)$. The homogeneity of E is conserved under the mean field time evolution $U^{\mathrm{mf}}(t)$, i.e.,

$$\langle \Omega_t^f, V\Omega_t^f \rangle_{\mathcal{H}_N} = \langle \Omega_0, V\Omega_0 \rangle_{\mathcal{H}_N}, \qquad \Omega_t^f = e^{-iH_N^f t}\Omega_0, \tag{11}$$

where $H_N^f = -\sum_{i=1}^{N} \Delta_{x_i}$ denotes the free Hamiltonian of the gas. The Schrödinger equation with Hamiltonian (6) defines therefore a self-consistent approximation. The reason why we call H^{mf} a mean field Hamiltonian is that to leading order, the potential V is replaced by its average value E.

In the rest of the article, we omit the subscripts $\mathcal{H}_{y_1,y_2}$, $\mathcal{H}_N$, or $\mathcal{H}_{y_1,y_2} \otimes \mathcal{H}_N$ on all scalar products and norms, since it is always clear from the argument on which space the scalar product or norm is meant.

1.2   Idea of the Proof

For deriving Theorem 1.1, we use Duhamel's expansion in order to decompose Ψ_t into different wave functions that correspond to different collision histories of the tracer particles. The main difficulty is to control the interaction with gas particles occupying momenta close to the Fermi edge. Our main ingredient here is the large shift in the energy and the thereby caused phase cancelation during the scattering with such particles. It turns out to be necessary but also sufficient to use a third-order expansion in the difference $H - H^{\mathrm{mf}}$. Let us stress again that $g = 1$. This prevents us from using a straightforward order-by-order expansion of the time evolution. Thus, after expanding to third order, we have to estimate an error term involving the whole time evolution $U(t)$. In order to convey the main ideas and techniques behind the proof, let us start by expanding

$$U(t)\Psi_0 - U^{\mathrm{mf}}(t)\Psi_0 = -i\int_0^t d\tau U^{\mathrm{mf}}(t-\tau)(H - H^{\mathrm{mf}})U^{\mathrm{mf}}(\tau)\Psi_0$$

$$-i\int_0^t d\tau_1 \Big(U(t-\tau) - U^{\mathrm{mf}}(t-\tau)\Big)(H - H^{\mathrm{mf}})U^{\mathrm{mf}}(\tau)\Psi_0, \tag{12}$$

which follows from expanding U around U^{mf} in terms of Duhamel's formula and then splitting $U = U^{\mathrm{mf}} + (U - U^{\mathrm{mf}})$. The first term on the r.h.s. contains deviations from the effective dynamics due to single particle–hole excitations. In order to present the main argument, let us first ignore the next-to-leading order energy correction $E_{re}(\varrho)$ as well as the effective potential $W_{\mathrm{eff},\varrho}$ in the following. Using some elementary algebra (only momenta inside the Fermi sphere can be annihilated and momenta outside the Fermi sphere created), one readily rewrites

$$(V - E)\Psi_0 = \frac{1}{L^2} \sum_{k=1}^{N} \sum_{l=N+1}^{\infty} \hat{v}(p_l - p_k) \left(\sum_{i=1}^{2} e^{i(p_l - p_k)y_i} \varphi_0 \right) \otimes a^*(p_l) a(p_k) \Omega_0,$$

(13)

where $a^*(p)$ and $a(p)$ denote the usual fermionic creation and annihilation operators, w.r.t. to a plane wave carrying momentum p. Abbreviating, $k_{kl}^{(i)}(\tau) = e^{i H_{y_1,y_2}^{\text{eff}} \tau} e^{i(p_l - p_k)y_i} e^{-i H_{y_1,y_2}^{\text{eff}} \tau}$, for $i = 1, 2$, it is also straightforward to arrive at

$$\left\| \int_0^t d\tau \, U^{\text{mf}}(-\tau)(V - E) U^{\text{mf}}(\tau) \Psi_0 \right\|^2$$

$$= \underbrace{\frac{1}{L^4} \sum_{k=1}^{N} \sum_{l=N+1}^{\infty} |\hat{v}(p_k - p_l)|^2 \left\| \sum_{i=1}^{2} \int_0^t d\tau e^{i(p_l^2 - p_k^2)\tau} k_{kl}^{(i)}(\tau) \varphi_0 \right\|^2}_{=\frac{1}{4}\|(V-E)\Omega_t^f\|^2}.$$

(14)

Due to the regularity of the potential v, it is unlikely that a single collision causes a large momentum transfer between φ_0 and Ω_0. This is reflected in the fact that the Fourier transform of a smooth and compactly supported function decays faster than any polynomial: for all $p \in \mathbb{N}$ there exists a constant D_p such that

$$|\hat{v}(p_k - p_l)| \leq \frac{D_p}{(1 + |p_k - p_l|)^p},$$

(15)

which follows directly from the Paley–Wiener Theorem, e.g., [3, Theorem XI.11]. At this point it is convenient to introduce the following notation. For $\varepsilon > 0$ we define $v^{\ell,\varepsilon}$ and $v^{s,\varepsilon}$ such that

$$\hat{v}^{\ell,\varepsilon}(p_k - p_l) = \theta(|p_k - p_l| - \varrho^\varepsilon) \hat{v}(p_k - p_l),$$

(16)

$$\hat{v}^{s,\varepsilon}(p_k - p_l) = \theta(\varrho^\varepsilon - |p_k - p_l|) \hat{v}(p_k - p_l).$$

(17)

The transition amplitude $|\hat{v}^{\ell,\varepsilon}(p_k - p_l)|^2$ is negligible for $\varrho \gg 1$ which can be inferred from (15). What remains to be bounded are the transitions in (14) with momentum transfer of order one, i.e.,

$$\frac{1}{L^4} \sum_{k=1}^{N} \sum_{l=N+1}^{\infty} |\hat{v}^{s,\varepsilon}(p_k - p_l)|^2 \left\| \sum_{i=1}^{2} \int_0^t d\tau e^{i(p_l^2 - p_k^2)\tau} k_{kl}^{(i)}(\tau) \varphi_0 \right\|^2.$$

(18)

The reason why this term vanishes as well is the oscillation of the integrand $e^{i(p_l^2 - p_k^2)\tau_1} k_{kl}^{(i)}(\tau_1) \varphi_0$. Outside a set of critical points of the phase for which $||p_l| - |p_k|| \leq \kappa(\varrho)$, for some appropriately small $\kappa(\varrho) \ll 1$, the energy shift grows rapidly: $p_l^2 - p_k^2 = (|p_l| + |p_k|)(|p_l| - |p_k|) \gtrsim \sqrt{\varrho}\kappa(\varrho) \gg 1$. By partial integration, one thus finds that

$$(18) \lesssim \frac{t^2}{L^4} \left[\sum_{k=1}^{N} \underbrace{\sum_{l=N+1}^{\infty}}_{\{ \text{ stationary points } \}} + \frac{1}{\varrho \kappa(\varrho)^2} \sum_{k=1}^{N} \sum_{l=N+1}^{\infty} \right] |\hat{v}^{s,\varepsilon}(p_k - p_l)|^2, \qquad (19)$$

which will be shown to vanish when ϱ tends to ∞, cf. Lemma 2.2. This result is the key ingredient to understand the proof of Theorem 1.1, and most of the contributions from the higher order Duhamel expansion in (12) can be estimated in an analogous way. However, there is one type of higher order processes for which the phase cancelation is not sufficiently strong.

This bring us to the next-to-leading order energy correction $W_{\text{eff},\varrho}(0)$ and the effective potential $W_{\text{eff},\varrho}(y_1 - y_2)$. It will be shown below that these two contributions which appear as

$$-i \int_0^t d\tau U^{\text{mf}}(t - \tau)(W_{\text{eff},\varrho}(0) + W_{\text{eff},\varrho}(y_1 - y_2))U^{\text{mf}}(\tau)\Psi_0 \qquad (20)$$

in first-order perturbation theory are canceled by terms from the second-order Duhamel expansion. Within the perturbation expansion, consider two immediate collisions with the same fermionic particle, where the second collision removes the particle–hole excitation which was caused in the first collision. This process leads to a contribution which cancels (20) up to a small error. In Sect. 2.5, this cancelation will be shown rigorously. Note that using an informal Dyson expansion, one can actually infer that the term $W_{\text{eff},\varrho}(0) + W_{\text{eff},\varrho}(y_1 - y_2)$ is due to processes from all orders of the expansion.

2 Proof of the Main Result

2.1 Notations and Definitions

In order to make the above heuristics more precise, let us start by introducing for $i \in \{1, 2\}$ the operators

$$k_{kl}^{(i)}(t) : \mathcal{H}_{y_1,y_2} \to \mathcal{H}_{y_1,y_2}, \qquad \varphi \mapsto k_{kl}^{(i)}(t)\varphi = e^{iH_{y_1,y_2}^{\text{eff}}t} e^{-i(p_k - p_l)y_i} e^{-iH_{y_1,y_2}^{\text{eff}}t}\varphi, \qquad (21)$$

$$g_{kl}^{(i)}(t) : \mathcal{H}_{y_1,y_2} \to \mathcal{H}_{y_1,y_2}, \qquad \varphi \mapsto g_{kl}^{(i)}(t)\varphi = e^{-i(p_k^2 - p_l^2)t} k_{kl}^{(i)}(t)\varphi, \qquad (22)$$

and

$$D(t) : \mathcal{H}_{y_1,y_2} \otimes \mathcal{H}_N \to \mathcal{H}_{y_1,y_2} \otimes \mathcal{H}_N, \qquad \Psi \mapsto D(t)\Psi = U(-t)U^{\mathrm{mf}}(t)\Psi. \tag{23}$$

We denote the Fourier transform of the potential v by $\hat{v}$, where $\hat{v}$ is defined such that

$$v(x) = \frac{1}{L^2} \sum_{k=1}^{\infty} \hat{v}(p_k)e^{ip_k x}. \tag{24}$$

Moreover, we use the following abbreviations:

- $\hat{v}_{kl} = \hat{v}(p_k - p_l)$,
- $\hat{v}_{kl}^{\ell,\varepsilon} = \theta\big(-\varrho^{\varepsilon} + |p_k - p_l|\big)\hat{v}_{kl}, \quad \hat{v}_{kl}^{s,\varepsilon} = \theta\big(\varrho^{\varepsilon} - |p_k - p_l|\big)\hat{v}_{kl}$,
- $E_k = p_k^2$,
- $\|\cdot\|_{\mathrm{TD}} = \lim_{\mathrm{TD}} \|\cdot\| = \lim_{N,L\to\infty,\varrho=const.} \|\cdot\|$,[1]
- $a^*(p_l)a(p_k)\Omega_0 = \Omega_0^{[l^*k]}$ and all kind of variations thereof. Here, $a^*(p_i)$ and $a(p_i)$ denote the fermionic creation and annihilation operators w.r.t. the mode $p_i \in (2\pi/L)\mathbb{Z}^2$.

2.2 Collision Histories

In the following, we will introduce wave functions $\Psi_1, ..., \Psi_4$ as well as $\Psi_{\mathrm{A}}, ..., \Psi_{\mathrm{F}}$ (see below), which are needed for the Duhamel expansion. As we explain hereafter, one can interpret $\Psi_{\mathrm{A}}, ..., \Psi_{\mathrm{F}}$ as different collision histories of the tracer particle. We set

$$\Psi_1^{(i,j)}(\tau_2, \tau_1) = \frac{1}{L^4} \sum_{k=1}^{N} \sum_{l=N+1}^{\infty} |\hat{v}_{kl}|^2 \Big(g_{lk}^{(i)}(\tau_2) g_{kl}^{(j)}(\tau_1)\varphi_0 \Big) \otimes \Omega_0, \tag{25}$$

$$\Psi_2^{(i,j)}(\tau_2, \tau_1) = \frac{1}{L^4} \sum_{k_1,k_2=1}^{N} \sum_{l=N+1}^{\infty} \hat{v}_{k_2 k_1} \hat{v}_{k_1 l} \Big(g_{k_2 k_1}^{(i)}(\tau_2) g_{k_1 l}^{(j)}(\tau_1)\varphi_0 \Big) \otimes \Omega_0^{[k_2 l^*]}, \tag{26}$$

$$\Psi_3^{(i,j)}(\tau_2, \tau_1) = \frac{1}{L^4} \sum_{k=1}^{N} \sum_{l_1,l_2=N+1}^{\infty} \hat{v}_{l_1 l_2} \hat{v}_{kl_1} \Big(g_{l_1 l_2}^{(i)}(\tau_2) g_{kl_1}^{(j)}(\tau_1)\varphi_0 \Big) \otimes \Omega_0^{[l_2^*k]}, \tag{27}$$

$$\Psi_4^{(i,j)}(\tau_2, \tau_1) = \frac{1}{L^4} \sum_{k_1,k_2=1}^{N} \sum_{l_1,l_2=N+1}^{\infty} \hat{v}_{k_2 l_2} \hat{v}_{k_1 l_1} \Big(g_{k_2 l_2}^{(i)}(\tau_2) g_{k_1 l_1}^{(j)}(\tau_1)\varphi_0 \Big) \otimes \Omega_0^{[l_2^* k_2 l_1^* k_1]}, \tag{28}$$

---

[1]Note that despite the chosen notation, $\|\cdot\|_{\mathrm{TD}}$ does not define a proper norm since $\|f\|_{\mathrm{TD}}$ may be zero for nonzero f.

which satisfy the equality (this is straightforward to verify)

$$U^{\text{mf}}(-\tau_2)(V-E)U^{\text{mf}}(\tau_2-\tau_1)(V-E)U^{\text{mf}}(\tau_1)\Psi_0 = \sum_{i,j=1}^{2} \left(\Psi_1^{(i,j)} + \Psi_2^{(i,j)} + \Psi_3^{(i,j)} + \Psi_4^{(i,j)} \right). \tag{29}$$

We further define

$$\Psi_A(t) = \int_0^t d\mu_2(\tau_2)U(-\tau_2)(W_{\text{eff},\varrho}(0) + W_{\text{eff},\varrho}(y_1 - y_2))U^{\text{mf}}(\tau_2-\tau_1)(V-E)U^{\text{mf}}(\tau_1)\Psi_0, \tag{30}$$

$$\Psi_B(t) = \int_0^t d\tau_1 U(-\tau_1)(W_{\text{eff},\varrho}(0) + W_{\text{eff},\varrho}(y_1 - y_2))U^{\text{mf}}(\tau_1)\Psi_0 + i\int_0^t d\mu_2(\tau)D(\tau_2) \sum_{i,j=1}^{2} \Psi_1^{(i,j)}(\tau_2,\tau_1), \tag{31}$$

$$\Psi_C(t) = \int_0^t d\mu_2(\tau)D(\tau_2) \sum_{i,j=1}^{2} \Psi_2^{(i,j)}(\tau_2,\tau_1), \tag{32}$$

$$\Psi_D(t) = \int_0^t d\mu_2(\tau)D(\tau_2) \sum_{i,j=1}^{2} \Psi_3^{(i,j)}(\tau_2,\tau_1), \tag{33}$$

$$\Psi_E(t) = \int_0^t d\mu_3(\tau)U(-\tau_3)\left(W_{\text{eff},\varrho}(0) + W_{\text{eff},\varrho}(y_1 - y_2)\right)U^{\text{mf}}(\tau_3) \sum_{i,j=1}^{2} \Psi_4^{(i,j)}(\tau_2,\tau_1), \tag{34}$$

$$\Psi_F(t) = \int_0^t d\mu_3(\tau)U(-\tau_3)(V-E)U^{\text{mf}}(\tau_3) \sum_{i,j=1}^{2} \Psi_4^{(i,j)}(\tau_2,\tau_1), \tag{35}$$

where we have introduced the shorthand notation

$$\int_0^t d\mu_n(\tau) = \int_0^t d\tau_1 \int_0^{\tau_1} d\tau_2 \dots \int_0^{\tau_{n-1}} d\tau_n. \tag{36}$$

Note that in comparison with [1, Section 2.2], the wave functions $\Psi_A(t)$, $\Psi_B(t)$, and $\Psi_E(t)$ read differently (the constant $E_{re}(\varrho)$ is replaced by $W_{\text{eff},\varrho}(0) + W_{\text{eff},\varrho}(y_1 - y_2)$), whereas $\Psi_C(t)$, $\Psi_D(t)$, and $\Psi_F(t)$ remain unchanged. The interesting difference is the appearance of $W_{\text{eff},\varrho}(y_1 - y_2)$ in $\Psi_B(t)$: Here, the effective potential, as well as the energy correction, needs to be included in order to cancel a nonvanishing contribution from $\Psi_1^{(i,j)}$.

The different wave functions $\Psi_X(t)$, $X \in \{A, B, C, D, E, F\}$ can be identified with the following collision histories of the tracer particles:

A: single collision which causes particle–hole excitations in the Fermi gas.
B: two collisions with the same fermionic particle, removing the particle–hole excitation which was caused in the first collision; the constant $W_{\text{eff},\varrho}(0)$, as well as the effective potential $W_{\text{eff},\varrho}(y)$, cancels the contribution in which the second collision follows immediately after the first one.

C: two collisions with the same fermionic particle; the second collision scatters the lifted particle into another momentum above the Fermi edge.

D: two collisions; the second collision scatters a particle from below the Fermi edge into the hole that was created in the first collision.

E: two collisions with two different particles; causing two particle–hole excitations.

F: three collisions; three particle–hole excitations but also all possible recollisions with the already scattered particles.

2.3 Proof of Theorem 1.1

In complete analogy to [1, Section 2.3], it can be shown that

$$\left\| U(t)\Psi_0 - U^{\text{mf}}(t)\Psi_0 \right\| \leq 2\left(\sqrt{\|\Psi_A(t)\|} + \sqrt{\|\Psi_B(t)\|} + \sqrt{\|\Psi_C(t)\|} \right.$$
$$\left. + \sqrt{\|\Psi_D(t)\|} + \sqrt{\|\Psi_E(t)\|} + \sqrt{\|\Psi_F(t)\|} \right). \quad (37)$$

The proof of the theorem is then completed by the following bounds for $\Psi_A, \ldots, \Psi_F$.

Lemma 2.1 *Let $\varepsilon > 0$ sufficiently small. Under the same assumptions as in Theorem 1.1, there exist positive constants C, C_ε such that for all $X \in \{A, B, C, D, E, F\}$ and any $t > 0$,*

$$\|\Psi_X(t)\|_{\mathcal{TD}} \leq C(1+t)^3 \left(\varrho^{-\frac{1}{4}+6\varepsilon} + C_\varepsilon \varrho^{\frac{1}{2}-\frac{1}{2\varepsilon}} \right). \quad (38)$$

2.4 Key Ingredients for the Proof of Lemma 2.1

Lemma 2.1 is analogous to [1, Lemma 2.1] where we only considered a single tracer particle coupled to the Fermi gas. In the present work, we do not carry out the full proof of Lemma 2.1, since to a large extent it is very similar to the proof of the single-tracer-particle case. We only state the key ingredients which go into the proof of the lemma and then elaborate on some details for the proof of (38) when $X = B$. As we already mentioned, this is the term which makes it necessary to include the $W_{\text{eff},\varrho}(0) + W_{\text{eff},\varrho}(y_1 - y_2)$ into the effective Hamiltonian. For the derivation of (38) when $X \in \{A, C, D, E, F\}$, one can use Lemmas 2.2 and 2.4 stated below, and then follow the strategy presented in [1] with only minor modifications.

In order to state the following lemmas, we need to introduce a suitable decomposition of the possible transitions in momentum space (the transitions appearing in $\Psi_1^{(i,j)}, \ldots, \Psi_4^{(i,j)}$). This is necessary in order to carry out the oscillating phase argument indicated below (18). To this end, let $\varepsilon > 0$, and define the two-dimensional index set

$$\mathfrak{S}^{\varepsilon}(N, \varrho) := \left\{ (k, l) : 1 \le k \le N, \ N + 1 \le l, \ |p_k - p_l| < \varrho^{\varepsilon} \right\} \subset \mathbb{N}^2, \quad (39)$$

and for $M \in \mathbb{N}$ the family of sets

$$\mathfrak{S}_n^{\varepsilon,M}(N, \varrho) := \left\{ (k, l) \in \mathfrak{S}^{\varepsilon}(N, \varrho) : \varrho^{-b_n} \le |p_l| - |p_k| < \varrho^{-b_{n+1}} \right\}, \quad 0 \le n \le M, \tag{40}$$

where

$$b_0 = \infty, \quad b_n = \frac{1}{2} - \frac{n-1}{M} \left(\frac{1}{2} + \varepsilon \right), \quad 1 \le n \le M. \tag{41}$$

For notational convenience, we omit the N-, ϱ-, ε-, and also the M-dependence in the notation: $\mathfrak{S} = \mathfrak{S}^{\varepsilon}(N, \varrho)$ and $\mathfrak{S}_n = \mathfrak{S}_n^{\varepsilon,M}(N, \varrho)$. The index set $\mathfrak{S}$ corresponds to the transitions that have to be controlled in (19), i.e., collisions with momentum transfer smaller than ϱ^{ε}. The sets of pairs of momenta $\{(p_k, p_l) \in (2\pi/L)^2 \mathbb{Z}^4 : (k, l) \in \mathfrak{S}_n\}$ are pairwise disjoint, and

$$\bigcup_{n=0}^{M} \left\{ (p_k, p_l) \in (2\pi/L)^2 \mathbb{Z}^4 : (k, l) \in \mathfrak{S}_n \right\} = \left\{ (p_k, p_l) \in (2\pi/L)^2 \mathbb{Z}^4 : (k, l) \in \mathfrak{S} \right\}. \tag{42}$$

The distance of modulus between the occupied momentum p_k and the new momentum state p_l increases in $\mathfrak{S}_n$ for increasing n:

$$|p_l| - |p_k| \ge \varrho^{-b_n} = \varrho^{-\frac{1}{2} + \frac{n-1}{M} \left(\frac{1}{2} + \varepsilon \right)} \quad \text{for } (k, l) \in \mathfrak{S}_n, \ 1 \le n \le M. \tag{43}$$

Hence, also the energy shift increases,

$$E_l - E_k = \left(|p_l| + |p_k| \right) \left(|p_l| - |p_k| \right) \ge k_F \varrho^{-b_n} = C \varrho^{\frac{n-1}{M} \left(\frac{1}{2} + \varepsilon \right)} \quad \text{for } (k, l) \in \mathfrak{S}_n, \tag{44}$$

$1 \le n \le M$. Note that ϱ^{-b_n} corresponds to the factor $\kappa(\varrho)$ in (19).

The key estimates that are used in order to prove Lemma 2.1 can now be summarized in the following two lemmas.

Lemma 2.2 *Assume $0 < \varepsilon < \frac{1}{2}$ and $M \in \mathbb{N}$. Let $v(x) \in C_0^{\infty}(\mathbb{T}^2) \cap C_0^{\infty}(\mathbb{R}^2)$ and $v^{\ell,\varepsilon}$, $v^{s,\varepsilon}$ defined as in (16), (17). Then there exist positive constants C, C_{ε} such that*

$$\lim_{TD} \frac{1}{L^4} \sum_{k=1}^{N} \sum_{l=N+1}^{\infty} |\hat{v}(p_k - p_l)| \le C\varrho^{\frac{1}{2}}, \tag{45}$$

$$\lim_{TD} \frac{1}{L^4} \sum_{k=1}^{N} \sum_{l=N+1}^{\infty} |\hat{v}^{\ell,\varepsilon}(p_k - p_l)| \le C_\varepsilon \varrho^{-1/\varepsilon}, \tag{46}$$

$$\lim_{TD} \frac{1}{L^4} \sum_{(k,l)\in\mathfrak{S}_0} \le C\varrho^{-\frac{1}{2}+\varepsilon}, \tag{47}$$

$$\lim_{TD} \frac{1}{L^4} \sum_{(k,l)\in\mathfrak{S}_n} \frac{1}{(E_l - E_k)^2} \le C\varrho^{-\frac{1}{2}+\varepsilon+\frac{1}{M}(\frac{1}{2}+\varepsilon)} \quad \text{for all } 1 \le n \le M, \tag{48}$$

$$|W_{\textit{eff},\varrho}(y)| \le C\varrho^{2\varepsilon} + C_\varepsilon \varrho^{-1/\varepsilon} \quad \text{for any } y \in \mathbb{R}. \tag{49}$$

Remark 2.3 The decomposition of $\mathfrak{S}$ into the sets $\mathfrak{S}_n$ is optimal in the sense that the r.h.s. of all estimates in (47) and (48) behaves asymptotically almost the same, namely, $\propto \varrho^{-\frac{1}{2}}$ for small ε and large M.

Except for (49), the lemma has been proven in [1, Section 2.2.]. The bounds in (45) and (46) depend mainly on the strong decay of $\hat{v}$, see (15). Let us also mention that the expression on the l.h.s. in (45) is equal to $\frac{1}{4}\|(V - E)\Omega_0\|^2$, and hence, it states that fluctuations in a dense ideal fermi gas are strongly suppressed compared to a bosonic or classical gas for which the variance of V would be at best proportional to the density ϱ (instead of $\sqrt{\varrho}$). In (47) and (48), one uses the definition of the $\mathfrak{S}_n$, the bound from (44), and then sums up (or integrates, after taking the thermodynamic limit) the remaining momentum space regions. The estimate in (49) follows directly from $|W_{\text{eff},\varrho}(y)| \le W_{\text{eff},\varrho}(0)$ and the bound for $E_{re}(\varrho) = W_{\text{eff},\varrho}(0)/2$, which has been derived as well in [1, Section 2.2].

Lemma 2.4 *Let $\varepsilon > 0$, $k_{kl}^{(i)}(t)$ as in (21), $g_{kl}^{(i)}(t)$ as in (22) and $\mathfrak{S} = \mathfrak{S}^\varepsilon(N, \varrho)$ as in (39). Given the same assumptions as in Theorem 1.1, the following bounds hold for all $\tau_1, \tau_2 \ge 0$ ($\chi_\mathfrak{S} : \mathbb{N}^2 \to \{0, 1\}$ denotes the characteristic function $\chi_\mathfrak{S}((k, l)) = 1$ when $(k, l) \in \mathfrak{S}$ and otherwise zero) and for $i, j \in \{1, 2\}$,*

$$\chi_\mathfrak{S}((k, l)) \, \|\partial_{\tau_1} k_{kl}^{(i)}(\tau_1)\varphi_0\| \le C\varrho^{2\varepsilon}(\varrho^{2\varepsilon} + C_\varepsilon \varrho^{-1/\varepsilon}.), \tag{50}$$

$$\chi_\mathfrak{S}((k, l)) \, \left(\|\partial_{\tau_2} k_{kl}(\tau_2) g_{kl}^{(i)}(\tau_1)\varphi_0\| + \|\partial_{\tau_2} k_{lk}(\tau_2) g_{kl}^{(i)}(\tau_1)\varphi_0\|\right) \le C\varrho^{2\varepsilon}(\varrho^{2\varepsilon} + C_\varepsilon \varrho^{-1/\varepsilon}.), \tag{51}$$

$$\chi_\mathfrak{S}((k, l))\chi_\mathfrak{S}((m, n)) \, \|\partial_{\tau_2} k_{lk}^{(i)}(\tau_2)\partial_{\tau_1} k_{mn}^{(j)}(\tau_1)\varphi_0\| \le C\varrho^{4\varepsilon}(\varrho^{2\varepsilon} + C_\varepsilon \varrho^{-1/\varepsilon}.), \tag{52}$$

$$\chi_\mathfrak{S}((k, l))\chi_\mathfrak{S}((m, n)) \, \|\partial_{\tau_1} k_{kl}^{(i)}(\tau_1) k_{mn}^{(j)}(\tau_1)\varphi_0\| \le C\varrho^{4\varepsilon}(\varrho^{2\varepsilon} + C_\varepsilon \varrho^{-1/\varepsilon}.) \tag{53}$$

This lemma tells us that integrating by parts in expressions like (18) is not expensive in terms of the ϱ-dependence, and hence we can really use the oscillation of the phase factor. The proof of the lemma follows from the definitions of $k_{kl}^{(j)}(t)$, $g_{kl}^{(j)}(t)$ and $\mathfrak{S}$, and then using regularity of $W \in C^2(\mathbb{R}^2)$ and the bound $\|\nabla^n W_{\text{eff},\varrho}\|_\infty \le \varrho^{2\varepsilon} + C_\varepsilon \varrho^{-1/\varepsilon}$, for $n = 1, 2$ which is obtained the same way as for $n = 0$. Note that W, $W_{\text{eff},\varrho}$, and its derivatives appear here due to the time derivative $\partial_{\tau_{1/2}}$.

2.5   Derivation of the Bound for $\|\Psi_B(t)\|_{TD}$

For $0 \le n \le M$ (M being some large integer), let

$$\Psi_{B}^{s,0}(t) = \frac{1}{L^4} \sum_{(k,l)\in\mathfrak{S}_0} |\hat{v}_{kl}^{s,\varepsilon}|^2 i \int_0^t d\mu_2(\tau) D(\tau_2) \sum_{i,j=1}^2 g_{lk}^{(i)}(\tau_2) g_{kl}^{(j)}(\tau_1) \Psi_0, \tag{54}$$

$$\Psi_{B,1}^{s,n}(t) = \frac{1}{L^4} \sum_{(k,l)\in\mathfrak{S}_n} |\hat{v}_{kl}^{s,\varepsilon}|^2 \int_0^t d\tau_2 D(\tau_2) \sum_{i,j=1}^2 \frac{g_{lk}^{(i)}(\tau_2) g_{kl}^{(j)}(t)}{(E_l - E_k)} \Psi_0, \tag{55}$$

$$\Psi_{B,2}^{s,n}(t) = \frac{1}{L^4} \sum_{(k,l)\in\mathfrak{S}_n} |\hat{v}_{kl}^{s,\varepsilon}|^2 \int_0^t d\tau_2 D(\tau_2) \sum_{i,j=1}^2 \frac{k_{lk}^{(i)}(\tau_2) k_{kl}^{(j)}(\tau_2)}{(E_l - E_k)} \Psi_0, \tag{56}$$

$$\Psi_{B,3}^{s,n}(t) = \frac{1}{L^4} \sum_{(k,l)\in\mathfrak{S}_n} |\hat{v}_{kl}^{s,\varepsilon}|^2 \int_0^t d\mu_2(\tau) D(\tau_2) \sum_{i,j=1}^2 \frac{g_{lk}^{(i)}(\tau_2) e^{i(E_l-E_k)\tau_1} \partial_{\tau_1} k_{kl}^{(j)}(\tau_1)}{(E_l - E_k)} \Psi_0, \tag{57}$$

$$\Psi_{B}^{\ell}(\tau) = \frac{1}{L^4} \sum_{l=1}^N \sum_{k=N+1}^{\infty} |\hat{v}_{kl}^{\ell,\varepsilon}|^2 \sum_{i,j=1}^2 g_{lk}^{(i)}(\tau_2) g_{kl}^{(j)}(\tau_1) \Psi_0. \tag{58}$$

Using that $|\hat{v}_{kl}|^2 = |\hat{v}_{kl}^{\ell,\varepsilon}|^2 + |\hat{v}_{kl}^{s,\varepsilon}|^2$, together with the decomposition defined in (40), and then a partial integration in the time variable, one verifies the identity

$$\Psi_B(t) = \int_0^t d\tau_1 U(-\tau_1)\Big(W_{\mathrm{eff},\varrho}(0) + W_{\mathrm{eff},\varrho}(y_1 - y_2)\Big) U^{\mathrm{mf}}(\tau_1)\Psi_0 - \sum_{n=1}^M \Psi_{B,2}^{s,n}(t)$$

$$+ \Psi_B^{s,0}(t) + \sum_{n=1}^M \Big[\Psi_{B,1}^{s,n}(t) - \Psi_{B,3}^{s,n}(t)\Big] + i \int_0^t d\mu_2(\tau) D(\tau_2) \Psi_B^{\ell}(\tau). \tag{59}$$

On the r.h.s. we have now identified the contribution in $\Psi_1^{(i,j)}$ which is canceled by the effective potential. To see this, we show that for any $M \ge 1$, the thermodynamic limit of the upper line is bounded in terms of

$$\left\| \int_0^t d\tau_1 U(-\tau_1)\Big(W_{\mathrm{eff},\varrho}(0) + W_{\mathrm{eff},\varrho}(y_1 - y_2)\Big) U^{\mathrm{mf}}(\tau_1)\Psi_0 - \sum_{n=1}^M \Psi_{B,2}^{s,n}(t) \right\|_{TD} \le C_\varepsilon \varrho^{-1/\varepsilon}. \tag{60}$$

Let us explain this in detail: In $\Psi_{B,2}^{s,n}(t)$, the factor $1/(E_l - E_k)$ is not sufficient to make this term small. This collision history corresponds to collisions and immediate recollisions between the tracer particles and the gas particles (instantaneous creation and annihilation of electron–hole pairs). As the particle–hole pair does not propagate in time, there is no additional phase factor that could be used here (in $\Psi_{B,1}^{s,n}$ and $\Psi_{B,3}^{s,n}$ such an additional phase factor is present which is the reason why these contributions will be small). Depending on the exact collision process, we obtain two distinct contributions: If the same tracer particle is causing the intermediate particle–hole excitation, the overall effect is given by the energy correction $W_{\mathrm{eff},\varrho}(0)$. The effective potential $W_{\mathrm{eff},\varrho}(y_1 - y_2)$ emerges if one tracer particle excites a gas particle, which subsequently interacts with the other tracer particle. To derive (60), we start by rewriting

$$D(\tau) = U(-\tau)U^{\mathrm{mf}}(\tau) = U(-\tau)e^{-i H_{y_1,y_2}^{\mathrm{eff}}\tau}e^{-i(-\sum_{i=1}^{N}\Delta_{x_i}+2\varrho\hat{v}(0)-W_{\mathrm{eff},\varrho}(0))\tau}.$$

Using

$$k_{lk}^{(i)}(\tau_2)k_{kl}^{(j)}(\tau_2) = e^{i H_{y_1,y_2}^{\mathrm{eff}}\tau_2}e^{i(p_k-p_l)(y_i-y_j)}e^{-i H_{y_1,y_2}^{\mathrm{eff}}\tau_2},$$

we obtain

$$D(\tau_2)\sum_{i,j=1}^{2}\frac{k_{lk}^{(i)}(\tau_2)k_{kl}^{(j)}(\tau_2)}{(E_l - E_k)}\Psi_0 = \sum_{i,j=1}^{2}U(-\tau_2)\frac{e^{i(p_k-p_l)(y_i-y_j)}}{(E_l - E_k)}U^{\mathrm{mf}}(\tau_2)\Psi_0$$

Abbreviating $\Gamma_{k,l}^{(i,j)}(t) = \int_0^t d\tau_2 U(-\tau_2)e^{i(p_k-p_l)(y_i-y_j)}U^{\mathrm{mf}}(\tau_2)\Psi_0$, it then follows that

$$\sum_{n=1}^{M}\Psi_{\mathrm{B},2}^{s,n}(t) = \sum_{i,j=1}^{2}\sum_{n=1}^{M}\frac{1}{L^4}\sum_{(k,l)\in\mathfrak{S}_n}\frac{|\hat{v}_{kl}^{s,\varepsilon}|^2}{(E_l - E_k)}\Gamma_{k,l}^{(i,j)}(t)$$

$$= \sum_{i,j=1}^{2}\frac{1}{L^4}\sum_{k=1}^{N}\sum_{l=N+1}^{\infty}\frac{|\hat{v}_{kl}|^2}{(E_l - E_k)}\theta\left(\varrho^\varepsilon - |p_k - p_l|\right)\theta\left(|p_l| - |p_k| - \varrho^{-\frac{1}{2}}\right)\Gamma_{k,l}^{(i,j)}(t).$$

$$(61)$$

We recall definition (3), that is

$$W_{\mathrm{eff},\varrho}(y) = 2\lim_{\mathrm{TD}}\frac{1}{L^4}\sum_{k=1}^{N}\sum_{l=N+1}^{\infty}\frac{|\hat{v}_{kl}|^2\cos((p_k - p_l)y)}{p_l^2 - p_k^2}\theta\left(|p_l| - |p_k| - \varrho^{-\frac{1}{2}}\right),$$

$$(62)$$

and introduce

$$\widetilde{W}_{\mathrm{eff},\varrho}^{\varepsilon}(y) = \frac{2}{L^4}\sum_{k=1}^{N}\sum_{l=N+1}^{\infty}\frac{|\hat{v}_{kl}|^2\cos((p_k - p_l)y)}{(E_l - E_k)}\theta\left(\varrho^\varepsilon - |p_k - p_l|\right)\theta\left(|p_l| - |p_k| - \varrho^{-\frac{1}{2}}\right).$$

$$(63)$$

It then follows that

$$\left\| \int_0^t d\tau_1 U(-\tau_1) \Big(W_{\text{eff},\varrho}(0) + W_{\text{eff},\varrho}(y_1 - y_2) \Big) U^{\text{mf}}(\tau_1)\Psi_0 - \sum_{n=1}^M \Psi_{B,2}^{s,n}(t) \right\|.$$

$$\leq \left\| \int_0^t d\tau U(-\tau) \left(W_{\text{eff},\varrho}(0) - \widetilde{W}_{\text{eff},\varrho}^\varepsilon(0) \right) U^{\text{mf}}(\tau)\Psi_0 \right\|$$

$$+ \left\| \int_0^t d\tau U(-\tau) \left(W_{\text{eff},\varrho}(y_1 - y_2) - \widetilde{W}_{\text{eff},\varrho}^\varepsilon(y_1 - y_2) \right) U^{\text{mf}}(\tau)\Psi_0 \right\|$$

$$\leq 2t \sup_{\tau \in [0,t]} \sup_{y \in \mathbb{T}^2} \left| W_{\text{eff},\varrho}(y) - \widetilde{W}_{\text{eff},\varrho}^\varepsilon(y) \right|. \tag{64}$$

For the remaining expression, we estimate

$$\lim_{\text{TD}} \left| W_{\text{eff},\varrho}(y) - \widetilde{W}_{\text{eff},\varrho}^\varepsilon(y) \right|$$

$$= 2 \left| \lim_{\text{TD}} \frac{1}{L^4} \sum_{k=1}^N \sum_{l=N+1}^\infty \frac{|\hat{v}_{kl}|^2 \cos((p_k - p_l)y)}{(E_l - E_k)} \theta\Big(-\varrho^\varepsilon + |p_k - p_l|\Big)\theta\Big(|p_l| - |p_k| - \varrho^{-\frac{1}{2}}\Big) \right|$$

$$\leq 2 \lim_{\text{TD}} \frac{1}{L^4} \sum_{k=1}^N \sum_{l=N+1}^\infty \frac{|\hat{v}_{kl}^{\ell,\varepsilon}|^2}{(E_l - E_k)} \theta\Big(|p_l| - |p_k| - \varrho^{-\frac{1}{2}}\Big)$$

$$\leq C \lim_{\text{TD}} \frac{1}{L^4} \sum_{k=1}^N \sum_{l=N+1}^\infty |\hat{v}_{kl}^{\ell,\varepsilon}|^2. \tag{65}$$

By (46), one thus obtains (60). Note that in the last step, we have used $E_l - E_k = (|p_l| - |p_k|)(|p_l| + |p_k|) \geq \varrho^{-\frac{1}{2}} k_F = C$ (since $|p_l| \geq k_F$).
It remains to estimate the other terms in (59). It follows with (48) that

$$\|\Psi_B^{s,0}(t)\|_{\text{TD}} \leq Ct^2 \sum_{(k,l) \in \mathfrak{S}_0} \leq Ct^2 \varrho^{-\frac{1}{2}+\varepsilon}, \tag{66}$$

as well as with (46),

$$\|\Psi_B^\ell(\tau)\|_{\text{TD}} \leq Ct^2 \lim_{\text{TD}} \frac{1}{L^4} \sum_{k_1=1}^N \sum_{l_1=N+1}^\infty |\hat{v}_{k_1 l_1}^{\ell,\varepsilon}|^2 \leq C_\varepsilon t^2 \varrho^{-1/\varepsilon}. \tag{67}$$

The bounds for the remaining wave functions are summarized in

Lemma 2.5 Let $\Psi_{B,1}^{s,n}(t)$ and $\Psi_{B,3}^{s,n}(t)$ as in (55) and (57). Then, under the same assumptions as in Theorem 1.1,

$$\|\Psi_{B,1}^{s,n}(t)\|_{\text{TD}} + \|\Psi_{B,3}^{s,n}(t)\|_{\text{TD}} \leq C(1+t)^2 \varrho^{-\frac{1}{2}+\varepsilon}\Big(\varrho^{\frac{1}{4}} + C_\varepsilon \varrho^{-1/\varepsilon}\Big)\varrho^{\frac{1}{M}(1+2\varepsilon)}, \tag{68}$$

holds for all $1 \leq n \leq M$ and any $t \geq 0$.

The proof of Lemma 2.5 follows from another partial integration in the time vari-

able, which gives an additional factor $1/(E_l - E_k)$. Moreover, one needs to use the estimate from (45) in order to control the time derivative of $D(t)$. The details can be found in [1, Section 2.4.3]. Taking the sum of all terms in (68), passing to the thermodynamic limit, and then choosing $M = \lfloor \ln \varrho \rfloor \leq \varrho^\varepsilon$, one finds

$$\sum_{n=1}^{M} \left(\|\Psi_{B,1}^{s,n}(t)\|_{TD} + \|\Psi_{B,3}^{s,n}(t)\|_{TD} \right) \leq C(1+t)^2 \left(\varrho^{-\frac{1}{4}+2\varepsilon} + C_\varepsilon \varrho^{-\frac{1}{2}+2\varepsilon-\frac{1}{\varepsilon}} \right). \quad (69)$$

This proves the bound for $\|\Psi_B(t)\|_{TD}$ in (38).

Acknowledgements M.J. and D.M. gratefully acknowledge financial support from the German Academic Scholarship Foundation. D.M. also acknowledges the support from Deutsche Forschungsgemeinschaft (DFG) through the Graduiertenkolleg 1838 "Spectral Theory and Dynamics of Quantum System".

References

1. Jeblick, M., Mitrouskas, D., Petrat, S., Pickl, P.: Free time evolution of a tracer particle coupled to a fermi gas in the high-density limit. Commun. Math. Phys. **356**(1), 143–187 (2017)
2. Bethe, H.A.: The electromagnetic shift of energy levels. Phys. Rev. **72**, 339–341 (1947)
3. Reed, M., Simon, B.: Methods of Modern Mathematical Physics II: Fourier Analysis, Self-adjointness. Academic Press, New York (1975)

Mean-Field Evolution of Fermions with Singular Interaction

Chiara Saffirio

Abstract We consider a system of N fermions in the mean-field regime interacting through an inverse power law potential $V(x) = |x|^{-\alpha}$, for $\alpha \in (0, 1]$. We prove the convergence of a solution of the many-body Schrödinger equation to a solution of the time-dependent Hartree–Fock equation in the sense of reduced density matrices, for a specific class of initial data, namely translation-invariant states. We stress the dependence on the singularity of the potential in the regularity of the initial data. The proof is an adaptation of Porta et al. (J. Stat. Phys. 166:1345–1364, 2017, [27]), where the case $\alpha = 1$ is treated.

1 Introduction

Fermionic mean-field regime. We consider a system of N particles obeying the Fermi statistics, whose state is represented by a wave function ψ_N lying in $L^2_a(\mathbb{R}^{3N})$, the space of square integrable functions antisymmetric in the exchange of particles. The Hamiltonian of the N particle system is given by

$$H_N^{\text{ext}} = \sum_{j=1}^{N} (-\varepsilon^2 \Delta_{x_j} + V_{\text{ext}}(x_j)) + \frac{1}{N} \sum_{i<j}^{N} V(x_i - x_j), \tag{1.1}$$

where V_{ext} is an external potential confining the system in a volume of order one. As $V_{\text{ext}} = 0$, the evolution in time of ψ_N is given by a solution to the Cauchy problem associated to the N-body Schrödinger equation, here denoted by $\psi_{N,t}$

C. Saffirio (✉)
Institute of Mathematics, University of Zurich, Winterthurerstrasse 190,
8057 Zurich, Switzerland
e-mail: chiara.saffirio@math.uzh.ch

© Springer Nature Switzerland AG 2018
D. Cadamuro et al. (eds.), *Macroscopic Limits of Quantum Systems*,
Springer Proceedings in Mathematics & Statistics 270,
https://doi.org/10.1007/978-3-030-01602-9_4

$$\begin{cases} i\varepsilon\partial_t\psi_{N,t} = \left[\sum_{j=1}^{N}(-\varepsilon^2\Delta_{x_j}) + \frac{1}{N}\sum_{i<j}^{N}V(x_i - x_j)\right]\psi_{N,t} \\ \\ \psi_{N,0} = \psi_N \in L_a^2(\mathbb{R}^{3N}). \end{cases} \tag{1.2}$$

In (1.2), the choice $\varepsilon = N^{-1/3}$ ensures the kinetic and the potential energy associated to (1.1) to be of comparable order, namely $O(N)$, for states initially confined in a volume of order one. This is due to the Fermi statistics that makes the kinetic energy of order $O(N^{5/3})$ (see Appendix A for more details). We underline that in this setting, in contrast with the bosonic case, the mean-field scaling for fermions comes coupled with a semiclassical limit (notice that here ε plays the role of $\hbar$) and it makes the analysis for fermions more complicated. Slightly different regimes have been analysed in [4–7, 16, 25, 26]. More precisely, [4, 25, 26] deal with states confined in a volume of order $O(N)$, while in [5–7, 16] the regime under consideration is such that the potential energy is sub-leading with respect to the potential energy.

Evolution of quasi-free states. To begin with, it is convenient to introduce the one-particle reduced density matrix of a wave function ψ_N as the nonnegative trace class operator

$$\gamma_N^{(1)} = N\mathrm{tr}_{2...N}|\psi_N\rangle\langle\psi_N|.$$

As we are interested in studying the dynamics of the system as $N \to \infty$, the choice of the initial data is crucial. Relevant initial states are given by the ground states of the Hamiltonian (1.1). At zero temperature the equilibrium of confined states is approximated by Slater determinants, i.e.,

$$\psi_{\mathrm{Slater}}(x_1, \ldots, x_N) = \frac{1}{\sqrt{N!}}\det(f_i(x_j))_{1\le i,j\le N} \tag{1.3}$$

where $\{f_j\}_{j=1,...,N}$ is an orthonormal system in the one-particle space $L^2(\mathbb{R}^3)$. A Slater determinant is a quasi-free state completely determined by its one-particle reduced density matrix. A simple computation shows that the one-particle reduced density associated to (1.3) is given by

$$\omega_N = \sum_{j=1}^{N}|f_j\rangle\langle f_j|,$$

the orthogonal projection on the N-dimensional linear space Span$\{f_1, \ldots, f_N\}$. The relevant initial data we want to consider minimize the Hartree–Fock energy functional

$$\mathcal{E}_{\mathrm{HF}}(\omega) = \mathrm{tr}\,(-\varepsilon^2\Delta + V_{\mathrm{ext}})\,\omega + \frac{1}{2N}\iint V(x-y)\,[\omega(x;x)\,\omega(y;y) - |\omega(x;y)|^2]\,dx\,dy. \tag{1.4}$$

As proved in [3, 18], the Hartree–Fock theory provides a good approximation of the ground state energy. It captures not only the leading order $O(N^{7/3})$ of the ground state as already established in the Thomas–Fermi theory (see [20, 21] for a review on the subject), but also errors smaller than $O(N^{5/3})$.

Convergence toward the Hartree–Fock dynamics. In [8, 13], it has been shown that the evolution of a Slater determinant approximating the ground state of the Hamiltonian (1.1) remains still close to a Slater determinant. Its evolved one-particle reduced density matrix associated to a Slater determinant is given by a solution to the time-dependent Hartree–Fock equation

$$i\varepsilon\partial_t\omega_{N,t} = [-\varepsilon^2\Delta + (V * \rho_t) - X_t, \, \omega_{N,t}], \tag{1.5}$$

that is the Euler–Lagrange equation of (1.4). For every $x \in \mathbb{R}^3$

$$\rho_t(x) = N^{-1}\omega_{N,t}(x; x)$$

is the density associated to the one-particle reduced density matrix $\omega_{N,t}$, $(V * \rho_t)$ represents the so-called direct term, while X_t is the exchange term defined through the operator kernel

$$X_t(x; y) = \frac{1}{N}V(x - y)\omega_{N,t}(x; y).$$

In [8], it has been proved that the Hartree–Fock approximation holds for initial states close to a Slater determinant with a semiclassical structure, namely

$$\omega_{N,t}(x; y) \simeq N\,\varphi\left(\frac{x - y}{\varepsilon}\right)\psi\left(\frac{x + y}{2}\right), \tag{1.6}$$

where ψ and φ determine, respectively, the density of particles and the momentum distribution. Heuristically, the integral kernel of $\omega_{N,t}$ varies in the direction $x - y$ on scales $O(\varepsilon)$ and in the direction $x + y$ on scales $O(1)$. This is precisely the structure that is expected to hold true in Slater determinants approximating equilibrium states and it reflects on the following structure:

$$\mathrm{tr}\,|[x, \omega_N]| \le CN\varepsilon \qquad \mathrm{tr}\,|[-i\varepsilon\nabla, \omega_N]| \le CN\varepsilon. \tag{1.7}$$

In [8], the propagation (global) in time of the bounds (1.7) has been shown, allowing for an approximation of the many-body Schrödinger equation by the Hartree–Fock dynamics on time scales of order one, while in [13] the approximation was proven only for short time intervals. The result in [8] has been extended to mixed states in [10] and to the case of fermions with semi-relativistic dispersion relation in [9]. A nice account of these results can be found in [12].

Mean-field in presence of singular interactions. When dealing with singular interactions $V(x) = 1/|x|^\alpha$, $\alpha \in (0, 1]$, the Hamiltonian takes the following form:

$$H_N = \sum_{i=1}^{N}(-\varepsilon^2 \Delta_{x_i}) + \frac{1}{N} \sum_{i<j}^{N} \frac{1}{|x-y|^\alpha}. \tag{1.8}$$

In particular, the case $\alpha = 1$ treated in [27] represents a system of N fermions interacting through a Coulomb potential, which describes, for instance, the dynamics of large atoms and molecules. In this case, the choice $\varepsilon = N^{-1/3}$ is justified by a rescaling of the space variables at a scale $O(N^{-1/3})$ (the typical distance of electrons from the nucleus) as suggested by the Thomas–Fermi theory (see [20, 21]). An analogous reasoning applies to the case of inverse power law potentials and, by appropriately scaling the time variable, it leads to

$$i\varepsilon \partial_t \psi_{N,t} = \left[\sum_{i=1}^{N}(-\varepsilon^2 \Delta_{x_i}) + \frac{1}{N} \sum_{i<j}^{N} \frac{1}{|x-y|^\alpha} \right] \psi_{N,t}. \tag{1.9}$$

More details on the rigorous justification of the mean-field scaling coupled to the semiclassical one in the Coulomb case can be found in [27].

We are now ready to introduce our main result:

Theorem 1.1. *Let ω_N be a sequence of orthogonal projections on $L^2(\mathbb{R}^3)$, with $\mathrm{tr}\,\omega_N = N$. Let $\omega_{N,t}$ be the solution of the Hartree–Fock equation (1.5) with initial data $\omega_{N,0} = \omega_N$. We assume that*

(i) $\mathrm{tr}\,(-\varepsilon^2 \Delta)\,\omega_N \leq CN$, for a constant $C > 0$ independent of N;
(ii) there exist $T > 0$, $p > 5/(3 - 2\alpha)$ and $C > 0$ such that

$$\sup_{t\in[0;T]} \sum_{i=1}^{3} \left[\|\rho_{|[x_i,\omega_{N,t}]|}\|_{L^1} + \|\rho_{|[x_i,\omega_{N,t}]|}\|_{L^p} \right] \leq CN\varepsilon, \tag{1.10}$$

where $\rho_{|[x_i,\omega_{N,t}]|}(z) = |[x_i, \omega_{N,t}]|(z; z)$ is the function obtained by considering the diagonal kernel of the operator $|[x_i, \omega_{N,t}]|$.

Let $\psi_N \in L_a^2(\mathbb{R}^{3N})$ be such that its one-particle reduced density matrix $\gamma_N^{(1)}$ satisfies

$$\mathrm{tr} \left| \gamma_N^{(1)} - \omega_N \right| \leq CN^\beta \tag{1.11}$$

for a constant $C > 0$ and an exponent $0 \leq \beta < 1$.

Consider the evolution $\psi_{N,t} = e^{-iH_N t/\varepsilon} \psi_N$, with the Hamiltonian (1.8) and let $\gamma_{N,t}^{(1)}$ be the corresponding one-particle reduced density matrix. Then, for every $\eta > 0$, there exists $C > 0$ such that

$$\sup_{t\in[0;T]} \left\| \gamma_{N,t}^{(1)} - \omega_{N,t} \right\|_{HS} \leq C \left[N^{\beta/2} + N^{(6-\alpha)/(6(3-\alpha))+\eta} \right] \tag{1.12}$$

and

$$\sup_{t\in[0;T]} \operatorname{tr} \left|\gamma_{N,t}^{(1)} - \omega_{N,t}\right| \leq C\left[N^\beta + N^{(15-4\alpha)/(6(3-\alpha))+\eta}\right]. \tag{1.13}$$

Remark 1.2. *Some remarks are in order.*

1. *The bounds* (1.12) *and* (1.13) *guarantee that the one-particle density matrix* $\gamma_{N,t}^{(1)}$ *is close to the solution of the Hartree–Fock equation in Hilbert–Schmidt and in trace norms as N is sufficiently large. Indeed, since* $\|\omega_{N,t}\|_{HS} = N^{1/2}$, $\|\gamma_{N,t}^{(1)}\|_{HS} = N^{1/2}$ *and* $\operatorname{tr} \omega_{N,t} = N$, $\operatorname{tr} \gamma_{N,t}^{(1)} = N$, *Eqs.* (1.12) *and* (1.13) *asserts that the difference between* $\gamma_{N,t}^{(1)}$ *and* $\omega_{N,t}$ *is smaller than the size of each component, thus the Hartree–Fock equation is a good approximation for the many-body evolution with singular interaction.*

2. *The exponent $0 \leq \beta < 1$ measures the initial number of excitations. In other words, β measures the number of particles that are not in the Slater determinant when $t = 0$.*

3. *The bounds are $N = \varepsilon^{-3}$ dependent. As already pointed out, it encodes the fact that the mean-field scaling is linked to a semiclassical limit in the fermionic setting. More precisely, as $N \to \infty$, the Wigner transform of a solution to the Hartree–Fock equation* (1.5) *converges (weakly) to a solution to the Vlasov equation. This statement has been proved in the case of regular interactions for pure states in [11]. Several results are available in the case of mixed states, see, for instance, [1, 2, 5, 17, 24, 29]. The papers [15, 22, 23] deal with singular interactions (here included the Coulomb case) for mixed states. For pure states, the semiclassical limit toward the Vlasov equation with singular interaction potential is an open problem.*

4. *Assumption (ii) is very strong. Indeed, we ask the bound to hold true for $\omega_{N,t}$ as $t \geq 0$. This is, in general, quite difficult to prove. In the context of regular interactions treated in [8], the authors deal with the propagation of the semiclassical structure* (1.7). *Grönwall inequality arguments employed in [8] to propagate the semiclassical structure* (1.7) *do not work here due to the singularity of the interaction. Of course, the less singular the potential is, the less integrability is required on $\rho_{|[x_i,\omega_{N,t}]|}$ and thus we expect condition* (1.10) *to be easier to verify. However, the problem of the derivation of the Hartree–Fock equation from a system of many interacting fermions is reduced to a PDE problem. Namely, it remains to prove that the bound in assumption (ii), if assumed at time $t = 0$, propagates at positive times. There is one special situation in which assumption (ii) is satisfied without any further requirement: the case of translation-invariant Slater determinants. This is, for instance, the case of a system of N fermions in a finite box of order one with periodic boundary conditions. In such a setting, Theorem 1.1 can be proved in the exact same way. Obviously, the Hartree–Fock dynamics becomes trivial, but it is interesting to see how the nontrivial dynamics given by the N-body Schrödinger equation can be approximated in the mean-field limit by a trivial one.*

5. *We underline that analogously to [8, 27] the exchange term does not play any role here and the exact same results hold true for $X_t = 0$. Moreover, as [9], we can extend the result to the case of fermions with semi-relativistic dispersion relation simply replacing hypothesis (i) by*

$$\text{tr} \left(\sqrt{-\varepsilon^2 \Delta + 1} \right) \omega_N \leq CN.$$

6. *Our result is not sensitive to the sign of the potential. Indeed, we can consider $V(x) = -1/|x|^\alpha$, though we believe assumption (ii) to be much harder to prove in this case.*

2   Second Quantization Formalism

Let us introduce the formalism of second quantization. We consider over $L^2(\mathbb{R}^3)$ the fermionic Fock space

$$\mathcal{F} = \bigoplus_{n \geq 0} L_a^2(\mathbb{R}^{3n}).$$

On $\mathcal{F}$, we define for every $f \in L^2(\mathbb{R}^3)$ the creation and annihilation operators in terms of operator valued distributions a_x^*, a_x, $x \in \mathbb{R}^3$,

$$a^*(f) = \int f(x) a_x^* \, dx, \quad a(f) = \int \bar{f}(x) a_x \, dx.$$

The second quantization of an operator O on $L^2(\mathbb{R}^3)$ is defined as

$$d\Gamma(O) = \int O(x; y) a_x^* a_y \, dx \, dy,$$

where $O(x; y)$ is the integral kernel of O. In particular, $O = 1$ corresponds to the second quantization of the number of particle operator $\mathcal{N}$

$$\mathcal{N} = \int a_x^* a_x \, dx.$$

Given a vector in the Fock space $\Psi \in \mathcal{F}$, the one-particle reduced density matrix associated with Ψ is the nonnegative trace class operator on $L^2(\mathbb{R}^3)$ whose kernel is given by

$$\gamma_\Psi^{(1)}(x; y) = \langle \Psi, a_y^* a_x \Psi \rangle. \tag{2.1}$$

Moreover, given $\Psi \in \mathcal{F}$ and a one-particle operator O on $L^2(\mathbb{R}^3)$, the expectation of the second quantization of O in the state Ψ is given by

$$\langle \Psi, d\Gamma(O)\, \Psi \rangle = \int O(x; y) \langle \Psi, a_x^* a_y\, \Psi \rangle \, dx \, dy = \text{tr } O\, \gamma_\Psi^{(1)}.$$

Notice that, according to this definition, the trace of the one-particle reduced density matrix (2.1) represents the expected number of particles in the state Ψ, namely

$$\text{tr } \gamma_\Psi^{(1)} = \langle \Psi, \mathcal{N}\Psi \rangle.$$

The next Lemma is taken from [8] and collects some useful bounds.

Lemma 2.1. *Let O be a bounded operator on $L^2(\mathbb{R}^3)$. Then, for every $\Psi \in \mathcal{F}$,*

$$\langle \Psi, d\Gamma(O)\, \Psi \rangle \leq \|O\| \langle \Psi, \mathcal{N}\Psi \rangle,$$
$$\|d\Gamma(O)\Psi\| \leq \|O\| \, \|\mathcal{N}\Psi\|.$$

Moreover, if O is a Hilbert–Schmidt operator,

$$\|d\Gamma(O)\Psi\| \leq \|O\|_{\text{HS}} \|\mathcal{N}^{1/2}\Psi\|,$$
$$\left\| \int O(x; y)\, a_x a_y\, \Psi \, dx \, dy \right\| \leq \|O\|_{\text{HS}} \|\mathcal{N}^{1/2}\Psi\|,$$
$$\left\| \int O(x; y)\, a_x^* a_y^*\, \Psi \, dx \, dy \right\| \leq \|O\|_{\text{HS}} \|\mathcal{N}^{1/2}\Psi\|.$$

If O is a trace class operator, we have

$$\|d\Gamma(O)\Psi\| \leq 2\text{tr } |O|,$$
$$\left\| \int O(x; y)\, a_x a_y\, \Psi \, dx \, dy \right\| \leq 2\text{tr } |O|,$$
$$\left\| \int O(x; y)\, a_x^* a_y^*\, \Psi \, dx \, dy \right\| \leq 2\text{tr } |O|.$$

We introduce the second quantization of the Hamiltonian (1.8) as the self-adjoint operator $\mathcal{H}_N$, whose restriction to the n-particle sector of the Fock space $\mathcal{F}$ is

$$\mathcal{H}_N|_{\mathcal{F}_n} = \sum_{j=1}^n -\varepsilon^2 \Delta_{x_j} + \frac{1}{N} \sum_{i<j}^n \frac{1}{|x_i - x_j|^\alpha}. \tag{2.2}$$

In particular, observe that the Hamiltonian $\mathcal{H}_N$ coincides with the Hamiltonian (1.8) when restricted to $\mathcal{F}_N$, the N-particle sector of the Fock space $\mathcal{F}$. Therefore, the dynamics of an initial data in $\mathcal{F}$ with N particles coincides with the evolution given by (1.2), where $V(x) = 1/|x|^\alpha$.

To define the second quantization of a Slater determinant (1.3), we introduce $\{f_j\}_{j=1,\dots,N}$, an orthonormal system in $L^2(\mathbb{R}^3)$, and the vacuum Ω. A Slater determinant on the Fock space $\mathcal{F}$ is given by

$$a^*(f_1)\ldots a^*(f_N)\Omega = \left\{0,\ldots,0,\frac{1}{\sqrt{N!}}\det(f_i(x_j))_{1\le i,j\le N},0,\ldots,0\right\}. \quad (2.3)$$

These states enjoy a remarkable structure: they can be obtained as the action of a Bogoliubov transformation on the vacuum Ω. More precisely, in our context a fermionic Bogoliubov transformation is a unitary linear map

$$\omega : L^2(\mathbb{R}^3) \oplus L^2(\mathbb{R}^3) \to L^2(\mathbb{R}^3) \oplus L^2(\mathbb{R}^3)$$

of the form

$$\omega = \begin{pmatrix} u & \bar{v} \\ v & \bar{u} \end{pmatrix}$$

where $u,\, v : L^2(\mathbb{R}^3) \to L^2(\mathbb{R}^3)$ are linear maps such that

$$u^*u + v^*v = 1,$$
$$u^*\bar{v} + v^*\bar{u} = 0,$$

and $\bar{v}$ denotes the complex conjugate of v, while v^* is the adjoint of v.

We say that the Bogoliubov transformation ω is implementable on the fermionic Fock space $\mathcal{F}$ if there exists a unitary operator $\mathfrak{R}_\omega : \mathcal{F} \to \mathcal{F}$ such that, for every $f,\, g \in L^2(\mathbb{R}^3)$,

$$\mathfrak{R}_\omega^*(a(f) + a^*(\bar{g}))\mathfrak{R}_\omega = a(\omega f) + a^*(\overline{\omega g}).$$

The Shale–Stinespring condition (see [28]) ensures that, if v is a Hilbert–Schmidt operator, then ω is indeed implementable. Thus, $\mathfrak{R}_\omega$ is the implementor of the Bogoliubov transformation ω.

We consider the one-particle reduced density matrix associated with the Slater determinant (2.3):

$$\omega_N = \sum_{i=1}^{N} |f_i\rangle\langle f_i|,$$

the orthogonal projection on the N-dimensional space given by $\mathrm{Span}\{f_1,\ldots,f_N\}$, where $\{f_i\}_{i=1,\ldots,N}$ are the orbitals of the Slater determinant. Moreover, there exists a unitary operator $\mathfrak{R}_{\omega_N}$, implementor of a Bogoliubov transformation, which generates a Slater determinant with orbitals $\{f_i\}_{i=1,\ldots,N}$, namely

$$\mathfrak{R}_{\omega_N}\Omega = a^*(f_1)\ldots a^*(f_N)\,\Omega, \quad (2.4)$$

such that for every $g \in L^2(\mathbb{R}^3)$

$$\mathfrak{R}_{\omega_N}^* a(g)\,\mathfrak{R}_{\omega_N} = a(u_N g) + a^*(\bar{v}_N\,\bar{g}) \quad (2.5)$$

where $u_N = 1 - \omega_N$, $v_N = \sum_{i=1}^{N} |\bar{f}_i\rangle\langle f_i|$. In other words, $\mathfrak{R}_{\omega_N}$ can be seen as the particle-hole transformation

$$\mathfrak{R}_{\omega_N} a^*(f_j) \mathfrak{R}^*_{\omega_N} = \begin{cases} a(f_j), & j \leq N \\ a^*(f_j), & j > N \end{cases}$$

where $\{f_j\}_{j=1,\dots,\infty}$ is the orthonormal basis of $L^2(\mathbb{R}^3)$ obtained by completing the orthonormal system $\{f_j\}_{j=1,\dots,N}$.

Equations (2.4) and (2.5) are convenient representations of Slater determinants and the main reason to look at the Fock space in this context. Indeed, $\mathfrak{R}_{\omega_N}$ describes fluctuations in the Slater determinant with reduced density ω_N: the Slater determinant is the new vacuum after the action of $\mathfrak{R}_{\omega_N}$; creation and annihilation operators act creating a particle outside the Slater determinant and annihilating a particle inside the Slater determinant.

3 Sketch of the Proof of Theorem 1.1

The proof of Theorem 1.1 is a direct consequence of the following Proposition (an adaptation of Theorem 2.2 in [27]).

Proposition 3.1. *Let ω_N be a sequence of orthogonal projections on $L^2(\mathbb{R}^3)$, with $\mathrm{tr}\,\omega_N = N$ and $\mathrm{tr}\,(-\varepsilon^2 \Delta)\,\omega_N \leq CN$. Let $\omega_{N,t}$ denote the solution of the Hartree–Fock equation (1.5) with initial data $\omega_{N,0} = \omega_N$. We assume that there exist $T > 0$, $p > 5/(3 - 2\alpha)$ and $C > 0$ such that*

$$\sup_{t\in[0;T]} \sum_{i=1}^{3} \left[\|\rho_{|[x_i,\omega_{N,t}]|}\|_{L^1} + \|\rho_{|[x_i,\omega_{N,t}]|}\|_{L^p} \right] \leq CN\varepsilon. \tag{3.1}$$

Let $\xi_N \in \mathcal{F}$ be a sequence with

$$\langle \xi_N, \mathcal{N}\xi_N \rangle \leq CN^{\beta}$$

for an exponent β, with $0 \leq \beta < 1$. We consider the evolution

$$\Psi_{N,t} = e^{-i\mathcal{H}_N t/\varepsilon} R_{\omega_N} \xi_N$$

and denote by $\gamma_{N,t}^{(1)}$ the one-particle reduced density of $\Psi_{N,t}$, as defined in (2.1). Then for any $\eta > 0$ small enough, there is a constant $C > 0$ such that

$$\sup_{t\in[0;T]} \left\| \gamma_{N,t}^{(1)} - \omega_{N,t} \right\|_{HS} \leq C \left[N^{\beta/2} + N^{1/2}\varepsilon^{(3-2\alpha-6\eta)/(2(3-\alpha))} \right]$$

and

$$\sup_{t \in [0;T]} \operatorname{tr} \left| \gamma_{N,t}^{(1)} - \omega_{N,t} \right| \leq C \left[N^\beta + N \varepsilon^{(3-2\alpha-6\eta)/(2(3-\alpha))} \right].$$

The proofs of Theorem 1.1 and Proposition 3.1 are immediate adaptations of Theorems 1.1 and 2.2 in [27] and can be found in Section 2 of [27]. The proof of Proposition 3.1 relies on the control on the growth of fluctuations established in Proposition 4.1. Therefore, one has to modify the proofs of Theorem 1.1 and Proposition 3.1 by using the estimates obtained in Proposition 4.1 instead of the one instead of the ones in Proposition 2.3 in [27].

To prove Proposition 4.1, we use a generalized Fefferman–de La Llave representation formula for the inverse power law potential $1/|x - y|^\alpha$ and the estimate, stated in the following Lemma, on the trace norm of the commutator between the multiplication operator given by the smooth function $\chi_{(r,z)}(x) = e^{|x-z|^2/r^2}$ and the solution $\omega_{N,t}$ to the Hartree–Fock equation (1.5):

Lemma 3.2 (Lemma 3.1 in [27]). *Let $\chi_{r,z}(x) = \exp(-x^2/r^2)$. Then, for all $0 < \delta < 1/2$ there exists $C > 0$ such that the pointwise bound*

$$\operatorname{tr} \left| [\chi_{(r,z)}, \omega_{N,t}] \right| \leq C r^{\frac{3}{2}-3\delta} \sum_{i=1}^{3} \|\rho_{|[x_i,\omega_{N,t}]|}\|_{L^1}^{\frac{1}{6}+\delta} \left(\rho_{|[x_i,\omega_{N,t}]|}^*(z) \right)^{\frac{5}{6}-\delta} \tag{3.2}$$

holds true. Here, $\varrho_{|[x_i,\omega_{N,t}]|}^*$ denotes the Hardy–Littlewood maximal function defined by

$$\rho_{|[x_i,\omega_{N,t}]|}^*(z) = \sup_{B:z \in B} \frac{1}{|B|} \int_B dx \, \rho_{|[x_i,\omega_{N,t}]|}(x) \tag{3.3}$$

with the supremum taken over all balls $B \in \mathbb{R}^3$ such that $z \in B$.

The proof of the above Lemma can be found in [27].

4   Control on the Growth of Fluctuations

We consider $\xi_N \in \mathcal{F}$ a vector in the fermionic Fock space such that the number of particles in the state ξ_N is bounded by a power of N, i.e.,

$$\langle \xi_N, \mathcal{N}\xi_N \rangle \leq C N^\beta$$

for some $\beta \in [0, 1)$ and a positive constant C. The time-evolution of ξ_N in the Fock space is given by the action of the semigroup generated by the Hamiltonian (2.2)

$$\Psi_{N,t} = e^{-i\mathcal{H}_N t/\varepsilon} \mathfrak{R}_{\omega_N} \xi_N.$$

In order to prove Proposition 3.1 (and thus Theorem 1.1) we compare the one-particle reduced density of $\Psi_{N,t}$ with the solution $\omega_{N,t}$ of the time-dependent Hartree–Fock equation (1.5). To this end, we introduce the *fluctuation dynamics*

$$\mathcal{U}_N(t) = \mathfrak{R}^*_{\omega_{N,t}} e^{-i\mathcal{H}_N t/\varepsilon} \mathfrak{R}_{\omega_N} \tag{4.1}$$

so that

$$\Psi_{N,t} = \mathfrak{R}_{\omega_{N,t}} \mathcal{U}_N(t) \xi_N,$$

where $\mathcal{U}_N(t)\xi_N \in \mathcal{F}$ describes the excitations of the Slater determinant at time $t > 0$. The following Proposition ensures the boundedness of the expectation of the number of excitations of the Slater determinant in the state $\Psi_{N,t}$.

Proposition 4.1. *Let ω_N be a fermionic operator such that $0 \leq \omega_N \leq 1$, $\omega_N^2 = \omega_N$ and* $\operatorname{tr} \omega_N = N$. *Assume that*

(i) $\operatorname{tr} (-\varepsilon^2 \Delta)\omega_N \leq CN$;
(ii) there exist a time $T > 0$ and a number $p > 5/(3 - 2\alpha)$ such that

$$\sup_{t \in [0,T]} \sum_{i=1}^3 \left[\|\rho_{|[x_i, \omega_{N,t}]|}\|_{L^1} + \|\rho_{|[x_i, \omega_{N,t}]|}\|_{L^p} \right] \leq C N \varepsilon$$

where C is a positive constant, possibly time dependent.

Let $\mathcal{U}_N(t)$ be the fluctuation dynamics defined in (4.1) and $\xi_N \in \mathcal{F}$, $\|\xi_N\| = 1$. Then, for every $\eta > 0$ small enough, there exists a positive constant C such that

$$\sup_{t \in [0,T]} \langle \mathcal{U}_N(t) \xi_N, \mathcal{N} \mathcal{U}_N(t) \xi_N \rangle \leq C \left[\langle \xi_N, \mathcal{N}\xi_N \rangle + N\varepsilon^{(3-2\alpha-6\eta)/(3-\alpha)} \right]. \tag{4.2}$$

Proof. To bound (4.2), we look for a Grönwall-type estimate on the quantity, which represents the expectation of excitations of the Slater determinant. We perform the time derivative of $\langle \mathcal{U}_N(t)\xi_N, \mathcal{N}\mathcal{U}_N(t)\xi_N \rangle$ and straightforward computations (see Proposition 3.3 in [8] for details) lead to

$$i\varepsilon \partial_t \langle \mathcal{U}_N(t)\xi_N, \mathcal{N}\mathcal{U}_N(t)\xi_N \rangle$$
$$= \frac{4i}{N} \mathfrak{Im} \iint \frac{1}{|x-y|^\alpha}$$
$$\left\{ \langle \mathcal{U}_N(t)\xi_N, a^*(u_{t,x}) a(\overline{v_{t,y}}) a(u_{t,y}) a(u_{t,x}) \mathcal{U}_N(t)\xi_N \rangle \right. \tag{4.3}$$
$$\langle \mathcal{U}_N(t)\xi_N, a^*(u_{t,y}) a^*(\overline{v_{t,y}}) a^*(\overline{v_{t,x}}) a(\overline{v_{t,x}}) \mathcal{U}_N(t)\xi_N \rangle$$
$$\left. \langle \mathcal{U}_N(t)\xi_N, a(\overline{v_{t,x}}) a(\overline{v_{t,y}}) a(u_{t,y}) a(u_{t,x}) \mathcal{U}_N(t)\xi_N \rangle \right\} dx \, dy$$
$$= I + II + III$$

where $u_{t,x}(z) = u_{N,t}(x; z)$, $v_{t,x}(z) = v_{N,t}(x; z)$,

$$I = \frac{4i}{N} \operatorname{Im} \iint \frac{1}{|x-y|^\alpha} \langle \mathcal{U}_N(t)\xi_N, \, a^*(u_{t,x}) a(\overline{v_{t,y}}) a(u_{t,y}) a(u_{t,x}) \mathcal{U}_N(t)\xi_N \rangle \, dx \, dy,$$

$$II = \frac{4i}{N} \operatorname{Im} \iint \frac{1}{|x-y|^\alpha} \langle \mathcal{U}_N(t)\xi_N, \, a^*(u_{t,y}) a^*(\overline{v_{t,y}}) a^*(\overline{v_{t,x}}) a(\overline{v_{t,x}}) \mathcal{U}_N(t)\xi_N \rangle \, dx \, dy,$$

$$III = \frac{4i}{N} \operatorname{Im} \iint \frac{1}{|x-y|^\alpha} \langle \mathcal{U}_N(t)\xi_N, \, a(\overline{v_{t,x}}) a(\overline{v_{t,y}}) a(u_{t,y}) a(u_{t,x}) \mathcal{U}_N(t)\xi_N \rangle \, dx \, dy.$$

To bound the r.h.s. in (4.3), we make use of a smooth version of a generalization of the Fefferman–de La Llave representation formula for radial potentials [14, 19]. In [19], an explicit expression for radial potentials which exhibit some decay at infinity is provided. In the context under consideration such a formula resumes in

$$\frac{1}{|x-y|^\alpha} = \frac{4}{\pi^2} \int_0^\infty \frac{dr}{r^{4+\alpha}} \int dz \, \chi_{(r,z)}(x) \, \chi_{(r,z)}(y) , \qquad (4.4)$$

where $\chi_{(r,z)}(x) = e^{-|x-z|^2/r^2}$. The advantage of this representation consists in the fact that the smooth part of the inverse power law potential represented by $\chi_{(r,z)}(\cdot)$ is decoupled from the singular part. In the formulation we are using, this representation is useful to isolate the commutator structure which is estimated in Lemma 3.2 (see Eq. (4.5) below).

Therefore, plugging (4.4) in I, we obtain

$$I \le \frac{C}{N} \int_0^\infty \frac{dr}{r^{4+\alpha}} \iiint \chi_{(r,z)}(x) \chi_{(r,z)}(y)$$

$$\langle \mathcal{U}_N(t)\xi_N, \, a^*(u_{t,x}) a(\overline{v}_{t,y}) a(u_{t,y}) a(u_{t,x}) \mathcal{U}_N(t)\xi_N \rangle \, dx \, dy \, dz$$

$$= \frac{C}{N} \int_0^\infty \frac{dr}{r^{4+\alpha}} \iint \chi_{(r,z)}(x) \langle \mathcal{U}_N(t)\xi_N, \, a^*(u_{t,x}) \, B_{r,z} \, a(u_{t,x}) \mathcal{U}_N(t)\xi_N \rangle \, dx \, dz$$

where

$$B_{r,z} = \int a(\overline{v_{t,y}})\chi_{(r,z)}(y)a(u_{t,y}) \, dy = \iint (\overline{v}_{N,t}\chi_{(r,z)}u_{N,t})(s_1; s_2) \, a_{s_1} a_{s_2} \, ds_1 \, ds_2.$$

Lemma 2.1 and the fact that u and v are orthogonal yield

$$\|B_{r,z}\| \le 2 \operatorname{tr} |\overline{v}_{N,t} \chi_{(r,z)} u_{N,t}| \le 2 \operatorname{tr} |[\chi_{(r,z)}, \omega_{N,t}]|. \qquad (4.5)$$

Therefore,

$$I \le \frac{C}{N} \int_0^\infty \frac{dr}{r^{4+\alpha}} \iint \chi_{(r,z)}(x) \operatorname{tr} |[\chi_{(r,z)}, \omega_{N,t}]| \, \|a(u_{t,x})\mathcal{U}_N(t)\xi_N\|^2 \, dx \, dz .$$

By Lemma 3.2, we get

$$|I| \leq C \frac{(N\varepsilon)^{\frac{1}{6}+\delta}}{N} \sum_{i=1}^{3} \int_0^\infty \frac{dr}{r^{\frac{5}{2}+\alpha+3\delta}} \int b_{r,i}(x) \, \|a(u_{t,x})\mathcal{U}_N(t)\xi_N\|^2 \, dx$$

$$\leq C \frac{(N\varepsilon)^{\frac{1}{6}+\delta}}{N} \sum_{i=1}^{3} \int_0^\infty \frac{dr}{r^{\frac{5}{2}+\alpha+3\delta}} \langle \mathcal{U}_N(t)\xi_N, \, d\Gamma(u_{N,t}b_{r,i}(x)u_{N,t})\mathcal{U}_N(t)\xi_N \rangle$$

where $d\Gamma(u_{N,t}b_{r,i}(x)u_{N,t})$ is the second quantization of the operator $u_{N,t}b_{r,i}(x)u_{N,t}$ with integral kernel

$$(u_{N,t}b_{r,i}(x)u_{N,t})(s_2; s_2) = \int u_{N,t}(s_1; x)b_{r,i}(x)u_{N,t}(x; s_2) \, ds_1 \, ds_2$$

and $b_{r,i}$ is defined as

$$b_{r,i}(x) = \int \chi_{(r,z)}(x) \left(\rho_{|[x_i, \omega_{N,t}]|}^*(z) \right)^{\frac{5}{6}-\delta} dz \, .$$

Notice that $\|u_{N,t}\| \leq 1$, thus Lemma 2.1 yields

$$|I| \leq C \frac{(N\varepsilon)^{\frac{1}{6}+\delta}}{N} \sum_{i=1}^{3} \int_0^\infty \frac{dr}{r^{\frac{5}{2}+\alpha+3\delta}} \|b_{r,i}\|_{L^\infty} \|\mathcal{N}^{1/2}\mathcal{U}_N(t)\xi_N\|^2 \, .$$

Hardy–Littlewood maximal inequality then implies

$$\|b_{r,i}\|_{L^\infty} \leq r^{\frac{3}{p}} \|\rho_{|[x_i, \omega_{N,t}]|}^*\|_{L^{(\frac{5}{6}-\delta)q}}^{\frac{5}{6}-\delta} \leq C r^{\frac{3}{p}} \|\rho_{|[x_i, \omega_{N,t}]|}\|_{L^{(\frac{5}{6}-\delta)q}}^{\frac{5}{6}-\delta} \tag{4.6}$$

where p, q are conjugated Hölder exponents coupled by the relation $\frac{1}{p} + \frac{1}{q} = 1$, with the constraint $\frac{6}{5-6\delta} < q < \infty$ in order to ensure $q > 1$ so that the last inequality in the r.h.s. of Eq. (4.6) holds true.

Now, we split the integral in the r variable into two parts: let $k > 0$ be a fixed positive number, then

$$|I| \leq C \frac{(N\varepsilon)^{\frac{1}{6}+\delta}}{N} \sum_{i=1}^{3} \left[\int_0^k \frac{dr}{r^{\frac{5}{2}+\alpha+3\delta-\frac{3}{p}}} \|\rho_{|[x_i, \omega_{N,t}]|}\|_{L^{(\frac{5}{6}-\delta)q}}^{\frac{5}{6}-\delta} \|\mathcal{N}^{1/2}\mathcal{U}_N(t)\xi_N\|^2 \right.$$

$$\left. + \int_k^\infty \frac{dr}{r^{\frac{5}{2}+\alpha+3\delta-\frac{3}{p'}}} \|\rho_{|[x_i, \omega_{N,t}]|}\|_{L^{(\frac{5}{6}-\delta)q'}}^{\frac{5}{6}-\delta} \|\mathcal{N}^{1/2}\mathcal{U}_N(t)\xi_N\|^2 \right]$$

where (p, q) and (p', q') are chosen to guarantee integrability of the integral in the r variable, namely

$$p < \frac{6}{3 + 2\alpha + 6\delta}, \qquad q > \frac{6}{3 - 2\alpha - 6\delta},$$

$$p' > \frac{6}{3 + 2\alpha + 6\delta}, \qquad q' < \frac{6}{3 - 2\alpha - 6\delta}.$$

With these choices, using hypothesis (ii) in Proposition 4.1, we obtain for every $t \in [0, T]$

$$|I| \leq C\varepsilon \|\mathcal{N}^{1/2}\mathcal{U}_N(t)\xi_N\|^2 = C\varepsilon \langle \mathcal{U}_N(t)\xi_N, \mathcal{N}\mathcal{U}_N(t)\xi_N \rangle. \tag{4.7}$$

The second term on the r.h.s. of (4.3) can be handled analogously

$$
\begin{aligned}
II &\leq \frac{C}{N} \iint \frac{1}{|x-y|^\alpha} \langle \mathcal{U}_N(t)\xi_N, a^*(\overline{v}_{t,x}) \, a^*(u_{t,x}) \, a^*(\overline{v}_{t,y}) \, a(\overline{v}_{t,x}) \mathcal{U}_N(t)\xi_N \rangle \, dx \, dy \\
&= \frac{C}{N} \int_0^\infty \frac{dr}{r^{4+\alpha}} \iiint \chi_{(r,z)}(x)\chi_{(r,z)}(y) \\
&\qquad\qquad \langle \mathcal{U}_N(t)\xi_N, a^*(\overline{v}_{t,x}) \, a^*(u_{t,x}) \, a^*(\overline{v}_{t,y}) \, a(\overline{v}_{t,x})\mathcal{U}_N(t)\xi_N \rangle \, dz \, dx \, dy \\
&= \frac{C}{N} \int dx \int_0^\infty \frac{dr}{r^{4+\alpha}} \int dz \chi_{(r,z)}(x) \langle \mathcal{U}_N(t)\xi_N, a^*(\overline{v}_{t,x}) \, B^*_{r,z} \, a(\overline{v}_{t,x})\mathcal{U}_N(t)\xi_N \rangle \\
&\leq C \frac{(N\varepsilon)^{\frac{1}{6}+\delta}}{N} \sum_{i=1}^3 \int_0^\infty \frac{dr}{r^{\frac{5}{2}+\alpha+3\delta}} \iint dx \, dz \, \chi_{(r,z)}(x) \left(\rho^*_{|[x_i,\omega_{N,t}]|}(z) \right)^{\frac{5}{6}-\delta} \\
&\qquad\qquad\qquad\qquad\qquad\qquad\qquad\qquad\qquad\qquad\qquad \|a(\overline{v}_{t,x})\mathcal{U}_N(t)\xi_N\|^2 \\
&\leq C \frac{(N\varepsilon)^{\frac{1}{6}+\delta}}{N} \sum_{i=1}^3 \int_0^\infty \frac{dr}{r^{\frac{5}{2}+\alpha|3\delta}} \langle \mathcal{U}_N(t)\xi_N, d\Gamma(\overline{v}_{N,t} \, b_{r,i}(x) \, \overline{v}_{N,t})\mathcal{U}_N(t)\xi_N \rangle \\
&\leq C \frac{(N\varepsilon)^{\frac{1}{6}+\delta}}{N} \sum_{i=1}^3 \int_0^\infty \frac{dr}{r^{\frac{5}{2}+\alpha+3\delta}} \|b_{r,i}\|_{L^\infty} \|\mathcal{N}^{1/2}\mathcal{U}_N(t)\xi_N\|^2
\end{aligned}
$$

and we conclude as in (4.7)

$$|II| \leq C\varepsilon \|\mathcal{N}^{1/2}\mathcal{U}_N(t)\xi_N\|^2 = C\varepsilon \langle \mathcal{U}_N(t)\xi_N, \mathcal{N}\mathcal{U}_N(t)\xi_N \rangle. \tag{4.8}$$

In order to close the Grönwall estimate, we need to bound III. This term, together with the initial quantity $\langle \xi_N, \mathcal{N}\xi_N \rangle$, determines the function of N which bounds the expectation of the number of fluctuations. To deal with III, we use again Eq. (4.4) to get

$$
\begin{aligned}
III &\leq \frac{C}{N} \int_0^\infty \frac{dr}{r^{4+\alpha}} \iiint \chi_{(r,z)}(x)\chi_{(r,z)}(y) \\
&\qquad\qquad \langle \mathcal{U}_N(t)\xi_N, a(\overline{v}_{t,x}) \, a(\overline{v}_{t,y}) \, a(u_{t,y}) \, a(u_{t,x})\mathcal{U}_N(t)\xi_N \rangle \, dz \, dx \, dy.
\end{aligned}
$$

We fix $k > 0$, to be chosen later as an ε dependent function, and divide III into two parts:

$$III_1 = \frac{C}{N} \int_0^k \frac{dr}{r^{4+\alpha}} \iiint \chi_{(r,z)}(x)\chi_{(r,z)}(y)$$

$$\langle \mathcal{U}_N(t)\xi_N, a(\overline{v}_{t,x})\,a(\overline{v}_{t,y})\,a(u_{t,y})\,a(u_{t,x})\mathcal{U}_N(t)\xi_N\rangle\,dz\,dx\,dy$$

$$III_2 = \frac{C}{N} \int_k^\infty \frac{dr}{r^{4+\alpha}} \iiint \chi_{(r,z)}(x)\chi_{(r,z)}(y)$$

$$\langle \mathcal{U}_N(t)\xi_N, a(\overline{v}_{t,x})\,a(\overline{v}_{t,y})\,a(u_{t,y})\,a(u_{t,x})\mathcal{U}_N(t)\xi_N\rangle\,dz\,dx\,dy.$$

As for the term III_1, we observe that the integral in the z variable cancels part of the singularity in the r variable by producing a factor r^3, i.e.,

$$\int \chi_{(r,z)}(x)\chi_{(r,z)}(y)\,dz = r^3\chi_{(\sqrt{2}r,x)}(y). \tag{4.9}$$

Plugging (4.9) into III_1, we obtain

$$III_1 = \frac{C}{N} \int_0^k \frac{dr}{r^{1+\alpha}} \int \langle \mathcal{U}_N(t)\xi_N, B_{\sqrt{2}r,x}\,a(\overline{v}_{t,x})\,a(u_{t,x})\mathcal{U}_N(t)\xi_N\rangle\,dx$$

$$\le \frac{C}{N} \int_0^k \frac{dr}{r^{1+\alpha}} \int \rho_t(x)^{1/2}\|B_{\sqrt{2}r,x}\|\,\|a(u_{t,x})\mathcal{U}_N(t)\xi_N\|\,dx$$

$$\le \frac{C}{N} \int_0^k \frac{dr}{r^{1+\alpha}} \int \rho_t(x)^{1/2}\mathrm{tr}\,|[\chi_{(\sqrt{2}r,x)}, \omega_{N,t}]|\,\|a(u_{t,x})\mathcal{U}_N(t)\xi_N\|\,dx,$$

where we have used

$$\|\overline{v}_{N,t}\|^2 = \omega_{N,t}(x;x) = \rho_t(x)$$

in the first inequality and (4.5) in the last inequality. Lemma 3.2, together with hypothesis (ii), yields

$$|III_1| \le C\frac{(N\varepsilon)^{\frac{1}{6}+\delta}}{N} \sum_{i=1}^3 \int_0^k dr\, r^{\frac{1}{2}-\alpha-3\delta} \int dx\, \rho_t(x)^{1/2} \sum_{i=1}^3 \|\rho_{|[x_i,\omega_{N,t}]}\|_{L^1}^{\frac{1}{6}+\delta} \left(\rho_{|[x_i,\omega_{N,t}]}^*(x)\right)^{\frac{5}{6}-\delta}$$

$$\le C\frac{(N\varepsilon)^{\frac{1}{6}+\delta}}{N} k^{\frac{3}{2}-\alpha-3\delta} \int \rho_t(x)^{1/2} \left(\rho_{|[x_i,\omega_{N,t}]}^*(x)\right)^{\frac{5}{6}-\delta} \|a(u_{t,x})\mathcal{U}_N(t)\xi_N\|\,dx$$

$$\tag{4.10}$$

Hölder inequality and hypothesis (ii) imply

$$|III_1| \le C\frac{(N\varepsilon)^{\frac{1}{6}+\delta}}{N} k^{\frac{3}{2}-\alpha-3\delta}\|\rho_t\|_{L^{\frac{5}{3}}}^{\frac{1}{2}} \sum_{i=1}^3 \|\rho_{|[x_i,\omega_{N,t}]}^*\|_{L^{\frac{25}{6}-5\delta}}^{\frac{5}{6}-\delta} \left(\int \|a(u_{t,x})\mathcal{U}_N(t)\xi_N\|^2\,dx\right)^{\frac{1}{2}}$$

$$\le C\frac{(N\varepsilon)}{N} k^{\frac{3}{2}-\alpha-3\delta}\|\rho_t\|_{L^{\frac{5}{3}}}^{\frac{1}{2}} \langle \mathcal{U}_N(t)\xi_N, d\Gamma(u_{N,t})\mathcal{U}_N(t)\xi_N\rangle$$

$$\le C\varepsilon\, k^{\frac{3}{2}-\alpha-3\delta}\|\rho_t\|_{L^{\frac{5}{3}}}^{\frac{1}{2}} \|\mathcal{N}^{1/2}\mathcal{U}_N(t)\xi_N\|^2.$$

Moreover, the $L^{5/3}$ norm of ρ_t can be bounded in terms of the initial data by standard kinetic energy inequalities that we report in Appendix A for completeness.

Thus by hypothesis (i), for every $t \in [0, T]$,

$$
\begin{aligned}
|III_1| &\leq C \sqrt{N} \, \varepsilon \, k^{\frac{3}{2}-\alpha-3\delta} \|\mathcal{N}^{1/2}\mathcal{U}_N(t)\xi_N\| \\
&\leq \varepsilon \|\mathcal{N}^{1/2}\mathcal{U}_N(t)\xi_N\|^2 + C\,N\,\varepsilon\,k^{3-2\alpha-6\delta}.
\end{aligned} \tag{4.11}
$$

As for the second term in III, we notice that

$$
\begin{aligned}
|III_2| &\leq \frac{C}{N} \int_k^\infty \frac{dr}{r^{4+\alpha}} \int \|B_{r,z}\|^2 \, dz \\
&\leq \frac{C}{N} \int_k^\infty \frac{dr}{r^{4+\alpha}} \int \mathrm{tr} \, |[\chi_{(r,z)}, \omega_{N,t}]| \, dz \\
&\leq C \frac{(N\varepsilon)^2}{N} \int_k^\infty \frac{dr}{r^{1+\alpha+6\delta}}
\end{aligned}
$$

The term on the r.h.s. is integrable for $\alpha > 0$ and $\delta \in (0, 1/2)$, thus

$$
|III_2| \leq C\,N\,\varepsilon^2\,k^{-\alpha-6\delta}. \tag{4.12}
$$

Combining the bounds (4.11) and (4.12), we obtain

$$
|III| \leq \varepsilon \|\mathcal{N}^{1/2}\mathcal{U}_N(t)\xi_N\|^2 + C\,N\,\varepsilon\,k^{3-2\alpha-6\delta} + C\,N\,\varepsilon^2\,k^{-\alpha-6\delta}.
$$

By optimizing in k, we get

$$
k = \varepsilon^{1/(3-\alpha)}
$$

and, therefore, III is bounded by

$$
|III_3| \leq \varepsilon \|\mathcal{N}^{1/2}\mathcal{U}_N(t)\xi_N\|^2 + C\,N\,\varepsilon^{3(2-\alpha-2\delta)/(3-\alpha)} \tag{4.13}
$$

Equations (4.7), (4.8) and (4.13) lead to a control on the growth of fluctuations quantified by the following Grönwall inequality

$$
\left| \frac{d}{dt} \langle \mathcal{U}_N(t)\xi_N, \mathcal{N}\mathcal{U}_N(t)\xi_N \rangle \right| \leq C \langle \mathcal{U}_N(t)\xi_N, \mathcal{N}\mathcal{U}_N(t)\xi_N \rangle + C\,N\,\varepsilon^{(3-2\alpha-6\delta)/(3-\alpha)}
$$

that implies, for every $t \in [0, T]$,

$$
\langle \mathcal{U}_N(t)\xi_N, \mathcal{N}\mathcal{U}_N(t)\xi_N \rangle \leq C \left[\langle \xi_N, \mathcal{N}\xi_N \rangle + N\varepsilon^{(3-2\alpha-6\delta)/(3-\alpha)} \right]
$$

and the Proposition is proved.   □

Acknowledgements The author is supported by the grant SNSF Ambizione PZ00P2_161287/1.

A Kinetic Energy Estimates

To bound the $L^{5/3}$ norm of the density ρ_t, we observe that the Lieb–Thirring inequality and the positivity of the interaction potential yield

$$\|\rho_t\|_{L^{5/3}}^{5/3} \leq \operatorname{tr}(-\Delta)\omega_{N,t} \leq \varepsilon^{-2}\mathcal{E}_{\mathrm{HF}}(\omega_{N,t})$$

where $\mathcal{E}_{\mathrm{HF}}$ is the Hartree–Fock energy functional defined in (1.4). Conservation of energy implies

$$\|\rho_t\|_{L^{5/3}}^{5/3} \leq \varepsilon^{-2}\mathcal{E}_{\mathrm{HF}}(\omega_{N,t}) = \varepsilon^{-2}\mathcal{E}_{\mathrm{HF}}(\omega_N).$$

To close the estimate using the assumption on the kinetic energy of the initial sequence $\operatorname{tr}(-\varepsilon^2\Delta)\omega_N \leq CN$, we observe that the potential energy can be bounded by the kinetic energy. Indeed, the Hardy–Littlewood–Sobolev inequality yields

$$\frac{1}{N}\int \frac{1}{|x-y|^\alpha}\rho_0(x)\rho_0(y)\,dx\,dy \leq \frac{C}{N}\|\rho_0\|_{L^{6/(5-\alpha)}}^2$$

By interpolation, using that $\frac{6}{6-\alpha} \in (1, \frac{5}{3})$ and $\|\rho_0\|_{L^1} = N$, we have

$$\frac{1}{N}\int \frac{1}{|x-y|^\alpha}\rho_0(x)\rho_0(y)\,dx\,dy \leq \frac{C}{N}\|\rho_0\|_{L^1}^{\frac{12-5\alpha}{6}}\|\rho_0\|_{L^{\frac{5}{3}}}^{\frac{5}{6}\alpha}$$

$$\leq CN^{1-\frac{5}{6}\alpha}\|\rho_0\|_{L^{\frac{5}{3}}}^{\frac{5}{6}\alpha}$$

$$\leq CN + N^{-\frac{2}{3}\alpha}\|\rho_0\|_{L^{\frac{5}{3}}}^{\frac{5}{3}\alpha},$$

where in the last line we have used Young's inequality $ab \leq \frac{a^p}{p} + \frac{b^q}{q}, p^{-1}+q^{-1}=1$, on the quantities $a = N^{1-\frac{\alpha}{2}}$ and $b = N^{-\frac{\alpha}{3}}\|\rho_0\|_{L^{\frac{5}{3}}}^{\frac{5}{6}\alpha}$ with $p = 2/(2-\alpha)$ and $q = 2/\alpha$. Thus, applying again the Lieb–Thirring inequality and recalling that $\varepsilon = N^{-1/3}$, we obtain

$$\frac{1}{N}\int \frac{1}{|x-y|^\alpha}\rho_0(x)\rho_0(y)\,dx\,dy \leq CN + C\operatorname{tr}(-\varepsilon^2\Delta)\omega_N$$

which gives the bound

$$\frac{1}{N}\int \frac{1}{|x-y|^\alpha}\rho_0(x)\rho_0(y)\,dx\,dy \leq CN$$

thanks to assumption (i).

References

1. Amour, L., Khodja, M., Nourrigat, J.: The semiclassical limit of the time dependent Hartree–Fock equation: the Weyl symbol of the solution. Anal. PDE **6**(7), 1649–1674 (2013)
2. Athanassoulis, A., Paul, T., Pezzotti, F., Pulvirenti, M.: Strong semiclassical approximation of Wigner functions for the Hartree dynamics. Rend. Lincei Mat. Appl. **22**, 525–552 (2011)
3. Bach, V.: Error bound for the Hartree–Fock energy of atoms and molecules. Commun. Math. Phys. **147**(3), 527–548 (1992)
4. Bach, V., Breteaux, S., Petrat, S., Pickl, P., Tzaneteas, T.: Kinetic energy estimates for the accuracy of the time-dependent Hartree–Fock approximation with Coulomb interaction. J. Math. Pures Appl. **105**(1), 1–30 (2016)
5. Bardos, C., Golse, F., Gottlieb, A.D., Mauser, N.J.: Mean-field dynamics of fermions and the time-dependent Hartree–Fock equation. J. Math. Pures Appl. **82**(6), 665–683 (2003)
6. Bardos, C., Golse, F., Gottlieb, A.D., Mauser, N.J.: Accuracy of the time-dependent Hatree–Fock approximation for uncorrelated initial states. J. Stat. Phys. **115**(3–4), 1037–1055 (2004)
7. Bardos, C., Ducomet, B., Golse, F., Mauser, N.J.: The TDHF approximation for Hamiltonians with m-particle interaction potentials. Commun. Math. Sci. **1**, 1–9 (2007)
8. Benedikter, N., Porta, M., Schlein, B.: Mean-field evolution of fermionic systems. Commun. Math. Phys. **331**(3), 1087–1131 (2014)
9. Benedikter, N., Porta, M., Schlein, B.: Mean-field dynamics of fermions with relativistic dispersion. J. Math. Phys. **55**, 021901 (2014) (10 pp.)
10. Benedikter, N., Jaksic, V., Porta, M., Saffirio, C., Schlein, B.: Mean-field evolution of fermionic mixed states. Commun. Pure Appl. Math. **69**(12), 2250–2303 (2016)
11. Benedikter, N., Porta, M., Saffirio, C., Schlein, B.: From the Hartree dynamics to the Vlasov equation. Arch. Ration. Mech. Anal. **221**(1), 273–334 (2016)
12. Benedikter, N., Porta, M., Schlein, B.: Effective Evolution Equations from Quantum Dynamics. Springer Briefs in Mathematical Physics, vol. 7. Springer, Berlin (2016)
13. Elgart, A., Erdős, L., Schlein, B., Yau, H.-T.: Nonlinear Hartree equation as the mean-field limit of weakly coupled fermions. J. Math. Pures Appl. (9) **83**(10), 1241–1273 (2004)
14. Fefferman, Ch.L., de la Llave, R.: Relativistic Stability of Matter - I. Rev. Mat. Iberoam. **2**(2), 119–213 (1986)
15. Figalli, A., Ligabò, M., Paul, T.: Semiclassical limit for mixed states with singular and rough potentials. Indiana Univ. Math. J. **61**(1), 193–222 (2013)
16. Fröhlich, J., Knowles, A.: A microscopic derivation of the time-dependent Hartree–Fock equation with Coulomb two-body interaction. J. Stat. Phys. **145**(1), 23–50 (2011)
17. Golse, F., Paul, T.: The Schrödinger equation in the mean-field and semiclassical regime. Arch. Ration. Mech. Anal. **223**, 57–94 (2017)
18. Graf, G.M., Solovej, J.P.: A correlation estimate with applications to quantum systems with Coulomb interactions. Rev. Math. Phys. **6**, 977–997 (1994)
19. Hainzl, C., Seiringer, R.: General decomposition of radial functions on $\mathbb{R}^n$ and applications to N-body quantum systems. Lett. Math. Phys. **61**(1), 75–84 (2002)
20. Lieb, E.H.: Thomas–Fermi and related theories of atoms and molecules. Rev. Mod. Phys. **53**(4), 603–641 (1981)
21. Lieb, E.H., Simon, B.: The Thomas–Fermi theory of atoms, molecules and solids. Adv. Math. **23**, 22–116 (1977)
22. Lions, P.-L., Paul, T.: Sur les mesures de Wigner. Rev. Mat. Iberoam. **9**, 553–618 (1993)
23. Markowich, P.A., Mauser, N.J.: The classical limit of a self-consistent quantum Vlasov equation. Math. Model. Methods Appl. Sci. **3**(1), 109–124 (1993)
24. Narnhofer, H., Sewell, G.L.: Vlasov hydrodynamics of a quantum mechanical model. Commun. Math. Phys. **79**(1), 9–24 (1981)
25. Petrat, S.: Hartree corrections in a mean-field limit for fermions with Coulomb interaction. J. Phys. A Math. Theor. (24), 244004 (2017) (19 pp.)
26. Petrat, S., Pickl, P.: A new method and a new scaling for deriving fermionic mean-field dynamics. Math. Phys. Anal. Geom. **19**(1), 51 pp. (2016)

27. Porta, M., Rademacher, S., Saffirio, C., Schlein, B.: Mean-field evolution of fermions with Coulomb interaction. J. Stat. Phys. **166**, 1345–1364 (2017)
28. Solovej, J.P.: Many Body Quantum Mechanics. Lecture Notes. Summer 2007. www.mathematik.uni-muenchen.de/~lowsorensen/Lehre/SoSe2013/MQM2/skript.pdf
29. Spohn, H.: On the Vlasov hierarchy. Math. Methods Appl. Sci. **3**(4), 445–455 (1981)

Recent Advances in the Theory of Bogoliubov Hamiltonians

Marcin Napiórkowski

Dedicated to Herbert Spohn on the occasion of his 70th birthday

Abstract Bosonic quadratic Hamiltonians, often called Bogoliubov Hamiltonians, play an important role in the theory of many-boson systems where they arise in a natural way as an approximation to the full many-body problem. In this note, we would like to give an overview of recent advances in the study of bosonic quadratic Hamiltonians. In particular, we relate the reported results to what can be called the time-dependent diagonalization problem.

1 Introduction

Bosonic quadratic Hamiltonians are operators that are formally given by expressions of the following form:

$$\mathbb{H} = \sum_{ij} h_{ij} a_i^* a_j + \frac{1}{2} \sum_{ij} k_{ij} a_i^* a_j^* + \frac{1}{2} \sum_{ij} \bar{k}_{ij} a_i a_j. \tag{1}$$

Here, h_{ij} is a self-adjoint matrix ($h_{ij} = h_{ij}^* = \bar{h}_{ij}^{\#}$), $\bar{k}_{ij}$ is the complex conjugate of the matrix k_{ij} and a_i^*/a_i are the bosonic creation/annihilation operators on the Fock space satisfying the canonical commutation relations

$$[a_i^*, a_j^*] = [a_i, a_j] = 0, \qquad [a_i, a_j^*] = \delta_{ij}.$$

M. Napiórkowski (✉)
Faculty of Physics, Department of Mathematical Methods in Physics,
University of Warsaw, Pasteura 5, 02-093 Warszawa, Poland
e-mail: marcin.napiorkowski@fuw.edu.pl

© Springer Nature Switzerland AG 2018
D. Cadamuro et al. (eds.), *Macroscopic Limits of Quantum Systems*,
Springer Proceedings in Mathematics & Statistics 270,
https://doi.org/10.1007/978-3-030-01602-9_5

101

If the sums in (1) are finite, which corresponds to the fact that the underlying one-particle Hilbert space is finite dimensional, then the operator $\mathbb{H}$ is well defined. However, if the one-particle Hilbert space is infinite dimensional, then the analysis of the operator (1) becomes more complicated. This includes even a proper definition of the formal expression given above. The goal of this note is to review recent results on $\mathbb{H}$ in the infinite-dimensional case.

Bogoliubov Hamiltonians are used in many different situations. For example, in the context of quantum field theory they appear as quantization of a scalar field with a mass-like perturbation [8]. In the context of many-body quantum mechanics, quadratic Hamiltonians arise in a natural way as an approximation to the full many-body problem. In fact, the name *Bogoliubov Hamiltonian* is related to such an approximation performed by Nikolay Bogoliubov in the famous 1947 paper *On the theory of superfluidity* [6]. A system of N bosons (confined in a box of size $V = L^3$ with periodic boundary conditions) that interact through a two-body potential v is described by the second quantized many-body Hamiltonian

$$H = \sum_{p \in \frac{2\pi}{L} \mathbb{Z}^3} p^2 a_p^* a_p + \frac{1}{2L^3} \sum_{p,q,k \in \frac{2\pi}{L} \mathbb{Z}^3} \hat{v}(k) a_{p+k}^* a_{q-k}^* a_q a_p. \tag{2}$$

As explained in detail in [31], Bogoliubov argued that at low energies, this system can be effectively described by a quadratic Hamiltonian

$$\mathbb{H}_{\text{Bog}} = \frac{N^2}{2V} + \sum_{p \neq 0} p^2 a_p^* a_p + \frac{N}{2V} \sum_{p \neq 0} \hat{v}(p)(2a_p^* a_p + a_p^* a_{-p}^* + a_p a_{-p}). \tag{3}$$

Finally, Bogoliubov noticed that introducing operators

$$b_p = c_p a_p + s_p a_{-p}^*, \tag{4}$$

with some appropriately chosen real c_p and s_p, one can *diagonalize* $\mathbb{H}_{\text{Bog}}$, that is rewrite it in the following form:

$$\mathbb{H}_{\text{Bog}} = E + \sum_p e(p) b_p^* b_p \tag{5}$$

where the constant E and function $e(p)$ depends on the parameters of the system (in particular, they depend on the interaction potential v). From (5), one can easily determine the spectrum of $\mathbb{H}_{\text{Bog}}$. Also, (5) provides a basis for the interpretation that at small energies, the collective behaviour of many particles can be interpreted as a system of non-interacting *quasiparticles* [11, 26].

The procedure that was applied by Bogoliubov to obtain (3) from (2) has since then been formulated in a more abstract setting and has become a standard tool in different subfields of condensed matter physics [12]. A rigorous justification of this approximation, which is, for example proving that the spectrum of $\mathbb{H}_{\text{Bog}}$ indeed approximates the spectrum of H, has only been achieved in some situations [4, 5, 10, 14, 20, 21, 27].

The other important idea introduced by Bogoliubov, which is the diagonalization of (3) leading to (5), is what we want to address in this review. Using the so-called Lie formula, it is easy to see that (4) can be written as

$$b_p = c_p a_p + s_p a^*_{-p} = e^{-X} a_p e^X$$

with $X = \sum_{p \neq 0} \beta_p (a^*_p a^*_{-p} - a_p a_{-p})$ for some appropriately chosen β_p. Since e^X is unitary, we see that (5) has been obtained from (3) by a unitary transformation on the Fock space. In particular, the b_p operators also obey the canonical commutation relations. The obvious questions arise: Can this be done for any quadratic bosonic Hamiltonian? If not, then under which conditions?

Before we can review the answers to these questions, we need to formulate them in a rigorous setting. This will be done in the next section. In Sect. 3, we will sketch the main ideas behind the proofs. These parts of this review are based on [25]. In Sect. 4, we will mention a result about time-dependent quadratic Hamiltonians. Finally, in Sect. 5, we will make a connection between the previous two sections in addressing the issue of what can be called a time-dependent diagonalization problem.

2 Main Result

2.1 Fock Space Formalism

Let us introduce the mathematical setting. Our one-body Hilbert space $\mathfrak{h}$ is a complex separable Hilbert space with inner product $\langle ., . \rangle$ which is linear in the second variable and anti-linear in the first. The bosonic Fock space is defined by

$$\mathcal{F}(\mathfrak{h}) := \bigoplus_{N=0}^{\infty} \bigotimes_{\text{sym}}^{N} \mathfrak{h} = \mathbb{C} \oplus \mathfrak{h} \oplus (\mathfrak{h} \otimes_s \mathfrak{h}) \oplus \cdots$$

Let $h > 0$ be a self-adjoint operator on $\mathfrak{h}$. Then, it can be lifted to the Fock space by

$$d\Gamma(h) := \bigoplus_{N=0}^{\infty} \left(\sum_{j=1}^{N} h_j \right) = 0 \oplus h \oplus (h \otimes 1 + 1 \otimes h) \oplus \cdots$$

A typical example is that $h = -\Delta + V(x)$ on $\mathfrak{h} = L^2(\mathbb{R}^d)$, where V is an external potential which serves to bind the particles. Another example is given by the identity operator. Then, $d\Gamma(1)$ is the particle number operator denoted by $\mathcal{N}$. The operator $d\Gamma(h)$ is well defined on the core

$$\bigcup_{M \geq 0} \overset{M}{\bigoplus_{n=0}} \overset{n}{\bigotimes_{\text{sym}}} D(h)$$

and it can be extended to a positive self-adjoint operator on Fock space by Friedrichs' extension.

To describe quadratic Hamiltonians on bosonic Fock space, we introduce the *creation* and *annihilation* operators. For any vector $f \in \mathfrak{h}$, the creation operator $a^*(f)$ and the annihilation operator $a(f)$ are defined by the following actions:

$$a^*(f_{N+1}) \left(\sum_{\sigma \in S_N} f_{\sigma(1)} \otimes \cdots \otimes f_{\sigma(N)} \right) = \frac{1}{\sqrt{N+1}} \sum_{\sigma \in S_{N+1}} f_{\sigma(1)} \otimes \cdots \otimes f_{\sigma(N+1)},$$

$$a(f_{N+1}) \left(\sum_{\sigma \in S_N} f_{\sigma(1)} \otimes \cdots \otimes f_{\sigma(N)} \right) = \sqrt{N} \sum_{\sigma \in S_N} \langle f_{N+1}, f_{\sigma(1)} \rangle f_{\sigma(2)} \otimes \cdots \otimes f_{\sigma(N)},$$

for all $f_1, \ldots, f_{N+1}$ in $\mathfrak{h}$, and all $N = 0, 1, 2, \ldots$. These operators satisfy the canonical commutation relations (CCRs)

$$[a(f), a(g)] = 0, \quad [a^*(f), a^*(g)] = 0, \quad [a(f), a^*(g)] = \langle f, g \rangle, \quad \forall f, g \in \mathfrak{h}. \quad (6)$$

In particular, for every $f \in \mathfrak{h}$ we have

$$a(f)|0\rangle = 0$$

where $|0\rangle = 1 \oplus 0 \oplus 0 \cdots$ is the Fock space vacuum. If $\{f_n\}_{n \geq 1} \subset D(h)$ is an arbitrary orthonormal basis for $\mathfrak{h}$, then it is common to denote the operators $a^*(f_i)$ and $a(f_i)$ by a_i^* and a_i, respectively. This explains the notation used in the introduction. In particular, the operators a_p used in the presentation of Bogoliubov's approach correspond to $a(V^{-1/2}e^{-ipx})$ (plane waves basis).

2.2 The Hamiltonian

In general, a quadratic Hamiltonian on Fock space is a linear operator which is quadratic in terms of creation and annihilation operators. For example, $d\Gamma(h)$ is a quadratic Hamiltonian because we can write

$$d\Gamma(h) = \sum_{m,n \geq 1} \langle f_m, h f_n \rangle a^*(f_m) a(f_n)$$

(the sum on the right side is independent of the choice of the basis). Thus, the matrix h_{ij} in (1) is given by $h_{ij} = \langle f_i, hf_j \rangle$. In this paper, we will consider a general quadratic operator of the following form:

$$\mathbb{H} = d\Gamma(h) + \frac{1}{2} \sum_{m,n \geq 1} \left(\langle J^*kf_m, f_n \rangle a(f_m)a(f_n) + \overline{\langle J^*kf_m, f_n \rangle} a^*(f_m)a^*(f_n) \right) \quad (7)$$

where $k : \mathfrak{h} \to \mathfrak{h}^*$ is an unbounded linear operator with $D(h) \subset D(k)$ (called *pairing operator*) and $J : \mathfrak{h} \to \mathfrak{h}^*$ is the anti-unitary operator[1] defined by

$$J(f)(g) = \langle f, g \rangle, \quad \forall f, g \in \mathfrak{h}.$$

Since $\mathbb{H}$ remains the same when k is replaced by $(k + Jk^*J)/2$, we will always assume without loss of generality that

$$k^* = J^*kJ^*. \quad (8)$$

In fact, the formula (7) is formal but $\mathbb{H}$ can be defined properly as a quadratic form as follows. For every normalized vector $\Psi \in \mathcal{F}(\mathfrak{h})$ with finite particle number expectation, namely $\langle \Psi, \mathcal{N}\Psi \rangle < \infty$, its one-particle density matrices $\gamma_\Psi : \mathfrak{h} \to \mathfrak{h}$ and $\alpha_\Psi : \mathfrak{h} \to \mathfrak{h}^*$ are linear operators defined by

$$\langle f, \gamma_\Psi g \rangle = \langle \Psi, a^*(g)a(f)\Psi \rangle, \quad \langle Jf, \alpha_\Psi g \rangle = \langle \Psi, a^*(g)a^*(f)\Psi \rangle, \quad \forall f, g \in \mathfrak{h}. \quad (9)$$

A formal calculation using (7) leads to the expression

$$\langle \Psi, \mathbb{H}\Psi \rangle = \mathrm{Tr}(h^{1/2}\gamma_\Psi h^{1/2}) + \Re\mathrm{Tr}(k^*\alpha_\Psi). \quad (10)$$

The formula (10) makes sense when $h^{1/2}\gamma_\Psi h^{1/2}$ and $k^*\alpha_\Psi$ are trace class operators. We will use (10) to define $\mathbb{H}$ as a quadratic form with a dense form domain described below.

Since $\mathfrak{h}$ is separable and $D(h)$ is dense in $\mathfrak{h}$, we can choose finite-dimensional subspaces $\{Q_n\}_{n=1}^\infty$ such that

$$Q_1 \subset Q_2 \subset \ldots \subset D(h) \quad \text{and} \quad \overline{\bigcup_{n \geq 1} Q_n} = \mathfrak{h}.$$

Then, it is straightforward to verify that

$$\mathcal{Q} := \bigcup_{M \geq 0} \bigcup_{n \geq 1} \left(\bigotimes_{\mathrm{sym}}^{M} Q_n \right) \quad (11)$$

---

[1]If $C : \mathfrak{h} \to \mathfrak{K}$ is anti-linear, then $C^* : \mathfrak{K} \to \mathfrak{h}$ is defined by $\langle C^*g, f \rangle_\mathfrak{h} = \langle Cf, g \rangle_\mathfrak{K}$ for all $f \in \mathfrak{h}, g \in \mathfrak{K}$. The anti-linear map C is an anti-unitary if $C^*C = 1_\mathfrak{h}$ and $CC^* = 1_\mathfrak{K}$.

is a dense subspace of $\mathcal{F}(\mathfrak{h})$. Moreover, for every normalized vector, $\Psi \in \mathcal{Q}$, γ_Ψ and α_Ψ are finite-rank operators with ranges in $D(h)$ and $JD(h)$, respectively. Note that $JD(h) \subset D(k^*)$ because $D(h) \subset D(k)$ and $k^* = J^*kJ^*$. Thus, $h^{1/2}\gamma_\Psi h^{1/2}$ and $k^*\alpha_\Psi$ are trace class and $\langle \Psi, \mathbb{H}\Psi \rangle$ is well defined by (10) for every normalized vector Ψ in $\mathcal{Q}$.

The first result we would like to report is as follows [25, Lemma 9]:

Proposition 1. *Assume that $h > 0$, $k^* = J^*kJ^*$, $kh^{-1}k^* \leq JhJ^*$ and $\mathrm{Tr}(kh^{-1}k^*) < \infty$. Then,*

$$\mathbb{H} \geq -\frac{1}{2}\mathrm{Tr}(kh^{-1}k^*)$$

as a quadratic form. Moreover, if $kh^{-1}k^* \leq \delta JhJ^*$ for some $0 \leq \delta < 1$, then

$$(1 + \sqrt{\delta})\mathrm{d}\Gamma(h) + \frac{\sqrt{\delta}}{2}\mathrm{Tr}(kh^{-1}k^*) \geq \mathbb{H} \geq (1 + \sqrt{\delta})\mathrm{d}\Gamma(h) - \frac{\sqrt{\delta}}{2}\mathrm{Tr}(kh^{-1}k^*)$$

as quadratic forms. Consequently, the quadratic form $\mathbb{H}$ defines a self-adjoint operator, still denoted by $\mathbb{H}$, such that $\inf \sigma(\mathbb{H}) \geq -\frac{1}{2}\mathrm{Tr}(kh^{-1}k^*)$.

Thus, the above proposition provides conditions under which the formal expressions (1)/(7) make sense. In the context of the earlier results (which will be reviewed after the statement of the main theorem), the main improvement is that we do not have to assume that k is a bounded operator.

2.3 Diagonalization

As mentioned in the introduction, a key feature of the quadratic Hamiltonians in Bogoliubov's theory is that they can be diagonalized to those of non-interacting systems by a special class of unitary operators which preserve the CCR algebra. It turns out that the diagonalization problem on Fock space can be associated to a diagonalization problem on $\mathfrak{h} \oplus \mathfrak{h}^*$ in a very natural way.

Since we will consider transformations on $\mathfrak{h} \oplus \mathfrak{h}^*$, it is convenient to introduce the generalized annihilation and creation operators

$$A(f \oplus Jg) = a(f) + a^*(g), \quad A^*(f \oplus Jg) = a^*(f) + a(g), \quad \forall f, g \in \mathfrak{h}. \tag{12}$$

They satisfy the conjugate and canonical commutation relations

$$A^*(F_1) = A(\mathcal{J}F_1), \quad \left[A(F_1), A^*(F_2)\right] = (F_1, \mathcal{S}F_2), \quad \forall F_1, F_2 \in \mathfrak{h} \oplus \mathfrak{h}^* \tag{13}$$

where we have introduced the block operators on $\mathfrak{h} \oplus \mathfrak{h}^*$

$$S = \begin{pmatrix} 1 & 0 \\ 0 & -1 \end{pmatrix}, \quad \mathcal{J} = \begin{pmatrix} 0 & J^* \\ J & 0 \end{pmatrix}. \tag{14}$$

Note that $S = S^{-1} = S^*$ is a unitary on $\mathfrak{h} \oplus \mathfrak{h}^*$ and $\mathcal{J} = \mathcal{J}^{-1} = \mathcal{J}^*$ is an anti-unitary.

We say that a bounded operator $\mathcal{V}$ on $\mathfrak{h} \oplus \mathfrak{h}^*$ is *unitarily implemented* by a unitary operator $\mathbb{U}_{\mathcal{V}}$ on Fock space if

$$\mathbb{U}_{\mathcal{V}} A(F) \mathbb{U}_{\mathcal{V}}^* = A(\mathcal{V}F), \quad \forall F \in \mathfrak{h} \oplus \mathfrak{h}^*. \tag{15}$$

It is easy to see that if (15) holds true, then the CCR (13) imply the following compatibility conditions:

$$\mathcal{J}\mathcal{V}\mathcal{J} = \mathcal{V}, \quad \mathcal{V}^* S \mathcal{V} = S = \mathcal{V} S \mathcal{V}^*. \tag{16}$$

Any bounded operator $\mathcal{V}$ on $\mathfrak{h} \oplus \mathfrak{h}^*$ satisfying (16) is called a *Bogoliubov transformation*.

The condition $\mathcal{J}\mathcal{V}\mathcal{J} = \mathcal{V}$ means that $\mathcal{V}$ has the following block form:

$$\mathcal{V} = \begin{pmatrix} U & J^*VJ^* \\ V & JUJ^* \end{pmatrix} \tag{17}$$

where $U : \mathfrak{h} \to \mathfrak{h}$ and $V : \mathfrak{h} \to \mathfrak{h}^*$ are linear-bounded operators. Under this form, the condition $\mathcal{V}^* S \mathcal{V} = S = \mathcal{V} S \mathcal{V}^*$ is equivalent to

$$U^*U = 1 + V^*V, \quad UU^* = 1 + J^*VV^*J, \quad V^*JU = U^*J^*V. \tag{18}$$

It is a fundamental result that a Bogoliubov transformation $\mathcal{V}$ of the form (17) is unitarily implementable if and only if it satisfies *Shale's condition* [28]

$$\|V\|_{\mathrm{HS}}^2 = \mathrm{Tr}(V^*V) < \infty. \tag{19}$$

Now, we come back to the problem of diagonalizing $\mathbb{H}$. Using the formal formula (7) and the assumption $k^* = J^*kJ^*$, we can write

$$\mathbb{H} = \mathbb{H}_{\mathcal{A}} - \frac{1}{2}\mathrm{Tr}(h) \tag{20}$$

where

$$\mathcal{A} := \begin{pmatrix} h & k^* \\ k & JhJ^* \end{pmatrix} \tag{21}$$

and

$$\mathbb{H}_{\mathcal{A}} := \frac{1}{2} \sum_{m,n \geq 1} \langle F_m, \mathcal{A} F_n \rangle A^*(F_m) A(F_n). \tag{22}$$

Here, $\{F_n\}_{n \geq 1}$ is an orthonormal basis for $\mathfrak{h} \oplus \mathfrak{h}^*$ and the definition $\mathbb{H}_{\mathcal{A}}$ is independent of the choice of the basis. In fact, $\mathbb{H}_{\mathcal{A}}$ is the Weyl quantization of $\mathcal{A}$. We refer to [9] for more details on this aspect of quadratic Hamiltonians. Finally, note that $\mathcal{J} \mathcal{A} \mathcal{J} = \mathcal{A}$ because of the symmetry condition $k^* = J^* k J^*$.

Now, let $\mathcal{V}$ be a Bogoliubov transformation on $\mathfrak{h} \oplus \mathfrak{h}^*$ which is implemented by a unitary operator $\mathbb{U}_{\mathcal{V}}$ on Fock space as in (15). Then, we can verify that $\mathbb{U}_{\mathcal{V}} \mathbb{H}_{\mathcal{A}} \mathbb{U}_{\mathcal{V}}^* = \mathbb{H}_{\mathcal{V} \mathcal{A} \mathcal{V}^*}$ and hence (20) is equivalent to

$$\mathbb{U}_{\mathcal{V}} \mathbb{H} \mathbb{U}_{\mathcal{V}}^* = \mathbb{H}_{\mathcal{V} \mathcal{A} \mathcal{V}^*} - \frac{1}{2} \mathrm{Tr}(h). \tag{23}$$

In particular, if $\mathcal{V} \mathcal{A} \mathcal{V}^*$ is *block diagonal*, namely

$$\mathcal{V} \mathcal{A} \mathcal{V}^* = \begin{pmatrix} \xi & 0 \\ 0 & J \xi J^* \end{pmatrix}$$

for some operator $\xi : \mathfrak{h} \to \mathfrak{h}$, then (23) reduces to

$$\mathbb{U}_{\mathcal{V}} \mathbb{H} \mathbb{U}_{\mathcal{V}}^* = d\Gamma(\xi) + \frac{1}{2} \mathrm{Tr}(\xi - h). \tag{24}$$

Note that all formulas (20), (23) and (24) are formal because h, ξ and $\xi - h$ may be not trace class. Nevertheless, the above heuristic argument suggests that the diagonalization problem on $\mathbb{H}$ can be reduced to the diagonalization problem on $\mathcal{A}$ by Bogoliubov transformations. This is summarized in the following two theorems [25, Theorems 1 and 2].

Theorem 2 (Diagonalization of bosonic block operators).
(i) (Existence). *Let $h : \mathfrak{h} \to \mathfrak{h}$ and $k : \mathfrak{h} \to \mathfrak{h}^*$ be (unbounded) linear operators satisfying $h = h^* > 0$, $k^* = J^* k J^*$ and $D(h) \subset D(k)$. Assume that the operator $G := h^{-1/2} J^* k h^{-1/2}$ is densely defined and extends to a bounded operator satisfying $\|G\| < 1$. Then, we can define the self-adjoint operator*

$$\mathcal{A} := \begin{pmatrix} h & k^* \\ k & J h J^* \end{pmatrix} > 0 \quad on \ \mathfrak{h} \oplus \mathfrak{h}^*$$

by Friedrichs' extension. This operator can be diagonalized by a bosonic Bogoliubov transformation $\mathcal{V}$ on $\mathfrak{h} \oplus \mathfrak{h}^*$ in the sense that

$$\mathcal{V} \mathcal{A} \mathcal{V}^* = \begin{pmatrix} \xi & 0 \\ 0 & J \xi J^* \end{pmatrix}$$

for a self-adjoint operator $\xi > 0$ on $\mathfrak{h}$. Moreover, we have

$$\|\mathcal{V}\| \leq \left(\frac{1 + \|G\|}{1 - \|G\|}\right)^{1/4}. \tag{25}$$

(ii) (Implementability). *Assume further that G is Hilbert–Schmidt. Then, $\mathcal{V}$ is unitarily implementable and, under the block form* (17),

$$\|V\|_{\mathrm{HS}} \leq \frac{2}{1 - \|G\|} \|G\|_{\mathrm{HS}}. \tag{26}$$

Next, we consider the diagonalization of quadratic Hamiltonians.

Theorem 3 (Diagonalization of quadratic Hamiltonians). *We keep all assumptions in* Theorem 2 *(that $\|G\| < 1$ and G is Hilbert–Schmidt) and assume further that $kh^{-1/2}$ is Hilbert–Schmidt. Then, the quadratic Hamiltonian $\mathbb{H}$, defined as a quadratic form by* (10), *is bounded from below and closable, and hence its closure defines a self-adjoint operator which we still denote by $\mathbb{H}$. Moreover, if $\mathbb{U}_\mathcal{V}$ is the unitary operator on Fock space implementing the Bogoliubov transformation $\mathcal{V}$ in* Theorem 2, *then*

$$\mathbb{U}_\mathcal{V} \mathbb{H} \mathbb{U}_\mathcal{V}^* = d\Gamma(\xi) + \inf \sigma(\mathbb{H}). \tag{27}$$

Finally, $\mathbb{H}$ has a unique ground state $\Psi_0 = \mathbb{U}_\mathcal{V}^*|0\rangle$ whose one-particle density matrices are $\gamma_{\Psi_0} = V^*V$ and $\alpha_{\Psi_0} = JU^*J^*V$ and

$$\inf \sigma(\mathbb{H}) = \mathrm{Tr}(h^{1/2}\gamma_{\Psi_0}h^{1/2}) + \Re\mathrm{Tr}(k^*\alpha_{\Psi_0}) \geq -\frac{1}{2}\|kh^{-1/2}\|_{\mathrm{HS}}^2. \tag{28}$$

In particular, $h^{1/2}\gamma_{\Psi_0}h^{1/2}$ and $k^*\alpha_{\Psi_0}$ are trace class.

2.4 Optimality of the Diagonalization Conditions

Having in mind the conditions in the statements above, let us consider the following example. Let h and k be multiplication operators on $\mathfrak{h} = L^2(\Omega, \mathbb{C})$, for some measure space Ω. Then J is simply complex conjugation and we can identify $\mathfrak{h}^* = \mathfrak{h}$ for simplicity. Assume that $h > 0$, but k is not necessarily real-valued. Then

$$\mathcal{A} := \begin{pmatrix} h & k \\ k & h \end{pmatrix} > 0 \quad \text{on } \mathfrak{h} \oplus \mathfrak{h}^*$$

if and only if $-1 < G < 1$ with $G := |k|h^{-1}$. In this case, $\mathcal{A}$ is diagonalized by the linear operator

$$\mathcal{V} := \sqrt{\frac{1}{2} + \frac{1}{2\sqrt{1 - G^2}}} \begin{pmatrix} 1 & \frac{-G}{1+\sqrt{1-G^2}} \\ \frac{-G}{1+\sqrt{1-G^2}} & 1 \end{pmatrix}$$

in the sense that

$$\mathcal{V}\mathcal{A}\mathcal{V}^* = \begin{pmatrix} \xi & 0 \\ 0 & \xi \end{pmatrix} \quad \text{with} \quad \xi := h\sqrt{1 - G^2} = \sqrt{h^2 - k^2} > 0.$$

It is straightforward to verify that $\mathcal{V}$ always satisfies the compatibility conditions (16). Moreover, $\mathcal{V}$ is bounded (and hence a Bogoliubov transformation) if and only if $\|G\| = \|kh^{-1}\| < 1$ and in this case

$$\|\mathcal{V}\| \sim (1 - \|G\|)^{-1/4} \tag{29}$$

(which means that the ratio between $\|\mathcal{V}\|$ and $(1 - \|G\|)^{-1/4}$ is bounded from above and below by universal positive constants). By Shale's condition (19), $\mathcal{V}$ is unitarily implementable if and only if kh^{-1} is Hilbert–Schmidt and in this case, under the conventional form: (17),

$$\|V\|_{\text{HS}} \sim (1 - \|G\|)^{-1/4}\|G\|_{\text{HS}}. \tag{30}$$

Finally, from (24) and the simple estimates

$$-\frac{1}{2}k^2h^{-1} \geq \xi - h = \sqrt{h^2 - k^2} - h \geq -k^2h^{-1}$$

we deduce that $\mathbb{H}$ is bounded from below if and only if $kh^{-1/2}$ is Hilbert–Schmidt and in this case

$$\inf \sigma(\mathbb{H}) \sim -\|kh^{-1/2}\|_{\text{HS}}^2. \tag{31}$$

Thus, in summary, in the above commutative example, we have the following *optimal* conditions:

- $\mathcal{A}$ is diagonalized by a Bogoliubov transformation $\mathcal{V}$ if and only if $\|kh^{-1}\| < 1$;
- $\mathcal{V}$ is unitarily implementable if and only if kh^{-1} is Hilbert–Schmidt;
- $\mathbb{H}$ is bounded from below if and only if $kh^{-1/2}$ is Hilbert–Schmidt.

If we now compare this with the conditions in Theorems 2 and 3, then condition $\|G\| < 1$ can be interpreted as a non-commutative analogue of the bound $\|kh^{-1}\| < 1$ in the commutative case. Furthermore, the implementability condition $\|G\|_{\text{HS}} < \infty$ can be interpreted as a non-commutative analogue of the condition $\|kh^{-1}\|_{\text{HS}} < \infty$ in the commutative case. Finally, the condition $\|kh^{-1/2}\|_{\text{HS}} < \infty$ is the same as in the commutative case. This necessary condition was proved by Bruneau and Dereziński in [7] when k is Hilbert–Schmidt. Note that in order to ensure that $\mathbb{H}$ is bounded from below, we do not really need the conditions $\|G\| < 1$ and $\|G\|_{HS} < \infty$ in Theorem 2, as stated in Proposition 1.

2.5 Comparison with Existing Results

Let us make some historical remarks. The physical model in Bogoliubov's 1947 paper [6] and described in the introduction corresponds to the case when dim $\mathfrak{h} = 2$ and $\mathcal{A}$ is a 2×2 real matrix which can be diagonalized explicitly (more precisely, in his case particles only come in pairs with momenta $\pm p$ and each pair can be diagonalized independently). In fact, when dim $\mathfrak{h}$ is finite, the diagonalization of $\mathcal{A}$ by symplectic matrices can be done by Williamson's Theorem [30]. We refer to Hörmander [17] for a complete discussion on the diagonalization problem in the finite-dimensional case.

In the 1950s and 1960s, Friedrichs [13] and Berezin [3] gave general diagonalization results in the case dim $\mathfrak{h} = +\infty$, assuming that h is bounded, k is Hilbert–Schmidt, and $\mathcal{A} \geq \mu > 0$ for a constant μ. Note that the gap condition $\mathcal{A} \geq \mu$ requires that $h \geq \mu > 0$.

In the present paper, we always assume that $\mathcal{A} > 0$ but we do not require a gap. In some cases, the weaker assumption $\mathcal{A} \geq 0$ might be also considered, but it is usually transferred back to the strict case $\mathcal{A} > 0$ by using an appropriate decomposition; see Kato and Mugibayashi [18] for a further discussion.

In many physical applications, it is important to consider unbounded operators. In the recent works on the excitation spectrum of interacting Bose gases, the diagonalization problem has been studied by Grech and Seiringer in [14] when h is a positive operator with compact resolvent and k is Hilbert–Schmidt, and then by Lewin, Nam, Serfaty and Solovej [20, Appendix A] when h is a general unbounded operator satisfying $h \geq \mu > 0$.

Very recently, in 2014, Bach and Bru [2] established for the first time the diagonalization problem when h is not bounded below away from zero. They assumed that $h > 0$, $\|kh^{-1}\| < 1$ and kh^{-s} is Hilbert–Schmidt for all $s \in [0, 1 + \varepsilon]$ for some $\varepsilon > 0$ (see conditions (A2) and (A5) in [2]).

3 Remarks About the Proof

Let us now make a few remarks about the proofs. All details can be found in [25].

To obtain the first bound in Proposition 1 we notice that since γ_Ψ and α_Ψ are finite-rank operators (when Ψ is a normalized vector in the domain $\mathcal{Q}$), we can use the cyclicity of the trace and the Cauchy–Schwarz inequality to write

$$|\text{Tr}(k^*\alpha_\Psi)| = |\text{Tr}(\alpha_\Psi k^*)| = |\text{Tr}((1 + J\gamma_\Psi J^*)^{-1/2}\alpha_\Psi h^{1/2}h^{-1/2}k^*(1 + J\gamma_\Psi J^*)^{1/2})|$$
$$\leq \|(1 + J\gamma_\Psi J^*)^{-1/2}\alpha_\Psi h^{1/2}\|_{\text{HS}}\|h^{-1/2}k^*(1 + J\gamma_\Psi J^*)^{1/2}\|_{\text{HS}}$$
$$= \left[\text{Tr}(h^{1/2}\alpha_\Psi^*(1 + J\gamma_\Psi J^*)^{-1}\alpha_\Psi h^{1/2})\right]^{1/2}$$
$$\times \left[\text{Tr}\left((1 + J\gamma_\Psi J^*)^{1/2}kh^{-1}k^*(1 + J\gamma_\Psi J^*)^{1/2}\right)\right]^{1/2}.$$

Using $\alpha_\Psi^*(1 + J\gamma_\Psi J^*)^{-1}\alpha_\Psi \leq \gamma_\Psi$ and $kh^{-1}k^* \leq JhJ^*$, we get

$$|\text{Tr}(k^*\alpha_\Psi)| \leq \left[\text{Tr}(h^{1/2}\gamma_\Psi h^{1/2})\right]^{1/2}\left[\text{Tr}(kh^{-1}k^*) + \text{Tr}(h^{1/2}\gamma_\Psi h^{1/2})\right]^{1/2}$$
$$\leq \text{Tr}(h^{1/2}\gamma_\Psi h^{1/2}) + \frac{1}{2}\text{Tr}(kh^{-1}k^*). \tag{32}$$

Here, in the last estimate, we have used $\sqrt{x(x + y)} \leq x + y/2$ for real numbers $x, y \geq 0$. Thus by definition (10), we get

$$\langle \Psi, \mathbb{H}\Psi \rangle = \text{Tr}(h^{1/2}\gamma_\Psi h^{1/2}) + \Re\text{Tr}(k^*\alpha_\Psi) \geq -\frac{1}{2}\text{Tr}(kh^{-1}k^*)$$

for $\Psi \in \mathcal{Q}$. Thus $\mathbb{H} \geq -(1/2)\text{Tr}(kh^{-1}k^*)$.

The other inequality in Proposition 1 follows (using the additional assumption) from a similar argument as in (32). This bound allows to conclude that the quadratic form $\mathbb{H}$ is not only bounded below but also closed and thus its closure defines a self-adjoint operator.

To prove Theorem 2 we employ a connection between the bosonic diagonalization problem and its fermionic analogue. Such kind of connection has been known for a long time; see Araki [1] for a heuristic discussion. To be precise, we will use the following diagonalization result for fermionic block operators.

Theorem 4 (Diagonalization of fermionic block operators). *Let B be a self-adjoint operator on $\mathfrak{h} \oplus \mathfrak{h}^*$ such that $JBJ = -B$ and such that $\dim \text{Ker}(B)$ is either even (possibly 0) or infinite. Then, there exists a unitary operator $\mathcal{U}$ on $\mathfrak{h} \oplus \mathfrak{h}^*$ such that $J\mathcal{U}J = \mathcal{U}$ and*

$$\mathcal{U}B\mathcal{U}^* = \begin{pmatrix} \xi & 0 \\ 0 & -J\xi J^* \end{pmatrix}$$

for some operator $\xi \geq 0$ on $\mathfrak{h}$. Moreover, if $\text{Ker}(B) = \{0\}$, then $\xi > 0$.

By applying Theorem 4 to $B = \mathcal{A}^{1/2}S\mathcal{A}^{1/2}$, with S given in (14), we can construct the Bogoliubov transformation $\mathcal{V}$ in Theorem 2 explicitly:

$$\mathcal{V} := \mathcal{U}|B|^{1/2}\mathcal{A}^{-1/2}.$$

This explicit construction is similar to the one used by Simon, Chaturvedi and Srinivasan [29] where they offered a simple proof of Williamsons' Theorem. The implementability of $\mathcal{V}$ is proved using a detailed study of $\mathcal{V}^*\mathcal{V} = \mathcal{A}^{-1/2}|B|\mathcal{A}^{-1/2}$.

To obtain the diagonalization result in Theorem 3, we recall from Proposition 1, that the quadratic form $\mathbb{H}$ defines a self-adjoint operator, still denoted by $\mathbb{H}$, such that

$$\inf \sigma(\mathbb{H}) \geq -\frac{1}{2}\mathrm{Tr}(kh^{-1}k^*).$$

Let $\mathcal{V}$ be as in Theorem 2. Let Ψ be a normalized vector in $\mathcal{Q}$ defined in (11). Consider $\Psi' := \mathbb{U}_{\mathcal{V}}^* \Psi$. It is straightforward to see that

$$\begin{pmatrix} \gamma_{\Psi'} & \alpha_{\Psi'}^* \\ \alpha_{\Psi'} & 1 + J\gamma_{\Psi'}J^* \end{pmatrix} = \mathcal{V}^* \begin{pmatrix} \gamma_{\Psi} & \alpha_{\Psi}^* \\ \alpha_{\Psi} & 1 + J\gamma_{\Psi}J^* \end{pmatrix} \mathcal{V}. \tag{33}$$

Moreover, we have

$$\mathcal{V}^* \begin{pmatrix} 0 & 0 \\ 0 & 1 \end{pmatrix} \mathcal{V} = \begin{pmatrix} X & Y^* \\ Y & 1 + JXJ^* \end{pmatrix} \tag{34}$$

with $X = V^*V$ and $Y = JV^*JU = JU^*J^*V$. From (33) and (34), we obtain

$$\begin{pmatrix} \gamma_{\Psi'} - X & \alpha_{\Psi'}^* - Y^* \\ \alpha_{\Psi'} - Y & J(\gamma_{\Psi'} - X)J^* \end{pmatrix} = \begin{pmatrix} \gamma_{\Psi'} & \alpha_{\Psi'}^* \\ \alpha_{\Psi'} & 1 + J\gamma_{\Psi'}J^* \end{pmatrix} - \begin{pmatrix} X & Y^* \\ Y & 1 + JXJ^* \end{pmatrix}$$

$$= \mathcal{V}^* \begin{pmatrix} \gamma_{\Psi} & \alpha_{\Psi}^* \\ \alpha_{\Psi} & 1 + J\gamma_{\Psi}J^* \end{pmatrix} \mathcal{V} - \mathcal{V}^* \begin{pmatrix} 0 & 0 \\ 0 & 1 \end{pmatrix} \mathcal{V}$$

$$= \mathcal{V}^* \begin{pmatrix} \gamma_{\Psi} & \alpha_{\Psi}^* \\ \alpha_{\Psi} & J\gamma_{\Psi}J^* \end{pmatrix} \mathcal{V}.$$

Recall that γ_{Ψ} and α_{Ψ} are finite-rank operators because $\Psi \in \mathcal{Q}$. Therefore, $\gamma_{\Psi'} - X$ and $\alpha_{\Psi'} - Y$ are also finite-rank operators. Using the cyclicity of the trace, we find that

$$\mathrm{Tr}(h^{1/2}(\gamma_{\Psi'} - X)h^{1/2}) + \Re\mathrm{Tr}(k^*(\alpha_{\Psi'} - Y))$$

$$= \frac{1}{2}\mathrm{Tr}\left[\mathcal{A}\begin{pmatrix} \gamma_{\Psi'} - X & \alpha_{\Psi'}^* - Y^* \\ \alpha_{\Psi'} - Y & J(\gamma_{\Psi'} - X)J^* \end{pmatrix}\right]$$

$$= \frac{1}{2}\mathrm{Tr}\left[\mathcal{A}\mathcal{V}^* \begin{pmatrix} \gamma_{\Psi} & \alpha_{\Psi}^* \\ \alpha_{\Psi} & J\gamma_{\Psi}J^* \end{pmatrix} \mathcal{V}\right] = \frac{1}{2}\mathrm{Tr}\left[\mathcal{V}\mathcal{A}\mathcal{V}^* \begin{pmatrix} \gamma_{\Psi} & \alpha_{\Psi}^* \\ \alpha_{\Psi} & J\gamma_{\Psi}J^* \end{pmatrix}\right]$$

$$= \frac{1}{2}\mathrm{Tr}\left[\begin{pmatrix} \xi & 0 \\ 0 & J\xi J^* \end{pmatrix} \begin{pmatrix} \gamma_{\Psi} & \alpha_{\Psi}^* \\ \alpha_{\Psi} & J\gamma_{\Psi}J^* \end{pmatrix}\right] = \mathrm{Tr}(\xi\gamma_{\Psi}) = \langle \Psi, \mathrm{d}\Gamma(\xi)\Psi\rangle.$$

Thus by the quadratic form expression (10), we have

$$\langle \Psi, \mathbb{U}_{\mathcal{V}}\mathbb{H}\mathbb{U}_{\mathcal{V}}^*\Psi\rangle = \langle \mathbb{U}_{\mathcal{V}}^*\Psi, \mathbb{H}\mathbb{U}_{\mathcal{V}}^*\Psi\rangle = \mathrm{Tr}(h^{1/2}\gamma_{\Psi'}h^{1/2}) + \Re\mathrm{Tr}(k^*\alpha_{\Psi'})$$

$$= \mathrm{Tr}(h^{1/2}(\gamma_{\Psi'} - X)h^{1/2}) + \Re\mathrm{Tr}(k^*(\alpha_{\Psi'} - Y))$$

$$\qquad + \mathrm{Tr}(h^{1/2}Xh^{1/2}) + \Re\mathrm{Tr}(k^*Y)$$

$$= \langle \Psi, \mathrm{d}\Gamma(\xi)\Psi\rangle + \mathrm{Tr}(h^{1/2}Xh^{1/2}) + \Re\mathrm{Tr}(k^*Y)$$

for all $\Psi \in \mathcal{Q}$. A more detailed analysis shows that $h^{1/2}Xh^{1/2}$ and k^*Y are trace class operators. Hence,

$$\mathbb{U}_\mathcal{V}\mathbb{H}\mathbb{U}_\mathcal{V}^* = d\Gamma(\xi) + \text{Tr}(h^{1/2}Xh^{1/2}) + \Re\text{Tr}(k^*Y).$$

Since $d\Gamma(\xi)$ has a unique ground state $|0\rangle$ with the ground state energy 0, we conclude that $\mathbb{H}$ has a unique ground state $\Psi_0 = \mathbb{U}_\mathcal{V}^*|0\rangle$ with the ground state energy

$$\inf \sigma(\mathbb{H}) = \text{Tr}(h^{1/2}Xh^{1/2}) + \Re\text{Tr}(k^*Y).$$

Finally, using (33) and (34) we find that $\gamma_{\Psi_0} = X$ and $\alpha_{\Psi_0} = Y$.

4   Time-Dependent Bogoliubov Hamiltonians

A crucial role in the theory of quadratic Hamiltonians is played by a class of states called *quasi-free* states. In fact, ground states of the quadratic Hamiltonians are quasi-free states (see [20, Theorem A.1]). Recall that a state Ψ in Fock space $\mathcal{F}(\mathfrak{h})$ is called a quasi-free state if it has finite particle number expectation and satisfies Wick's Theorem:

$$\langle \Psi, a^\#(f_1)a^\#(f_2)\cdots a^\#(f_{2n-1})\Psi\rangle = 0, \tag{35}$$

$$\langle \Psi, a^\#(f_1)a^\#(f_2)\cdots a^\#(f_{2n})\Psi\rangle = \sum_{\sigma\in P_{2n}} \prod_{j=1}^n \langle \Psi, a^\#(f_{\sigma(2j-1)})a^\#(f_{\sigma(2j)})\Psi\rangle \tag{36}$$

for all n and for all $f_1, \ldots, f_n \in \mathfrak{h}$, where $a^\#$ is either the creation or annihilation operator and P_{2n} is the set of pairings,

$$P_{2n} = \{\sigma \in S(2n) \mid \sigma(2j-1) < \min\{\sigma(2j), \sigma(2j+1)\} \text{ for all } j\}.$$

It is clear that if Ψ is a quasi-free, then the projection $|\Psi\rangle\langle\Psi|$ is determined completely by its one-particle density matrices γ_Ψ and α_Ψ. States implemented by a Bogoliubov transformation such as $\Psi_0 = \mathbb{U}_\mathcal{V}^*|0\rangle$ in Theorem 3, are called *pure quasi-free* states.

Since quasi-free states are so closely related to quadratic Hamiltonians, a natural question one can ask is whether this property (of being a quasi-free state) is preserved under time evolution generated by a Bogoliubov Hamiltonian. In the case of time-independent quadratic Hamiltonians, the answer is positive and the proof is rather straightforward. In the case of time-dependent Hamiltonians, the same remains true, but the arguments are different and we will sketch them below (for more details see [22], also see [23, 24] for further applications).

Let $\Phi(t)$ be a state in the Fock space $\mathcal{F}$ (which could, in general, depend on time). Assume $\Phi(t)$ satisfies the Bogoliubov equation

$$i\partial_t \Phi(t) = \mathbb{H}(t)\Phi(t), \quad \Phi(t=0) = \Phi(0). \tag{37}$$

Here, $\mathbb{H}(t)$ is a quadraric, time-dependent Hamiltonian in the Fock space:

$$\mathbb{H}(t) = d\Gamma(h(t)) + \frac{1}{2}\iint_{\mathbb{R}^3\times\mathbb{R}^3}\left(K_2(t,x,y)a_x^*a_y^* + \overline{K_2(t,x,y)}a_x a_y\right)dx\,dy.$$

Let us focus on the special situation when $h(t)$ can be decomposed as

$$h(t) = -\Delta + h_1(t).$$

For simplicity, assume furthermore that $h_1(t)$ and K_2 are bounded uniformly in time. We then have the following.

Proposition 5 (Bogoliubov equation). *(i) If $\Phi(0)$ belongs to the quadratic form domain $\mathcal{Q}(d\Gamma(1-\Delta))$, then Eq. (37) has a unique global solution in $\mathcal{Q}(d\Gamma(1-\Delta))$. Moreover, the pair of density matrices $(\gamma_{\Phi(t)}, \alpha_{\Phi(t)})$ is the unique solution to*

$$\begin{cases} i\partial_t\gamma = h\gamma - \gamma h + K_2\alpha - \alpha^* K_2^*, \\ i\partial_t\alpha = h\alpha + \alpha h^{\mathrm{T}} + K_2 + K_2\gamma^{\mathrm{T}} + \gamma K_2, \\ \gamma(t=0) = \gamma_{\Phi(0)}, \quad \alpha(t=0) = \alpha_{\Phi(0)}. \end{cases} \tag{38}$$

(ii) If we assume further that $\Phi(0)$ is a quasi-free state in $\mathcal{F}$, then $\Phi(t)$ is also a quasi-free state for all $t > 0$.

Let us sketch the proof. Recall that $\gamma_{\Phi(t)} : \mathfrak{h} \to \mathfrak{h}$, $\alpha_{\Phi(t)} : \overline{\mathfrak{h}} \equiv \mathfrak{h}^* \to \mathfrak{h}$ are operators with kernels $\gamma_{\Phi(t)}(x,y) = \langle \Phi(t), a_y^* a_x \Phi(t)\rangle$, $\alpha_{\Phi(t)}(x,y) = \langle \Phi(t), a_x a_y \Phi(t)\rangle$ and $K_2 : \overline{\mathfrak{h}} \equiv \mathfrak{h}^* \to \mathfrak{h}$ is an operator with kernel $K_2(t,x,y)$.

For existence and uniqueness of $\Phi(t)$, we refer to [19, Theorem 7]. To derive (38), we use (37) to compute

$$i\partial_t\gamma_{\Phi(t)}(x',y') = i\partial_t\langle\Phi(t), a_{y'}^* a_x \Phi(t)\rangle = \langle\Phi(t), [a_{y'}^* a_x, \mathbb{H}(t)]\Phi(t)\rangle$$

$$= \iint h(t,x,y)\Big(\delta(x'-x)\gamma_{\Phi(t)}(y,y') - \delta(y'-y)\gamma_{\Phi(t)}(x',x)\Big)dxdy$$

$$+ \frac{1}{2}\iint k(t,x,y)\Big(\delta(x'-x)\alpha_{\Phi(t)}^*(y,y') + \delta(x'-y)\alpha_{\Phi(t)}^*(y',x)\Big)dxdy$$

$$- \frac{1}{2}\iint k^*(t,x,y)\Big(\delta(y'-y)\alpha_{\Phi(t)}(x,x') + \delta(y'-x)\alpha_{\Phi(t)}(y,x')\Big)dxdy$$

$$= \Big(h(t)\gamma_{\Phi(t)} - \gamma_{\Phi(t)}h(t) + K_2(t)\alpha_{\Phi(t)}^* - \alpha_{\Phi(t)}K_2^*(t)\Big)(x',y').$$

This is the first equation in (38). The second equation is proved similarly.

Let us now discuss the second statement. We will write $(\gamma, \alpha) = (\gamma_{\Phi(t)}, \alpha_{\Phi(t)})$ for short. Let us introduce

$$X := \gamma + \gamma^2 - \alpha\alpha^*, \quad Y := \gamma\alpha - \alpha\gamma^{\mathrm{T}}.$$

It is a general fact (see e.g. [22, Lemma 8]) that $\Phi(t)$ is a quasi-free state if and only if $X(t) = 0$ and $Y(t) = 0$. In particular, we have $X(0) = 0$ and $Y(0) = 0$ by the assumption on $\Phi(0)$. Using (38), it is straightforward to see that

$$i\partial_t X = hX - Xh + K_2 Y^* - Y K_2^*,$$
$$i\partial_t X^2 = (i\partial_t X)X + X(i\partial_t X) = hX^2 - X^2 h + (K_2 Y^* - Y K_2^*)X + X(K_2 Y^* - Y K_2^*).$$

Then, we take the trace and use $\mathrm{Tr}(hX^2 - X^2 h) = 0$ (hX^2 and $X^2 h$ may be not trace class but we can introduce a cut-off; see [22] for details). We find that

$$\|X(t)\|_{\mathrm{HS}}^2 \leq 4 \int_0^t \|K_2(s)\| \cdot \|X(s)\|_{\mathrm{HS}} \cdot \|Y(s)\|_{\mathrm{HS}} \, \mathrm{d}s.$$

We also obtain a similar bound for $\|Y(t)\|_{\mathrm{HS}}$. Summing these estimates and using the fact that $\|K_2(t)\|$ is bounded uniformly in time, we conclude by Grönwall's inequality that $X(t) = 0$, $Y(t) = 0$ for all $t > 0$.

A similar argument can be used to show the uniqueness of solutions to (38).

5 Time-Dependent Diagonalization Problem

In this section, we would like to make a remark how Eqs. (37) and (38) are related to what can be called the *time-dependent diagonalization problem*.

Assume we are given an effective (since quadratic rather than many-body, as explained in the introduction) evolution

$$i\partial_t \Phi(t) = \mathbb{H}(t)\Phi(t), \quad \Phi(t = 0) = \Phi(0).$$

Here, $\mathbb{H}(t)$ is a quadraric, time-dependent Hamiltonian in the Fock space. As stated in Proposition 5, if $\Phi(0)$ is a pure quasi-free state, then $\Phi(t)$ is also a pure quasi-free state for $t > 0$. Thus

$$\Phi(t) = \mathbb{U}_\mathcal{V}^*(t)|0\rangle \tag{39}$$

with, now, the Bogoliubov transformation being time-dependent. Inserting (39) into (37), we obtain

$$\left(i\partial_t \mathbb{U}_\mathcal{V}^*(t) - \mathbb{H}(t)\mathbb{U}_\mathcal{V}^*(t)\right)|0\rangle = 0$$

which implies

$$\big(\mathbb{U}_V(t)(i\partial_t \mathbb{U}_V^*(t)) - \mathbb{U}_V(t)\mathbb{H}(t)\mathbb{U}_V^*(t)\big)\,|0\rangle = 0. \tag{40}$$

It follows from (40) that

$$\mathbb{U}_V(t)(i\partial_t \mathbb{U}_V^*(t)) - \mathbb{U}_V(t)\mathbb{H}(t)\mathbb{U}_V^*(t) = d\Gamma(\xi(t)) \tag{41}$$

for some operator $\xi(t) : \mathfrak{h} \to \mathfrak{h}$. This argument can be inverted and one can ask the following question: Given a (time-dependent) quadratic Hamiltonian $\mathbb{H}(t)$, does there exist a unitarily implemented Bogoliubov transformation $\mathbb{U}_V(t)$ such that the operator

$$\mathbb{U}_V(t)(i\partial_t \mathbb{U}_V^*(t)) - \mathbb{U}_V(t)\mathbb{H}(t)\mathbb{U}_V^*(t)$$

is diagonal? This question we shall call the *time-dependent diagonalization problem.*

The above argument together with Proposition 5 implies the following.

Theorem 6 (Time-dependent diagonalization problem). *Let $\Phi(t)$ be a pure quasi-free state and and $\mathbb{U}_V^*(t)$ the corresponding unitarily implemented Bogoliubov transformation such that $\Phi(t) = \mathbb{U}_V^*(t)|0\rangle$. Let $(\gamma_{\Phi(t)}, \alpha_{\Phi(t)})$ be the pair of density matrices associated to $\Phi(t)$. Let $\mathbb{H}(t)$ be the quadratic operator defined in (37). Then, under the assumptions of Proposition 5, we have that*

$$\mathbb{U}_V(t)(i\partial_t \mathbb{U}_V^*(t)) - \mathbb{U}_V(t)\mathbb{H}(t)\mathbb{U}_V^*(t) = d\Gamma(\xi(t))$$

for some operator $\xi(t) : \mathfrak{h} \to \mathfrak{h}$ if and only if

$$\begin{cases} i\partial_t \gamma = h\gamma - \gamma h + K_2\alpha - \alpha^* K_2^*, \\ i\partial_t \alpha = h\alpha + \alpha h^{\mathrm{T}} + K_2 + K_2\gamma^{\mathrm{T}} + \gamma K_2. \end{cases}$$

We would like to stress that formulating the time-dependent diagonalization problem in the language of quasi-free states yields a system of coupled linear equations for the reduced density matrices. Since solving linear equations is easier than nonlinear ones, one could claim that this is the right approach.

Indeed, a different approach would be to formulate this question in a more direct manner. As mentioned in the introduction, it is well known [3, 7] that pure quasi-free states can be represented in the explicit form:

$$\mathbb{U}_{V_k}^*(t)|0\rangle = \exp\left(i\chi_N(t) + \iint \left[\overline{k(t,x,y)}a_x a_y - k(t,x,y)a_x^* a_y^*\right] dxdy\right)|0\rangle \tag{42}$$

where $\chi_N(t) \in \mathbb{R}$ is a phase factor. One can now insert (42) into the left-hand side of (41) and demand that all terms involving $a_x^* a_y^*$ and $a_x a_y$ disappear. This condition will determine an equation for the kernel $k(t,x,y)$. To our knowledge, this approach was used for the first time in [15]. The resulting equation will be however highly nonlinear. This can be easily seen from the relations

$$U_{\mathcal{V}_k}^*(t)a(f)U_{\mathcal{V}_k}(t) = a(\cosh(2k)f) + a^*(\sinh(2k)\overline{f})$$
$$U_{\mathcal{V}_k}^*(t)a^*(f)U_{\mathcal{V}_k}(t) = a^*(\cosh(2k)f) + a(\sinh(2k)\overline{f})$$
(43)

for any $f \in \mathfrak{h}$. Here, $\cosh(2k)$ and $\sinh(2k)$ denote the linear operators on $\mathfrak{h}$ is given by

$$\cosh(2k) := \sum_{n \geq 0} \frac{1}{(2n)!}((2k)(\overline{2k}))^n, \quad \sinh(2k) := \sum_{n \geq 0} \frac{1}{(2n+1)!}((2k)(2\overline{k}))^n(2k)$$

where the products of k have to be understood as products of operators (given by the kernels $k(t, x, y)$). In fact, this relation shows that the natural variables for the diagonalization problem are $\cosh(2k)$ and $\sinh(2k)$, but the resulting equations are still nonlinear in those variables. It has been then observed in [16] that when expressed in new variables, these equations can be rewritten as linear equations that are equivalent to (38).

Adapting ideas from [16] to our setting, we shall now present an alternative derivation of (37). To this end, we will first rewrite the time-dependent diagonalization problem (41) on the classical level, that is before Weyl quantization. According to (22), the classical counterpart of the term $U_{\mathcal{V}}(t)\mathbb{H}(t)U_{\mathcal{V}}^*(t)$ is given by $\mathcal{V}\mathcal{A}\mathcal{V}^*$ where $\mathcal{V}$ and $\mathcal{A}$ are defined in (17) and (21), respectively, (and, now, being time-dependent and with the off-diagonal term in $\mathcal{A}$ being K_2 rather than k according to the notation in (37)). Note that with $U_{\mathcal{V}_k}^*(t)$ given by (42), it follows from the formula for Weyl quantization (22) that

$$U_{\mathcal{V}_k}^*(t) = e^{\mathbb{H}_{\mathcal{K}}}$$

where

$$\mathcal{K} = \begin{pmatrix} 0 & -k \\ \overline{k} & 0 \end{pmatrix}.$$

It follows that

$$U_{\mathcal{V}_k}(t)\partial_t U_{\mathcal{V}_k}^*(t) = e^{-\mathbb{H}_{\mathcal{K}}}\partial_t e^{\mathbb{H}_{\mathcal{K}}}.$$

Since

$$e^{-\mathbb{H}_{\mathcal{K}(t)}}\partial_t e^{\mathbb{H}_{\mathcal{K}(t)}} = \lim_{s \to 0} e^{-\mathbb{H}_{\mathcal{K}(t)}}\frac{1}{s}\left(e^{\mathbb{H}_{\mathcal{K}(t+s)}} - e^{\mathbb{H}_{\mathcal{K}(t)}}\right)$$

$$= \lim_{s \to 0}\frac{1}{s}\int_0^1 d\lambda \frac{d}{d\lambda}\left(e^{-\mathbb{H}_{\mathcal{K}(t)}\lambda}e^{\mathbb{H}_{\mathcal{K}(t+s)}\lambda}\right)$$

$$= \lim_{s \to 0}\int_0^1 d\lambda e^{-\mathbb{H}_{\mathcal{K}(t)}\lambda}\frac{-\mathbb{H}_{\mathcal{K}}(t) + \mathbb{H}_{\mathcal{K}}(t+s)}{s}e^{\mathbb{H}_{\mathcal{K}(t+s)}\lambda}$$

$$= \int_0^1 d\lambda e^{-\mathbb{H}_{\mathcal{K}(t)}\lambda}\left(\partial_t \mathbb{H}_{\mathcal{K}}(t)\right)e^{\mathbb{H}_{\mathcal{K}(t)}\lambda}.$$

using the so-called Lie formula, we see that

$$e^{-\mathbb{H}_{\mathcal{K}}(t)}\partial_t e^{\mathbb{H}_{\mathcal{K}}(t)} = \int_0^1 d\lambda \left(\partial_t \mathbb{H}_{\mathcal{K}}(t) - [\mathbb{H}_{\mathcal{K}}(t), \partial_t \mathbb{H}_{\mathcal{K}}(t)] + \cdots\right).$$

Since the Weyl quantization is a Lie algebra isomorphism of antihermitian matrices we deduce that

$$e^{-\mathbb{H}_{\mathcal{K}}(t)}\partial_t e^{\mathbb{H}_{\mathcal{K}}(t)} = \mathbb{H}_{e^{-\mathcal{K}}\partial_t e^{\mathcal{K}}}.$$

We already saw that

$$e^{-\mathbb{H}_{\mathcal{K}}(t)}\mathbb{H}_A e^{\mathbb{H}_{\mathcal{K}}(t)} = \mathbb{H}_{e^{-\mathcal{K}}Ae^{\mathcal{K}}}.$$

All this implies that on the classical level, the time-dependent diagonalization problem is equivalent to

$$\mathcal{V}_k \widetilde{\mathcal{A}} \mathcal{V}_k^*$$

being block diagonal, with $\mathcal{V}_k = e^{-\mathcal{K}}$ and $\widetilde{\mathcal{A}}$ is given (21) where h is replaced by $\widetilde{h} = i\partial_t + h$.

Now, recall that by general properties of Bogoliubov transformations, we have $\mathcal{V}_k^{-1} = S\mathcal{V}_k^* S$. Since $\mathcal{V}_k \widetilde{\mathcal{A}} \mathcal{V}_k^* = \mathcal{V}_k \widetilde{\mathcal{A}} S \mathcal{V}_k^{-1} S$ is block diagonal, $\mathcal{V}_k \widetilde{\mathcal{A}} S \mathcal{V}_k^{-1}$ is also block diagonal. Therefore, $[\mathcal{V}_k \widetilde{\mathcal{A}} S \mathcal{V}_k^{-1}, S] = 0$, and hence $[\widetilde{\mathcal{A}} S, \mathcal{V}_k^{-1} S \mathcal{V}_k] = 0$. Using $\mathcal{V}_k^{-1} = S\mathcal{V}_k^* S$ again we have

$$\mathcal{V}_k^{-1} S \mathcal{V}_k = S \mathcal{V}_k^* \mathcal{V}_k = \begin{pmatrix} 1 + 2X_k & 2Y_k^* \\ -2Y_k & -(1 + 2JX_k J^*) \end{pmatrix} \qquad (44)$$

(recall (34) for the definition of X_k and Y_k). Therefore,

$$
\begin{aligned}
0 = [\mathcal{A}S, \mathcal{V}_k^{-1} S \mathcal{V}_k] &= \begin{pmatrix} \widetilde{h} & -K_2^* \\ K_2 & -J\widetilde{h}J^* \end{pmatrix} \begin{pmatrix} 1 + 2X_k & 2Y_k^* \\ -2Y_k & -(1 + 2JX_k J^*) \end{pmatrix} \\
&\quad - \begin{pmatrix} 1 + 2X_k & 2Y_k^* \\ -2Y_k & -(1 + 2JX_k J^*) \end{pmatrix} \begin{pmatrix} \widetilde{h} & -K_2^* \\ K_2 & -J\widetilde{h}J^* \end{pmatrix} \\
&= 2 \begin{pmatrix} \widetilde{h}X_k - X_k\widetilde{h} + K_2^*Y_k - Y_k^*K_2 & \widetilde{h}Y_k^* + Y_k^*J\widetilde{h}J^* + K_2^*JX_kJ^* + X_kK_2^* + K_2^* \\ J\widetilde{h}J^*Y_k + Y_k\widetilde{h} + K_2X_k + JX_kJ^*K_2 + K_2 & J\widetilde{h}X_kJ^* - JX_k\widetilde{h}J^* + K_2Y_k^* - Y_kK_2^* \end{pmatrix}.
\end{aligned}
$$

Finally, just as at the end of Sect. 3, we identify γ with X_k and α with Y_k. All this, together with the fact that in the setting described in Sect. 4 the operator J simply yields a complex conjugation, leads to (38).

Acknowledgements The support of the National Science Centre (NCN) project Nr. 2016/21/D/ST1/02430 is gratefully acknowledged.

References

1. Araki, H.: On the diagonalization of a bilinear Hamiltonian by a Bogoliubov transformation. Publ. RIMS, Kyoto Univ. Ser. A **4**, 387–412 (1968)
2. Bach, V., Bru, J.-B.: Diagonalizing quadratic Bosonic operators by non-autonomous flow equation. Mem. Am. Math. Soc. **240**(1138) (2016)
3. Berezin, F.: The Method of Second Quantization. Pure and Applied Physics. A Series of Monographs and Textbooks. Academic Press, New York (1966)
4. Boccato, C., Brennecke, C., Cenatiempo, S., Schlein, B.: The excitation spectrum of Bose gases interacting through singular potentials. J. Eur. Math. Soc. (in press). Preprint arXiv:1704.04819
5. Boccato, C., Brennecke, C., Cenatiempo, S., Schlein, B.: Bogoliubov Theory in the Gross–Pitaevskii Limit. Preprint arXiv:1801.01389
6. Bogoliubov, N.N.: On the theory of superfluidity. J. Phys. (USSR) **11**, 23 (1947)
7. Bruneau, L., Dereziński, J.: Bogoliubov Hamiltonians and one-parameter groups of Bogoliubov transformations. J. Math. Phys. **48**, 022101 (2007)
8. Dereziński, J.: Quantum fields with classical perturbations. J. Math. Phys. **55**, 075201 (2014)
9. Dereziński, J.: Bosonic quadratic Hamiltonians. J. Math. Phys. **58**, 121101 (2017)
10. Dereziński, J., Napiórkowski, N.: Excitation spectrum of interacting bosons in the mean-field infinite-volume limit. Annales Henri Poincaré **15**, 2409–2439 (2014). Erratum: Annales Henri Poincaré **16**, 1709–1711 (2015)
11. Dereziński, J., Meissner, K.A., Napiórkowski, M.: On the energy-momentum spectrum of a homogeneous Fermi gas. Annales Henri Poincaré **14**(1), 1–36 (2013)
12. Dereziński, J., Napiórkowski, N., Solovej, J.P.: On the minimization of Hamiltonians over pure Gaussian states. In: Siedentop, H. (ed.) Complex Quantum Systems. Analysis of Large Coulomb Systems. Lecture Notes Series Institute for Mathematical Sciences, vol. 24, pp. 151–161. National University of Singapore (2013)
13. Friedrichs, K.O.: Mathematical Aspects of the Quantum Theory of Field. Interscience Publishers Inc., New York (1953)
14. Grech, P., Seiringer, R.: The excitation spectrum for weakly interacting Bosons in a trap. Commun. Math. Phys. **322**, 559–591 (2013)
15. Grillakis, M.G., Machedon, M., Margetis, D.: Second-order corrections to mean field evolution of weakly interacting Bosons. I. Commun. Math. Phys. **294**, 273–301 (2010)
16. Grillakis, M., Machedon, M.: Pair excitations and the mean field approximation of interacting Bosons, I. Commun. Math. Phys. **324**, 601–636 (2013)
17. Hörmander, L.: Sympletic classification of quadratic forms, and general Mehler formulas. Math. Z. **219**, 413–449 (1995)
18. Kato, Y., Mugibayashi, N.: Friedrichs–Berezin transformation and its application to the spectral analysis of the BCS reduced hamiltonian. Prog. Theor. Phys. **38**, 813–831 (1967)
19. Lewin, M., Nam, P.T., Schlein, B.: Fluctuations around Hartree states in the mean-field regime. Am. J. Math. **137**, 1613–1650 (2015)
20. Lewin, M., Nam, P.T., Serfaty, S., Solovej, J.P.: Bogoliubov spectrum of interacting Bose gases. Commun. Pure Appl. Math. **68**, 413–471 (2015)
21. Nam, P.T., Seiringer, R.: Collective excitations of Bose gases in the mean-field regime. Arch. Ration. Mech. Anal. **215**, 381–417 (2015)
22. Nam, P.T., Napiórkowski, M.: Bogoliubov correction to the mean-field dynamics of interacting bosons. Adv. Theor. Math. Phys. **21**(3), 683–738 (2017)
23. Nam, P.T., Napiórkowski, M.: A note on the validity of Bogoliubov correction to mean-field dynamics. Journal de Mathmatiques Pures et Appliques **108**(5), 662–688 (2017)
24. Nam, P.T., Napiórkowski, M.: Norm approximation for many-body quantum dynamics: focusing case in low dimensions. preprint arXiv:1710.09684
25. Nam, P.T., Napiórkowski, M., Solovej, J.P.: Diagonalization of Bosonic quadratic Hamiltonians by Bogoliubov transformations. J. Funct. Anal. **270**(11), 4340–4368 (2016)
26. Napiórkowski, M.: Excitation spectrum and quasiparticles in quantum gases. A rigorous approach. Ph.D. thesis, University of Warsaw (2014)

27. Seiringer, R.: The excitation spectrum for weakly interacting bosons. Commun. Math. Phys. **306**, 565–578 (2011)
28. Shale, D.: Linear symmetries of free Boson fields. Trans. Am. Math. Soc. **103**, 149–167 (1962)
29. Simon, R., Chaturvedi, S., Srinivasan, V.: Congruences and canonical forms for a positive matrix: Application to the Schweinler–Wigner extremum principle. J. Math. Phys. **40**, 3632 (1999)
30. Williamson, J.: On the algebraic problem concerning the normal forms of linear dynamical systems. Am. J. Math. **58**, 141–163 (1936)
31. Zagrebnov, V.A., Bru, J.-B.: The Bogoliubov model of weakly imperfect Bose gas. Phys. Rep. **350**, 291–434 (2001)

Multi-scale Analysis in the Occupation Numbers of Particle States: An Application to Three-Modes Bogoliubov Hamiltonians

Alessandro Pizzo

Abstract In this paper, we describe a multi-scale technique introduced in Pizzo (Bose particles in a box I. A convergent expansion of the ground state of a three-modes Bogoliubov Hamiltonian, [32]) to study many-body quantum systems. The method is based on the Feshbach–Schur map and the scales are represented by occupation numbers of particle states. The main purpose of this method is to implement singular perturbation theory to deal with *large field problems*. A simple model to apply our method is the *three-modes* (including the zero mode) Bogoliubov Hamiltonian that here we consider for a sufficiently small ratio between the kinetic energy and the Fourier component of the (positive type) potential corresponding to the two nonzero modes. In space dimension $d \geq 3$, for an arbitrarily large box and at *fixed*, large particle density ρ (i.e., ρ is independent of the size of the box), this method provides the construction of the ground state vector of the system and its expansion, up to any desired precision, in terms of the bare operators and the ground state energy. In the mean field limiting regime (i.e., at fixed box volume $|\Lambda|$ and for a number of particles, N, sufficiently large), this method provides the same results in any dimension $d \geq 1$. Furthermore, in the mean field limit, we can replace the ground state energy with the Bogoliubov energy in the expansion of the ground state vector.

1 Introduction: Interacting Bose Gas in a Box

This paper is about a novel multi-scale technique, appeared recently in [32–34], to study many-body quantum systems. Here, we outline the strategy used to derive some of the results established in [32]. For proofs and details, we refer the reader to the original reference [32].

A. Pizzo (✉)
Dipartimento di Matematica, Università di Roma "Tor Vergata",
Via della Ricerca Scientifica 1, 00133 Roma, Italy
e-mail: pizzo@mat.uniroma2.it

© Springer Nature Switzerland AG 2018
D. Cadamuro et al. (eds.), *Macroscopic Limits of Quantum Systems*,
Springer Proceedings in Mathematics & Statistics 270,
https://doi.org/10.1007/978-3-030-01602-9_6

123

The multi-scale technique is based on the Feshbach–Schur map and the scales are represented by occupation numbers of particle states. The main purpose of this method is controlling perturbation theory in presence of so-called *large field problems*; see Remark 3.3 later on.

The *three-modes* (including the zero mode) "Bogoliubov Hamiltonian" (defined below) is a convenient, simple model where our method provides an algorithm to construct the ground state vector of the system. It is obtained from the Schrödinger Hamiltonian which describes the dynamics of an interacting gas of Bose atoms ("particles") in a box of side L, see (1.1). Starting from this Hamiltonian, the interaction is then restricted to particles with wave vectors/modes taking just three values, namely the wave vectors/modes[1] $k_j = \pm \frac{2\pi}{L} j$ and $k_0 = 0$ (see definitions (1.2)). The three-modes Bogoliubov Hamiltonian is defined in detail in the section below.

The main result of the paper is an algorithm to compute the ground state vector (and, in principle, any eigenvector) of the three-modes system up to any arbitrary precision, for a number of particles, N, large but fixed. The expansions are in fact not in $\frac{1}{N}$ but in a parameter—not small—related to the ratio between the kinetic energy of the modes $\pm j$ and the Fourier component of the pair potential corresponding to mode j.

1.1 Definition of the Model

Consider the Hamiltonian describing a gas of (spinless) nonrelativistic Bose particles that, at zero temperature, are constrained to a d-dimensional box of side $L > 1$ with $d \geq 1$. The particles interact through a pair potential with a coupling constant proportional to the inverse of the particle density ρ. The rigorous description of this system has many intriguing mathematical aspects not completely clarified yet. In spite of remarkable contributions also in recent years, some important problems are still open to date, in particular, in connection to the thermodynamic limit and the exact structure of the ground state vector. Amongst the rigorous results concerning the spectrum and the ground state of this system (and of similar models) in various regimes (mean field, Gross–Pitaevskii, and thermodynamic limit), we mention [2–8, 11–20, 22–31, 35, 36]. More details about these references can be found in the original paper [32].

Though the number of particles, N, is fixed under time evolution, we use the formalism of second quantization that turns out to be very convenient to implement the technique. The Hamiltonian corresponding to the pair potential $\phi(x - y)$ and to the coupling constant $\frac{1}{\rho} = \frac{|\Lambda|}{N} =: \lambda > 0$ is

$$H := \int \frac{1}{2m} (\nabla a^*)(\nabla a)(x) dx + \frac{\lambda}{2} \int \int a^*(x) a^*(y) \phi(x - y) a(y) a(x) dx dy$$

$$(1.1)$$

---

[1]We use the word mode both for j and $\frac{2\pi}{L} j$

where reference to the integration domain $\Lambda := \{x \in \mathbb{R}^d \mid |x_i| \le \frac{L}{2}, \, i = 1, 2, \ldots, d\}$ is omitted, periodic boundary conditions are assumed, and dx is Lebesgue measure in d dimensions. Concerning units, we have set $\hbar$ equal to 1. Here, the operators $a^*(x)$, $a(x)$ are the usual operator-valued distributions on the bosonic Fock space

$$\mathcal{F} := \Gamma \left(L^2 (\Lambda, \mathbb{C}; dx) \right)$$

that satisfy the canonical commutation relations (CCR)

$$[a^\#(x), a^\#(y)] = 0, \qquad [a(x), a^*(y)] = \delta(x - y)\mathbb{1}_\mathcal{F},$$

with $a^\# := a$ or a^*. In terms of the field modes, they read

$$a(x) = \sum_{\mathbf{j} \in \mathbb{Z}^d} \frac{a_\mathbf{j} e^{i k_\mathbf{j} \cdot x}}{|\Lambda|^{\frac{1}{2}}}, \quad a^*(x) = \sum_{\mathbf{j} \in \mathbb{Z}^d} \frac{a_\mathbf{j}^* e^{-i k_\mathbf{j} \cdot x}}{|\Lambda|^{\frac{1}{2}}}, \tag{1.2}$$

where $k_\mathbf{j} := \frac{2\pi}{L} \mathbf{j}, \mathbf{j} = (j_1, j_2, \ldots, j_d), j_1, j_2, \ldots, j_d \in \mathbb{Z}$, and $|\Lambda| = L^d$, with CCR

$$[a_\mathbf{j}^\#, a_{\mathbf{j}'}^\#] = 0, \qquad [a_\mathbf{j}, a_{\mathbf{j}'}^*] = \delta_{\mathbf{j}, \mathbf{j}'}. \tag{1.3}$$

The unique (up to a phase) vacuum vector of $\mathcal{F}$ is denoted by Ω ($\|\Omega\| = 1$). The operator H is meant to be restricted to the subspace $\mathcal{F}^N$ of vectors with exactly N particles.

Upon implementing the volume integrations in (1.1), and keeping only the terms that are quadratic in the operators a_0, a_0^*, one gets the complete (particles number preserving) Bogoliubov Hamiltonian

$$H^{Bog} := \frac{1}{2} \sum_{\mathbf{j} \in \mathbb{Z}^d \setminus \{0\}} \hat{H}_\mathbf{j}^{Bog} \tag{1.4}$$

plus a function of the number operator $\sum_{\mathbf{j} \in \mathbb{Z}^d} a_\mathbf{j}^* a_\mathbf{j}$, where

$$\hat{H}_\mathbf{j}^{Bog} := \sum_{\mathbf{j}' = \pm \mathbf{j}} (k_{\mathbf{j}'}^2 + \phi_{\mathbf{j}'} \frac{a_0^* a_0}{N}) a_{\mathbf{j}'}^* a_{\mathbf{j}'} + \phi_\mathbf{j} \frac{a_0^* a_0^* a_\mathbf{j} a_{-\mathbf{j}}}{N} + \phi_{-\mathbf{j}} \frac{a_0 a_0 a_\mathbf{j}^* a_{-\mathbf{j}}^*}{N} \tag{1.5}$$

with $\{\phi_\mathbf{j} ; \mathbf{j} \in \mathbb{Z}^d\}$ the Fourier components of the pair potential $\phi(z)$. Indeed, the Hamiltonian in (1.4) with the operators a_0, a_0^* replaced with c-numbers[2] has been proposed in [9] to study the spectrum of the Hamiltonian H in (1.1) for large $N = \rho |\Lambda|$, based on the expectation that, for the system in its ground state, the mode corresponding to $\mathbf{0}$ is the only one macroscopically occupied in the limit $N \to \infty$,

---

[2]Concerning the validity in the thermodynamic limit of the replacement of $a_0^* a_0$ with its average we refer the reader to Appendix D of [28].

and that the fluctuations of $a_0^* a_0$ around its average are negligible. This is also the reason to neglect from the beginning the terms which are cubic or quartic in the operators a_j, a_j^*, $\mathbf{j} \neq \mathbf{0}$, and are not reabsorbed in the function of the number operator $\sum_{j \in \mathbb{Z}^d} a_j^* a_j$ mentioned after (1.4). This approximation has been proven to be correct at fixed $|\Lambda|$ to identify, in the limit $N \to \infty$, the spectrum of H with the spectrum predicted by Bogoliubov using the (quadratic) Hamiltonian described above; see also Remark 1.2. On the contrary, the omitted terms are not negligible in the (still poorly understood) thermodynamic limit. In scaling regimes of a dilute gas like the Gross–Pitaevskii limit, the validity of the Bogoliubov conjecture for the spectrum of the Hamiltonian proven recently in [8] requires a careful analysis of the so-called cubic and quartic terms.

As it is apparent from the definition in (1.4), H^{Bog} can be seen as a collection of three-modes systems. In fact, the three-modes system represents the main building block in the construction of the ground state of the Bogoliubov Hamiltonian (see (1.4)) and of the complete Hamiltonian (see (1.1)) in the mean field limiting regime; see [33, 34], respectively.

In this paper, for a given couple of modes $\pm \mathbf{j}_*$, we study the Hamiltonian

$$H_{\mathbf{j}_*}^{Bog} := \sum_{\mathbf{j} \in \mathbb{Z}^d \setminus \{\pm \mathbf{j}_*\}} k_{\mathbf{j}}^2 a_{\mathbf{j}}^* a_{\mathbf{j}} + \sum_{\mathbf{j} = \pm \mathbf{j}_*} (k_{\mathbf{j}}^2 + \phi_{\mathbf{j}} \frac{a_0^* a_0}{N}) a_{\mathbf{j}}^* a_{\mathbf{j}} + \phi_{\mathbf{j}_*} \frac{a_0^* a_0^* a_{\mathbf{j}_*} a_{-\mathbf{j}_*}}{N} + \phi_{\mathbf{j}_*} \frac{a_0 a_0 a_{\mathbf{j}_*}^* a_{-\mathbf{j}_*}^*}{N} \quad (1.6)$$

$$=: \sum_{\mathbf{j} \in \mathbb{Z}^d \setminus \{\pm \mathbf{j}_*\}} k_{\mathbf{j}}^2 a_{\mathbf{j}}^* a_{\mathbf{j}} + \hat{H}_{\mathbf{j}_*}^{Bog} \quad (1.7)$$

where $\phi_{\mathbf{j}_*} = \phi_{-\mathbf{j}_*} > 0$. Rather than the operator $\hat{H}_{\mathbf{j}_*}^{Bog}$, the latter (i.e., $H_{\mathbf{j}_*}^{Bog}$) is more convenient for later use (see [33, 34]) where by an iterative procedure we add the interaction of a new couple of modes at a time. A crucial role in our expansions and estimates is played by

$$\frac{k_{\mathbf{j}_*}^2}{\phi_{\mathbf{j}_*}} =: \epsilon_{\mathbf{j}_*}, \quad (1.8)$$

that at fixed $\mathbf{j}_*$ becomes arbitrarily small as $L \to \infty$. In the definition below, we report the assumptions on the pair potential ϕ considered in the papers [32–34].

Definition 1.1. The pair potential $\phi(x - y)$ is a bounded, real-valued function that is periodic, i.e., $\phi(z) = \phi(z + \mathbf{j}L)$ for $\mathbf{j} \in \mathbb{Z}^d$, and satisfies the following conditions:

1. $\phi(z) = \frac{1}{|\Lambda|} \sum_{\mathbf{j} \in \mathbb{Z}^d} \phi_{\mathbf{j}} e^{i k_{\mathbf{j}} z}$ is an even function, in consequence $\phi_{\mathbf{j}} = \phi_{-\mathbf{j}}$.
2. $\phi(z)$ is of positive type, i.e., the Fourier components $\phi_{\mathbf{j}}$ are nonnegative.
3. The pair interaction has a fixed but arbitrarily large ultraviolet cutoff (i.e., the nonzero Fourier components $\phi_{\mathbf{j}}$ form a finite set) with the requirements below to be satisfied:
   (3.1) (Strong Interaction Potential Assumption) The ratio $\epsilon_{\mathbf{j}}$ between the kinetic energy of the modes $\pm \mathbf{j} \neq \mathbf{0} = (0, \ldots, 0)$ and the corresponding Fourier component $\phi_{\mathbf{j}}(\neq 0)$ of the potential is sufficiently small.
   (3.2) For some $1 > \mu > 0$, $\theta > 0$ and all nonzero $\phi_{\mathbf{j}}$

$$\frac{\phi_{\mathbf{j}}}{\Delta_0} \frac{N^{\mu}}{N(N - N^{\mu})} < \frac{1}{2} \ , \quad \frac{1}{N^{\mu}} \leq O((\sqrt{\epsilon_{\mathbf{j}}})^{1+\theta}), \tag{1.9}$$

where $\Delta_0 = \min \left\{ k_{\mathbf{j}}^2 \mid \mathbf{j} \in \mathbb{Z}^d \setminus \{\mathbf{0}\} \right\}$ and N is the number of particles in the box.

We point out that assumption (3.2) in Definition 1.1 is always fulfilled in the mean field limiting regime, hence in the regime considered in [33, 34] for the Bogoliubov and the complete Hamiltonian, respectively. With regard to the model treated in the present paper, i.e., the *three modes* Bogoliubov Hamiltonian in (1.6) for arbitrary values of L and ρ, assumption (3.2) plays a nontrivial role in the last step of the Feshbach–Schur flow; more comments in Sect. 3.2.

1.2 Main Result

The main result of the paper [32] is the expansion, up to any arbitrary precision, of the ground state vector, ψ_{gs}, of the three-modes Hamiltonian (see (1.6)) at a sufficiently large but *fixed* number of particles N, in terms of the bare operators and the groundstate energy z_*, as described in the theorem below:

Theorem 1.1. *Let d be larger than or equal to 3, $\epsilon_{\mathbf{j}}$, sufficiently small, and $\rho = \frac{N}{L^d}$ sufficiently large independently of L. Then, $\forall \zeta > 0$ the following estimate holds true:*

$$\|\psi_{gs} - (\psi_{gs})_{\zeta}\| \leq \zeta \tag{1.10}$$

where $(\psi_{gs})_{\zeta}$ is a ζ-dependent finite sum of vectors obtained by applying suitable ζ-dependent finite products of the bare operators $\frac{1}{\hat{H}_{\mathbf{j}_*}^0 - z_*}$ and $W_{\mathbf{j}_*}$, $W_{\mathbf{j}_*}^*$ to the vector $\eta = \frac{a_0^* \dots a_0^*}{\sqrt{N}} \Omega$, where

$$\hat{H}_{\mathbf{j}_*}^0 := (k_{\mathbf{j}_*}^2 + \phi_{\mathbf{j}_*} \frac{a_0^* a_0}{N}) a_{\mathbf{j}_*}^* a_{\mathbf{j}_*} + (k_{\mathbf{j}_*}^2 + \phi_{\mathbf{j}_*} \frac{a_0^* a_0}{N}) a_{-\mathbf{j}_*}^* a_{-\mathbf{j}_*}, \tag{1.11}$$

$$\phi_{\mathbf{j}_*} \frac{a_0^* a_0^* a_{\mathbf{j}_*} a_{-\mathbf{j}_*}}{N} =: W_{\mathbf{j}_*} \ , \quad \phi_{\mathbf{j}_*} \frac{a_0 a_0 a_{\mathbf{j}_*}^* a_{-\mathbf{j}_*}^*}{N} =: W_{\mathbf{j}_*}^*. \tag{1.12}$$

In the limit $N \to \infty$ at fixed L, the ground state energy z_* can be replaced with

$$E_{\mathbf{j}_*}^{Bog} := -\left[k_{\mathbf{j}_*}^2 + \phi_{\mathbf{j}_*} - \sqrt{(k_{\mathbf{j}_*}^2)^2 + 2\phi_{\mathbf{j}_*} k_{\mathbf{j}_*}^2} \right]. \tag{1.13}$$

Furthermore, in the mean field limiting regime (i.e., at fixed box volume $|\Lambda|$ and for a number of particles, N, sufficiently large), the expansion of the ground state wave function in (1.10) holds in any dimension $d \geq 1$; see also Remark 3.9.

Remark 1.2. We recall that in the mean field limit, i.e., at L fixed and for $N \to \infty$, the ground state energy of the complete Hamiltonian in (1.1) tends to $E^{Bog} := \frac{1}{2} \sum_{\mathbf{j} \in \mathbb{Z}^d \setminus \{0\}} E_{\mathbf{j}}^{Bog}$, and, more importantly, the spectrum of H tends to the spectrum of H^{Bog}; see [15, 19, 21, 29, 35]. In some regimes, the expected asymptotic expression ("Lee–Yuang–Yang formula") has been rigorously obtained also in the thermodynamic limit (for details see [10, 18, 36]).

As compared to previous works on the mean field limit of the Hamiltonian in (1.1), the main purpose of the expansion stated in Theorem 1.1—and likewise of the analysis in [33, 34]—is to provide a detailed control of the ground state vector not restricted to the two-point correlation function (which is enough to derive Bose–Einstein condensation). The ultimate goal is to provide the expansion of relevant physical quantities that can be controlled at large but finite particle density ρ uniformly in the box size when (also) *large field* problems appear; see Remark 3.3. Of course, the results obtained in [32] are restricted to amputated Hamiltonians and therefore concern only marginally the *infrared problem* associated with Bose–Einstein condensation in the thermodynamic limit, but can help to shed light on some of the features of the ground state vector that become relevant at fixed density ρ and $|\Lambda| \to \infty$. Concerning the thermodynamic limit, we also mention [5] which contains an overview of their long-term program to see symmetry breaking in a weakly interacting many-boson system on a three-dimensional lattice at low temperature, including their strategy to control the large field contributions to the partition function of the gas.

2 Motivations and Features of the Strategy

We know that, at fixed volume $|\Lambda|$, the expectation value of the number operator[3] $\sum_{\mathbf{j} \in \mathbb{Z}^d \setminus \{0\}} a_{\mathbf{j}}^* a_{\mathbf{j}}$ in the ground state of the Hamiltonian (1.1) remains bounded in the mean field limit (i.e., $\lambda = \frac{|\Lambda|}{N}$ and $N \to \infty$); see the papers by Seiringer and by Lewin et al.; cf [35], [21]. Starting from this fact, one might think of a multi-scale procedure leading to an effective Hamiltonian for spectral values in a neighborhood of the ground state energy. An obvious candidate for such an effective Hamiltonian is (a multiple of) the orthogonal projection onto the state where all the particles are in the zero mode.

The Feshbach–Schur map is a very useful tool to construct effective Hamiltonians. We recall that, given a (separable) Hilbert space $\mathcal{H}$ and two projections $\mathscr{P}$ and $\overline{\mathscr{P}}$, with $\mathscr{P} + \overline{\mathscr{P}} = \mathbb{1}_{\mathcal{H}}$, the *Feshbach–Schur map*, $\mathscr{F}$, associated with $(\mathscr{P}, \overline{\mathscr{P}})$ maps every pair (K, z) of a closed operator K defined on the entire range of $\mathscr{P}$ and a complex number z, with the property that $\overline{\mathscr{P}}(K - z)\overline{\mathscr{P}}$ is bounded-invertible on $\overline{\mathscr{P}}\mathcal{H}$, to the operator $\mathscr{F}(K - z)$ acting on the subspace $\mathscr{P}\mathcal{H} \subset \mathcal{H}$ formally given by

---

[3]The operator $\sum_{\mathbf{j} \in \mathbb{Z}^d \setminus \{0\}} a_{\mathbf{j}}^* a_{\mathbf{j}}$ counts the number of particles in the nonzero modes states.

$$\mathscr{F}(K-z) := \mathscr{P}(K-z)\mathscr{P} - \mathscr{P}K\overline{\mathscr{P}}\frac{1}{\overline{\mathscr{P}}(K-z)\overline{\mathscr{P}}}\overline{\mathscr{P}}K\mathscr{P}. \qquad (2.1)$$

The Feshbach map $\mathscr{F}$ is *"isospectral"* in the sense that if $\mathscr{F}$ is defined on (K, z) then z belongs to the point spectrum of K if and only if 0 is in the point spectrum of $\mathscr{F}(K-z)$; moreover, z belongs to the resolvent set of K if and only if $\mathscr{F}(K-z)$ is bounded-invertible on $\mathscr{P}\mathcal{H}$ (see [1]).

The use of the Feshbach–Schur map for the spectral analysis of quantum field theory systems started with the seminal work by Bach, Fröhlich, and Sigal, followed by refinements of the technique and variants; see the references in [32]. In those papers, the use of the Feshbach–Schur map is in the spirit of the functional integral renormalization group, and the projections $(\mathscr{P}, \overline{\mathscr{P}})$ are directly related to energy subspaces of the free Hamiltonian. However, as a mathematical tool, the Feshbach map enjoys an enormous flexibility due to the freedom in the choice of the couple of projections $\mathscr{P}, \overline{\mathscr{P}}$. The effectiveness of the choice depends on the features of the Hamiltonian.

In the system that we study the total number of particles is conserved under time evolution. The effective Hamiltonian that we want to construct suggests to relate the Feshbach–Schur projections $(\mathscr{P}, \overline{\mathscr{P}})$ to subspaces of states with definite number of particles in the modes labeled by $\left\{\frac{2\pi}{L}\mathbf{j}; \mathbf{j} \in \mathbb{Z}^d\right\}$. More precisely, consider the eigenspace of $\sum_{\mathbf{j}=\pm\mathbf{j}_*} a_{\mathbf{j}}^* a_{\mathbf{j}}$ corresponding to the eigenvalue i, i.e., the subspace of states containing i particles in the modes associated with $\pm\frac{2\pi}{L}\mathbf{j}_*$. Observe that the interaction part of the second quantized Hamiltonian in (1.1) can connect two eigenspaces corresponding to distinct eigenvalues, i and i', only if $i - i' = \pm1, \pm2$. The selection rules of the interaction Hamiltonian with respect to the occupation numbers of the particle states associated with the modes $\left\{\frac{2\pi}{L}\mathbf{j}; \mathbf{j} \in \mathbb{Z}^d\right\}$ suggest to construct a flow of Feshbach–Schur maps associated with projections onto such eigenspaces with decreasing eigenvalue i.

In [32–34], we consider the system in the *strong interaction potential regime*: by this we mean that the ratio between each nonzero Fourier component of the potential, $\phi_{\mathbf{j}}$, and the corresponding kinetic energy, $k_{\mathbf{j}}^2$, must be sufficiently large. For a (positive definite) potential $\phi \in L^1$ such that $\int \phi(z)dz > 0$, this is precisely the regime that is relevant in the thermodynamic limit because at fixed $\mathbf{j}$ the ratio $\phi_{\mathbf{j}}/(k_{\mathbf{j}})^2$ diverges like L^2, being $k_{\mathbf{j}} := \frac{2\pi}{L}\mathbf{j}$ and L the side of the box.

2.1 Feshbach–Schur Projections

For the three-modes system described by the Hamiltonian $H_{\mathbf{j}_*}^{Bog}$ where $\epsilon_{\mathbf{j}_*} := \frac{k_{\mathbf{j}_*}^2}{\phi_{\mathbf{j}_*}}$ is sufficiently small, we consider the Feshbach–Schur flow associated with the sequence of projections defined below. Without loss of generality, we can suppose that $H_{\mathbf{j}_*}^{Bog}$

is restricted to $\mathcal{F}^N$ with N even. Then, for i even number, $0 \leq i \leq N-2$, we define $\overline{\mathscr{P}^{(i)}} := Q_{\mathbf{j}_*}^{(>i+1)}$, $\mathscr{P}^{(i)} := Q_{\mathbf{j}_*}^{(i,i+1)}$ where:

- $Q_{\mathbf{j}_*}^{(0,1)} :=$ the projection (in $\mathcal{F}^N$) onto the subspace generated by vectors with $N-0 = N$ or $N-1$ particles in the modes $\mathbf{j}_*$ and $-\mathbf{j}_*$, i.e., the operator $a_{\mathbf{j}_*}^* a_{\mathbf{j}_*} + a_{-\mathbf{j}_*}^* a_{-\mathbf{j}_*}$ has eigenvalues N and $N-1$ when restricted to $Q_{\mathbf{j}_*}^{(0,1)} \mathcal{F}^N$,
- $Q_{\mathbf{j}_*}^{(>1)} :=$ the projection onto the orthogonal complement of $Q_{\mathbf{j}_*}^{(0,1)} \mathcal{F}^N$ in $\mathcal{F}^N$,

and, iteratively, up to $i = N-2$

- $Q_{\mathbf{j}_*}^{(i,i+1)} :=$ the projection onto the subspace of $Q_{\mathbf{j}_*}^{(>i-1)} \mathcal{F}^N$ spanned by the vectors with $N-i$ or $N-i-1$ particles in the modes $\mathbf{j}_*$ and $-\mathbf{j}_*$,
- $Q_{\mathbf{j}_*}^{(>i+1)} :=$ the projection onto the orthogonal complement of $Q_{\mathbf{j}_*}^{(i,i+1)} Q_{\mathbf{j}_*}^{(>i-1)} \mathcal{F}^N$ in $Q_{\mathbf{j}_*}^{(>i-1)} \mathcal{F}^N$.

Hence, we can write

$$Q_{\mathbf{j}_*}^{(>i+1)} + Q_{\mathbf{j}_*}^{(i,i+1)} = Q_{\mathbf{j}_*}^{(>i-1)}. \tag{2.2}$$

As a final couple of Feshbach–Schur projections, we employ

$$\mathscr{P}^{(N)} := |\eta\rangle\langle\eta| \quad , \quad \overline{\mathscr{P}^{(N)}} := Q_{\mathbf{j}_*}^{(>N-1)} - |\eta\rangle\langle\eta|,$$

where $\eta := \frac{1}{\sqrt{N!}} a_0^* \ldots a_0^* \Omega$, i.e., η is the state where all the N particles are in the zero-mode state.

The Feshbach–Schur flow is implemented starting from $H_{\mathbf{j}_*}^{Bog} - z$ and using the formula in (2.1) iteratively by means of the projections $\overline{\mathscr{P}^{(i)}} := Q_{\mathbf{j}_*}^{(>i+1)}$ and $\mathscr{P}^{(i)} := Q_{\mathbf{j}_*}^{(i,i+1)}$, up to the final step $i = N$. At each step i the corresponding Feshbach–Schur Hamiltonian is an operator on $Q_{\mathbf{j}_*}^{(>i+1)} \mathcal{F}^N$, and the iteration is at least formally consistent due to the identity in (2.2). The final Feshbach–Schur Hamiltonian is the finite rank operator $f(z)|\eta\rangle\langle\eta|$. In the range $(-\infty, z_{max})$ of the spectral parameter z where the flow can be implemented, we show that there exists a unique z_* such that $f(z_*) = 0$. Then, z_* is an eigenvalue (the ground state energy) of the original Hamiltonian $H_{\mathbf{j}_*}^{Bog}$ due to the isospectrality that holds at each step of the Feshbach–Schur flow. Feshbach–Schur theory provides also an algorithm (cf Sect. 3.3) to reconstruct the eigenvector of the original Hamiltonian $H_{\mathbf{j}_*}^{Bog}$ associated with the eigenvalue z_* starting from the eigenvector (i.e., η) with eigenvalue zero of the final Feshbach–Schur Hamiltonian $f(z_*)|\eta\rangle\langle\eta|$.

2.2 Outline of the Feshbach–Schur Flow

We shall iterate the Feshbach–Schur map starting from $i = 0$ up to $i = N - 2$ with i even, using the projections $\mathscr{P}^{(i)}$ and $\overline{\mathscr{P}^{(i)}}$ for the i-th step[4] of the iteration where

$$\mathscr{P}^{(i)} := Q_{\mathbf{j_*}}^{(>i+1)} \quad , \quad \overline{\mathscr{P}^{(i)}} := Q_{\mathbf{j_*}}^{(i,i+1)}. \tag{2.3}$$

We denote by $\mathscr{F}^{(i)}$ the Feshbach–Schur map at the i-th step. We start applying $\mathscr{F}^{(0)}$ to $H_{\mathbf{j_*}}^{Bog} - z$ and compute

$$\mathscr{K}_{\mathbf{j_*}}^{Bog\,(0)}(z) \tag{2.4}$$

$$:= \mathscr{F}^{(0)}(H_{\mathbf{j_*}}^{Bog} - z) \tag{2.5}$$

$$= Q_{\mathbf{j_*}}^{(>1)}(H_{\mathbf{j_*}}^{Bog} - z)Q_{\mathbf{j_*}}^{(>1)} - Q_{\mathbf{j_*}}^{(>1)} H_{\mathbf{j_*}}^{Bog} Q_{\mathbf{j_*}}^{(0,1)} \frac{1}{Q_{\mathbf{j_*}}^{(0,1)}(H_{\mathbf{j_*}}^{Bog} - z)Q_{\mathbf{j_*}}^{(0,1)}} Q_{\mathbf{j_*}}^{(0,1)} H_{\mathbf{j_*}}^{Bog} Q_{\mathbf{j_*}}^{(>1)} \tag{2.6}$$

$$= Q_{\mathbf{j_*}}^{(>1)}(H_{\mathbf{j_*}}^{Bog} - z)Q_{\mathbf{j_*}}^{(>1)} - Q_{\mathbf{j_*}}^{(>1)} W_{\mathbf{j_*}} Q_{\mathbf{j_*}}^{(0,1)} \frac{1}{Q_{\mathbf{j_*}}^{(0,1)}(H_{\mathbf{j_*}}^{Bog} - z)Q_{\mathbf{j_*}}^{(0,1)}} Q_{\mathbf{j_*}}^{(0,1)} W_{\mathbf{j_*}}^* Q_{\mathbf{j_*}}^{(>1)}. \tag{2.7}$$

Then, we iteratively define

$$\mathscr{K}_{\mathbf{j_*}}^{Bog\,(i)}(z) := \mathscr{F}^{(i)}(\mathscr{K}_{\mathbf{j_*}}^{Bog\,(i-2)}(z)), \quad i = 0, \ldots, N - 2 \quad \text{with } i \text{ even}, \tag{2.8}$$

where $\mathscr{K}_{\mathbf{j_*}}^{Bog\,(-2)}(z) \equiv H_{\mathbf{j_*}}^{Bog} - z$.

Notice that, for l and l' even numbers, $Q_{\mathbf{j_*}}^{(l,l+1)} W_{\mathbf{j_*}} Q_{\mathbf{j_*}}^{(l',l'+1)} \neq 0$ only if $l - l' = 2$ and $Q_{\mathbf{j_*}}^{(l,l+1)} W_{\mathbf{j_*}}^* Q_{\mathbf{j_*}}^{(l',l'+1)} \neq 0$ only if $l - l' = -2$. This implies

$$Q_{\mathbf{j_*}}^{(>3)} \mathscr{K}_{\mathbf{j_*}}^{Bog\,(0)}(z) Q_{\mathbf{j_*}}^{(2,3)} = Q_{\mathbf{j_*}}^{(>3)} W_{\mathbf{j_*}} Q_{\mathbf{j_*}}^{(2,3)} \tag{2.9}$$

and

$$Q_{\mathbf{j_*}}^{(>3)} \mathscr{K}_{\mathbf{j_*}}^{Bog\,(0)}(z) Q_{\mathbf{j_*}}^{(>3)} = Q_{\mathbf{j_*}}^{(>3)} H_{\mathbf{j_*}}^{Bog} Q_{\mathbf{j_*}}^{(>3)}. \tag{2.10}$$

Hence, a straightforward calculation shows that

$$\mathscr{K}_{\mathbf{j_*}}^{Bog\,(2)}(z) \tag{2.11}$$

$$= Q_{\mathbf{j_*}}^{(>3)}(H_{\mathbf{j_*}}^{Bog} - z)Q_{\mathbf{j_*}}^{(>3)} \tag{2.12}$$

$$- Q_{\mathbf{j_*}}^{(>3)} W_{\mathbf{j_*}} Q_{\mathbf{j_*}}^{(2,3)} \frac{1}{Q_{\mathbf{j_*}}^{(2,3)}(H_{\mathbf{j_*}}^{Bog} - W_{\mathbf{j_*}} Q_{\mathbf{j_*}}^{(0,1)} \frac{1}{Q_{\mathbf{j_*}}^{(0,1)}(H_{\mathbf{j_*}}^{Bog} - z)Q_{\mathbf{j_*}}^{(0,1)}} Q_{\mathbf{j_*}}^{(0,1)} W_{\mathbf{j_*}}^* - z)Q_{\mathbf{j_*}}^{(2,3)}} Q_{\mathbf{j_*}}^{(2,3)} W_{\mathbf{j_*}}^* Q_{\mathbf{j_*}}^{(>3)}. \tag{2.13}$$

Assuming that the use of the expansion below is legitimate

---

[4]We use this notation though the number of steps is in fact $1 + i/2$ being i an even number.

$$Q_{\mathbf{j}_*}^{(2,3)} \frac{1}{Q_{\mathbf{j}_*}^{(2,3)}(H_{\mathbf{j}_*}^{Bog} - W_{\mathbf{j}_*} Q_{\mathbf{j}_*}^{(0,1)} \frac{1}{Q_{\mathbf{j}_*}^{(0,1)}(H_{\mathbf{j}_*}^{Bog} - z)Q_{\mathbf{j}_*}^{(0,1)}} Q_{\mathbf{j}_*}^{(0,1)} W_{\mathbf{j}_*}^* - z)Q_{\mathbf{j}_*}^{(2,3)}} Q_{\mathbf{j}_*}^{(2,3)} \tag{2.14}$$

$$= Q_{\mathbf{j}_*}^{(2,3)} \sum_{l_2=0}^{\infty} \frac{1}{Q_{\mathbf{j}_*}^{(2,3)}(H_{\mathbf{j}_*}^{Bog} - z)Q_{\mathbf{j}_*}^{(2,3)}} \times \tag{2.15}$$

$$\times \left[Q_{\mathbf{j}_*}^{(2,3)} W_{\mathbf{j}_*} Q_{\mathbf{j}_*}^{(0,1)} \frac{1}{Q_{\mathbf{j}_*}^{(0,1)}(H_{\mathbf{j}_*}^{Bog} - z)Q_{\mathbf{j}_*}^{(0,1)}} Q_{\mathbf{j}_*}^{(0,1)} W_{\mathbf{j}_*}^* Q_{\mathbf{j}_*}^{(2,3)} \frac{1}{Q_{\mathbf{j}_*}^{(2,3)}(H_{\mathbf{j}_*}^{Bog} - z)Q_{\mathbf{j}_*}^{(2,3)}} \right]^{l_2} Q_{\mathbf{j}_*}^{(2,3)} ,$$

and introducing the notation

$$W_{\mathbf{j}_* ; i,i'} := Q_{\mathbf{j}_*}^{(i,i+1)} W_{\mathbf{j}_*} Q_{\mathbf{j}_*}^{(i',i'+1)} \quad , \quad W_{\mathbf{j}_* ; i,i'}^* := Q_{\mathbf{j}_*}^{(i,i+1)} W_{\mathbf{j}_*}^* Q_{\mathbf{j}_*}^{(i',i'+1)} ,$$

we can write

$$\mathscr{K}_{\mathbf{j}_*}^{Bog\,(2)}(z) \tag{2.16}$$

$$= Q_{\mathbf{j}_*}^{(>3)}(H_{\mathbf{j}_*}^{Bog} - z)Q_{\mathbf{j}_*}^{(>3)} \tag{2.17}$$

$$- \sum_{l_2=0}^{\infty} Q_{\mathbf{j}_*}^{(>3)} W_{\mathbf{j}_*} Q_{\mathbf{j}_*}^{(2,3)} \frac{1}{Q_{\mathbf{j}_*}^{(2,3)}(H_{\mathbf{j}_*}^{Bog} - z)Q_{\mathbf{j}_*}^{(2,3)}} \times \tag{2.18}$$

$$\times \left[W_{\mathbf{j}_* ; 2,0} \frac{1}{Q_{\mathbf{j}_*}^{(0,1)}(H_{\mathbf{j}_*}^{Bog} - z)Q_{\mathbf{j}_*}^{(0,1)}} W_{\mathbf{j}_* ; 0,2}^* \frac{1}{Q_{\mathbf{j}_*}^{(2,3)}(H_{\mathbf{j}_*}^{Bog} - z)Q_{\mathbf{j}_*}^{(2,3)}} \right]^{l_2} Q_{\mathbf{j}_*}^{(2,3)} W_{\mathbf{j}_*}^* Q_{\mathbf{j}_*}^{(>3)} .$$

With the definition

$$R_{\mathbf{j}_* ; i,i}^{Bog}(z) := Q_{\mathbf{j}_*}^{(i,i+1)} \frac{1}{Q_{\mathbf{j}_*}^{(i,i+1)}(H_{\mathbf{j}_*}^{Bog} - z)Q_{\mathbf{j}_*}^{(i,i+1)}} Q_{\mathbf{j}_*}^{(i,i+1)} , \tag{2.19}$$

for $4 \le i \le N - 2$ we get

$$\mathscr{K}_{\mathbf{j}_*}^{Bog\,(i)}(z) \tag{2.20}$$

$$= Q_{\mathbf{j}_*}^{(>i+1)}(H_{\mathbf{j}_*}^{Bog} - z)Q_{\mathbf{j}_*}^{(>i+1)} \tag{2.21}$$

$$- \sum_{l_i=0}^{\infty} Q_{\mathbf{j}_*}^{(>i+1)} W_{\mathbf{j}_*} R_{\mathbf{j}_* ; i,i}^{Bog}(z) \left[W_{i,i-2} R_{\mathbf{j}_* ; i-2,i-2}^{Bog}(z) \times \tag{2.22}$$

$$\times \sum_{l_{i-2}=0}^{\infty} \left[W_{\mathbf{j}_* ; i-2,i-4} \dots W_{\mathbf{j}_* ; i-4,i-2}^* R_{\mathbf{j}_* ; i-2,i-2}^{Bog}(z) \right]^{l_{i-2}} W_{\mathbf{j}_* ; i-2,i}^* R_{\mathbf{j}_* ; i,i}^{Bog}(z) \right]^{l_i} W_{\mathbf{j}_*}^* Q_{\mathbf{j}_*}^{(>i+1)} \tag{2.23}$$

where i is an even number and the expression corresponding to ... in (2.23) is made precise in Theorem 3.1 reported below.

3 Statement of the Results and Role of the Assumptions

In the following sections, we specify the results that are obtained in [32] for the three-modes Bogoliubov Hamiltonian.

3.1 Feshbach–Schur Map up to the $N-2$ Step

For the implementation up to the $N-2$-th step of the Feshbach–Schur map outlined in Sect. 2.2, we require $\frac{1}{N} \leq \epsilon_{\mathbf{j}_*}^{\nu}$ for some $\nu > \frac{11}{8}$ and $\epsilon_{\mathbf{j}_*}$ sufficiently small.

The precise form of the Feshbach–Schur operators is determined in Theorem 3.1 of [32] that we report below

Theorem 3.1. *Let $\epsilon_{\mathbf{j}_*}$ and $E_{\mathbf{j}_*}^{Bog}$ defined in (1.8) and (1.13) respectively. Then, for*

$$z \leq E_{\mathbf{j}_*}^{Bog} + (\delta - 1)\phi_{\mathbf{j}_*}\sqrt{\epsilon_{\mathbf{j}_*}^2 + 2\epsilon_{\mathbf{j}_*}}(< 0) \tag{3.1}$$

with $\delta = 1 + \sqrt{\epsilon_{\mathbf{j}_*}}$, $\frac{1}{N} \leq \epsilon_{\mathbf{j}_*}^{\nu}$ for some $\nu > \frac{11}{8}$, and $\epsilon_{\mathbf{j}_*}$ sufficiently small, the operators $\mathscr{K}_{\mathbf{j}_*}^{Bog\,(i)}(z)$ defined in (2.8), where $0 \leq i \leq N-2$ and even, are well defined.[5] For $i = 0$, $\mathscr{K}_{\mathbf{j}_*}^{Bog\,(0)}(z)$ is given in (2.7). For $i = 2, 4, 6, \ldots, N-2$, they correspond to

$$\mathscr{K}_{\mathbf{j}_*}^{Bog\,(i)}(z) = Q_{\mathbf{j}_*}^{(>i+1)}(H_{\mathbf{j}_*}^{Bog} - z)Q_{\mathbf{j}_*}^{(>i+1)} \tag{3.2}$$

$$-Q_{\mathbf{j}_*}^{(>i+1)}W_{\mathbf{j}_*}\,R_{\mathbf{j}_*\,;\,i,i}^{Bog}(z)\sum_{l_i=0}^{\infty}\left[\Gamma_{\mathbf{j}_*\,;\,i,i}^{Bog}(z)R_{\mathbf{j}_*\,;\,i,i}^{Bog}(z)\right]^{l_i}W_{\mathbf{j}_*}^*Q_{\mathbf{j}_*}^{(>i+1)}$$

where:

•

$$\Gamma_{\mathbf{j}_*\,;\,2,2}^{Bog}(z) := W_{\mathbf{j}_*\,;\,2,0}\,R_{\mathbf{j}_*\,;\,0,0}^{Bog}(z)W_{\mathbf{j}_*\,;\,0,2}^* \tag{3.3}$$

• *for $N-2 \geq i \geq 4$,*

$$\Gamma_{\mathbf{j}_*\,;\,i,i}^{Bog}(z) := W_{\mathbf{j}_*\,;\,i,i-2}\,R_{\mathbf{j}_*\,;\,i-2,i-2}^{Bog}(z)\sum_{l_{i-2}=0}^{\infty}\left[\Gamma_{\mathbf{j}_*\,;\,i-2,i-2}^{Bog}(z)R_{\mathbf{j}_*\,;\,i-2,i-2}^{Bog}(z)\right]^{l_{i-2}}W_{\mathbf{j}_*\,;\,i-2,i}^* \tag{3.4}$$

$$= W_{\mathbf{j}_*\,;\,i,i-2}\,(R_{\mathbf{j}_*\,;\,i-2,i-2}^{Bog}(z))^{\frac{1}{2}}\sum_{l_{i-2}=0}^{\infty}\left[(R_{\mathbf{j}_*\,;\,i-2,i-2}^{Bog}(z))^{\frac{1}{2}}\Gamma_{\mathbf{j}_*\,;\,i-2,i-2}^{Bog}(z)(R_{\mathbf{j}_*\,;\,i-2,i-2}^{Bog}(z))^{\frac{1}{2}}\right]^{l_{i-2}} \times (3.5)$$

$$\times (R_{\mathbf{j}_*\,;\,i-2,i-2}^{Bog}(z))^{\frac{1}{2}}W_{\mathbf{j}_*\,;\,i-2,i}^* \,.$$

Notice that the flow can be implemented for z in a neighborhood of the Bogoliubov energy $E_{\mathbf{j}_*}^{Bog}$ (see (1.13)). As it is well known, $E_{\mathbf{j}_*}^{Bog}$ corresponds to the ground state energy of $H_{\mathbf{j}_*}^{Bog}$ in the limit $N \to \infty$; furthermore, for $\delta = 2$, the r-h-s of (3.1) corresponds to the first excited energy level in the mean field limit (see [35]).

Remark 3.2. We observe that:

(i) For space dimension $d \geq 3$, and at fixed particle density $\rho = \frac{N}{L^d}$, recalling that $\epsilon_{\mathbf{j}_*}$ goes to zero like L^{-2}, the bound $\frac{1}{N} \leq \epsilon_{\mathbf{j}_*}^{\nu}$ is fulfilled, for $(\frac{11}{8} <)\nu < \frac{3}{2}$, if the box is sufficiently large;

---

[5] $\mathscr{K}_{\mathbf{j}_*}^{Bog\,(i)}(z)$ is self-adjoint on the domain of the Hamiltonian $Q_{\mathbf{j}_*}^{(>i+1)}(H_{\mathbf{j}_*}^{Bog} - z)Q_{\mathbf{j}_*}^{(>i+1)}$.

(ii) For $d = 1, 2$, if at fixed $\mathbf{j}_*$ and $\phi_{\mathbf{j}_*}$, the box size tends to infinity, the particle density ρ must be suitably divergent to ensure the bound $\frac{1}{N} \leq \epsilon_{\mathbf{j}_*}^{\nu}$;

(iii) The bound $\frac{1}{N} \leq \epsilon_{\mathbf{j}_*}^{\nu}$ holds trivially in the mean field limiting regime where the size of the box is kept fixed and the number of particles, N, can be arbitrarily large irrespective of the box size.

fThe result of Theorem 3.1 follows from an upper bound to the norm of

$$\check{\Gamma}_{\mathbf{j}_*;i,i}^{Bog} := \sum_{l_i=0}^{\infty} [(R_{\mathbf{j}_*;i,i}^{Bog}(z))^{\frac{1}{2}} \Gamma_{\mathbf{j}_*;i,i}^{Bog} (R_{\mathbf{j}_*;i,i}^{Bog}(z))^{\frac{1}{2}}]^{l_i} . \tag{3.6}$$

To this end, under the same assumptions of Theorem 3.1, in Lemma 3.2 of [32], we prove the key estimate (for z as in (3.1))

$$\left\| [R_{\mathbf{j}_*;N-2l,N-2l}^{Bog}(z)]^{\frac{1}{2}} W_{\mathbf{j}_*;N-2l,N-2l+2}^* [R_{\mathbf{j}_*;N-2l+2,N-2l+2}^{Bog}(z)]^{\frac{1}{2}} \right\| \tag{3.7}$$

$$\leq \frac{1}{2 \left[1 + a_{\epsilon_{\mathbf{j}_*}} - \frac{2b_{\epsilon_{\mathbf{j}_*}}^{(\delta)}}{2l-1} - \frac{1-c_{\epsilon_{\mathbf{j}_*}}^{(\delta)}}{(2l-1)^2} \right]^{\frac{1}{2}}} \tag{3.8}$$

where

$$a_{\epsilon_{\mathbf{j}_*}} := 2\epsilon_{\mathbf{j}_*} + O(\epsilon_{\mathbf{j}_*}^{\nu}) , \tag{3.9}$$

$$b_{\epsilon_{\mathbf{j}_*}}^{(\delta)} := (1 + \epsilon_{\mathbf{j}_*})\delta \, \chi_{[0,2)}(\delta)\sqrt{\epsilon_{\mathbf{j}_*}^2 + 2\epsilon_{\mathbf{j}_*}} \tag{3.10}$$

and

$$c_{\epsilon_{\mathbf{j}_*}}^{(\delta)} := -(1 - \delta^2 \, \chi_{[0,2)}(\delta))(\epsilon_{\mathbf{j}_*}^2 + 2\epsilon_{\mathbf{j}_*}) \tag{3.11}$$

with $\chi_{[0,2)}$ the characteristic function of the interval $[0, 2)$. This allows us to control a nontrivial numerical sequence associated to the norm of $\check{\Gamma}_{\mathbf{j}_*;i,i}^{Bog}$ and yielding the following upper bound provided z is in the interval specified in (3.1):

$$\|\check{\Gamma}_{\mathbf{j}_*;N-2l,N-2l}^{Bog}(z)\| \leq \frac{2}{\left[1 + \sqrt{\eta a_{\epsilon_{\mathbf{j}_*}}} - \frac{b_{\epsilon_{\mathbf{j}_*}}/\sqrt{\eta a_{\epsilon_{\mathbf{j}_*}}}}{2l-\epsilon_{\mathbf{j}_*}^{\Theta}} \right]} \tag{3.12}$$

with $\eta = 1 - \sqrt{\epsilon_{\mathbf{j}_*}}$, where $\Theta := \min\{2(\nu - \frac{11}{8}); \frac{1}{4}\}$, and $b_{\epsilon_{\mathbf{j}_*}} = b_{\epsilon_{\mathbf{j}_*}}^{(\delta)}|_{\delta=1+\sqrt{\epsilon_{\mathbf{j}_*}}}$.

Remark 3.3. We stress that the detailed form of the bound in (3.7) as a function of $\epsilon_{\mathbf{j}_*}$ and l is crucial to control the norm on the l-h-s of (3.12) for arbitrarily small values of $\epsilon_{\mathbf{j}_*}$. This is in view of the *large field problem* that appears as $\epsilon_{\mathbf{j}_*} \to 0$. Indeed, in this limit the effective operator in formula (3.2) is constructed in a *large field region*, i.e., in a region where the field is such that the Hamiltonian interaction cannot be controlled by the kinetic term, hence to implement a perturbation scheme, the Hamiltonian must be split so that the "free part" (see (1.11)) includes a positive, normal ordered term coming from the interaction. However, this would not be enough to derive a convergent expansion without the multi-scale analysis in the particles number, by means of the projections $\mathscr{P}^{(i)}$, $\overline{\mathscr{P}^{(i)}}$.

Remark 3.4. The u.v. (ultraviolet) cutoff mentioned in Definition 1.1 implies an upper bound on $\epsilon_{\mathbf{j}_*}$. The results in Theorem 1.1 can be easily derived in a regime where the kinetic energy $k_{\mathbf{j}}^2$ is large with respect to the potential $\phi_{\mathbf{j}_*}$, making use of the usual perturbative expansion where the free part (i.e., $\sum_{\mathbf{j} \in \mathbb{Z}^d} k_{\mathbf{j}}^2 a_{\mathbf{j}}^* a_{\mathbf{j}}$) of the Hamiltonian dominates the interaction. In the intermediate regime where the kinetic energy and the potential terms are comparable, the result is surely true though the key estimate in (3.7) is derived assuming $\epsilon_{\mathbf{j}_*}$ sufficiently small, but an analogous (and probably even better) estimate should be derived when $\epsilon_{\mathbf{j}_*}$ is of order 1, by properly taking all the terms into account. The substantial complications in the control of the perturbative scheme appear as $\epsilon_{\mathbf{j}_*}$ gets ever smaller; see Remark 3.3.

In the mean field limiting regime (i.e., N arbitrarily large at fixed L), for the expansion of the ground state vector of the Hamiltonian in (1.1) analogous to the one in Theorem 1.1, the removal of the u.v. cutoff is expected to be just a technical complication in dealing with the intermediate regime corresponding to ratios $\epsilon_{\mathbf{j}}$ of order 1.

3.2 Last Step of the Feshbach–Schur Flow

For the step from $i = N - 2$ to $i = N$, we consider the projections $\mathscr{P}^{(N)} := \mathscr{P}_\eta := |\eta\rangle\langle\eta|$ and $\overline{\mathscr{P}^{(N)}} := \overline{\mathscr{P}_\eta}$ such that

$$\mathscr{P}^{(N)} + \overline{\mathscr{P}^{(N)}} = Q_{\mathbf{j}_*}^{(>N-1)}. \tag{3.13}$$

Formally, we get

$$\mathscr{K}_{\mathbf{j}_*}^{Bog\,(N)}(z) \tag{3.14}$$

$$:= \mathscr{F}^{(N)}(\mathscr{K}_{\mathbf{j}_*}^{Bog\,(N-2)}(z)) \tag{3.15}$$

$$= \mathscr{P}_\eta (H_{\mathbf{j}_*}^{Bog} - z)\mathscr{P}_\eta \tag{3.16}$$

$$- \mathscr{P}_\eta W_{\mathbf{j}_*} R_{\mathbf{j}_*;N-2,N-2}^{Bog}(z) \sum_{l_{N-2}=0}^{\infty} [\Gamma_{\mathbf{j}_*;N-2,N-2}^{Bog}(z) R_{\mathbf{j}_*;N-2,N-2}^{Bog}(z)]^{l_{N-2}} W_{\mathbf{j}_*}^* \mathscr{P}_\eta$$

$$- \mathscr{P}_\eta W_{\mathbf{j}_*} \overline{\mathscr{P}_\eta} \frac{1}{\overline{\mathscr{P}_\eta}\mathscr{K}_{\mathbf{j}_*}^{Bog\,(N-2)}(z)\overline{\mathscr{P}_\eta}} \overline{\mathscr{P}_\eta} W_{\mathbf{j}_*}^* \mathscr{P}_\eta \tag{3.17}$$

$$= \left(-z - \langle \eta, \ W_{\mathbf{j}_*} R_{\mathbf{j}_*;N-2,N-2}^{Bog}(z) \sum_{l_{N-2}=0}^{\infty} [\Gamma_{\mathbf{j}_*;N-2,N-2}^{Bog}(z) R_{\mathbf{j}_*;N-2,N-2}^{Bog}(z)]^{l_{N-2}} W_{\mathbf{j}_*}^* \eta \rangle \right)\mathscr{P}_\eta \tag{3.18}$$

$$- \mathscr{P}_\eta W_{\mathbf{j}_*} \overline{\mathscr{P}_\eta} \frac{1}{\overline{\mathscr{P}_\eta}\mathscr{K}_{\mathbf{j}_*}^{Bog\,(N-2)}(z)\overline{\mathscr{P}_\eta}} \overline{\mathscr{P}_\eta} W_{\mathbf{j}_*}^* \mathscr{P}_\eta \tag{3.19}$$

where the last term (i.e., the one in (3.19)) is in fact zero if the expression is well defined, since $\overline{\mathscr{P}}_\eta W_{\mathbf{j}_*}^* \mathscr{P}_\eta = 0$ by definition of $\mathscr{P}_\eta$ and $\overline{\mathscr{P}}_\eta$.

Then, the procedure in [32] consists in studying

$$f_{\mathbf{j}_*}(z) := -z - \langle \eta, \, W_{\mathbf{j}_*} \, R_{\mathbf{j}_*;\,N-2,N-2}^{Bog}(z) \sum_{l_{N-2}=0}^{\infty} [\Gamma_{\mathbf{j}_*;\,N-2,N-2}^{Bog}(z) R_{\mathbf{j}_*;\,N-2,N-2}^{Bog}(z)]^{l_{N-2}} \, W_{\mathbf{j}_*}^* \eta \rangle \tag{3.20}$$

to prove the existence of a unique solution z_* of the (fixed point) equation $f_{\mathbf{j}_*}(z) = 0$. In addition to the standing assumptions (see the beginning of Sect. 3.1) needed to implement the flow up to the step $N - 2$, in order to establish the existence of z_*, we require that

$$\epsilon_{\mathbf{j}_*}^2 + \frac{\epsilon_{\mathbf{j}_*}}{N^\gamma} + \frac{1}{N} \leq k_\gamma \epsilon_{\mathbf{j}_*} \sqrt{\epsilon_{\mathbf{j}_*}} \quad , \quad \frac{1}{N^{1-\gamma}} \leq k_\gamma \epsilon_{\mathbf{j}_*} , \tag{3.21}$$

for some $0 < \gamma < 1$ and some constant $k_\gamma > 0$ sufficiently small. Under these further assumptions, we prove that there is a unique z_* in the interval

$$z < E_{\mathbf{j}_*}^{Bog} + (\frac{2\sqrt{2}+3}{6})\sqrt{\epsilon_{\mathbf{j}_*}} \phi_{\mathbf{j}_*} \sqrt{\epsilon_{\mathbf{j}_*}^2 + 2\epsilon_{\mathbf{j}_*}} . \tag{3.22}$$

As explained in the remark below, at fixed particle density ρ, (3.21) yields a lower bound on the space dimension d if we want to implement the flow for an arbitrarily large box.

Remark 3.5. For a box of arbitrarily large side length $L (< \infty)$, the existence of z_* is proven if $\rho \geq \rho_0 (L/L_0)^{3-d}$ where ρ_0 is sufficiently large and $L_0 = 1$. Consequently, for $d \geq 3$, it is enough to require ρ be sufficiently large but *independent* of L and the result holds for a finite box of arbitrarily large volume $|\Lambda|$. In the mean field limiting regime, the existence of the point z_* is established for any space dimension $d \geq 1$.

Next (see Lemma 4.4 of [32]), using the information on the existence of the solution to the fixed point equation in the interval displayed in (3.22), we prove the invertibility of

$$\overline{\mathscr{P}}_\eta \mathscr{K}_{\mathbf{j}_*}^{Bog\,(N-2)}(z) \overline{\mathscr{P}}_\eta$$

in the space $\overline{\mathscr{P}}_\eta \mathcal{F}_N$, for z in the interval

$$z < \min \left\{ z_* + \frac{\Delta_0}{2} \, ; \, E_{\mathbf{j}_*}^{Bog} + \sqrt{\epsilon_{\mathbf{j}_*}} \phi_{\mathbf{j}_*} \sqrt{\epsilon_{\mathbf{j}_*}^2 + 2\epsilon_{\mathbf{j}_*}} \right\} \tag{3.23}$$

where $\Delta_0 := \min \left\{ k_{\mathbf{j}}^2 \, | \, \mathbf{j} \in \mathbb{Z}^d \setminus \{0\} \right\}$. This readily implies that, for z in (3.23), the Feshbach–Schur operator $\mathscr{K}_{\mathbf{j}_*}^{Bog\,(N)}(z)$ is well defined and has the form

$$\mathscr{K}_{\mathbf{j}_*}^{Bog\,(N)}(z) = f_{\mathbf{j}_*}(z) |\eta\rangle \langle \eta| . \tag{3.24}$$

Remark 3.6. For the last step of the Feshbach–Schur flow, Condition (3.2) in Definition 1.1 is also needed. In this regard, we observe that

(i) At fixed particle density and for $d \geq 2$, Condition (3.2) is fulfilled if L is sufficiently large.
(ii) Condition (3.2) is trivially fulfilled for any dimension $d \geq 1$ in the mean field limiting regime.

3.3 Construction and Expansion of the Ground State Vector

In all cases where the existence of z_* is proven, we can verify that z_* is the ground state energy of $H_{\mathbf{j}_*}^{Bog}$ and construct the ground state vector of the system. To this purpose, we recall a well known result (see e.g. [1]) that concerns the construction of eigenvectors. Assuming that the Feshbach–Schur map is well defined on a pair (K, z_*), for some $z_* \in \mathbb{C}$, and that the kernel of

$$\mathscr{F}(K - z_*) := \mathscr{P}(K - z_*)\mathscr{P} - \mathscr{P}K\overline{\mathscr{P}}\frac{1}{\overline{\mathscr{P}}(K - z_*)\overline{\mathscr{P}}}\overline{\mathscr{P}}K\mathscr{P} \tag{3.25}$$

is nontrivial, with $\psi_* \neq 0$ belonging to this kernel, then

$$\left[\mathscr{P} - \frac{1}{\overline{\mathscr{P}}(K - z_*)\overline{\mathscr{P}}}\overline{\mathscr{P}}K\mathscr{P}\right]\psi_* \tag{3.26}$$

is an eigenvector of K with eigenvalue z_*. The dimensions of the kernels of $K - z_*$ and of $\mathscr{F}(K - z_*)$ coincide.

We observe that the isospectrality property that we have mentioned in Sect. 2— see the comment to the definition in (2.1)—holds up to the last step. Hence, if $\mathscr{K}_{\mathbf{j}_*}^{Bog\,(N)}(z_*)\eta = 0$ then also the Hamiltonian $\mathscr{K}_{\mathbf{j}_*}^{Bog\,(N-2)}(z_*)$ has eigenvalue zero and, from formula (3.26), the corresponding eigenvector is

$$\left[\mathscr{P}_\eta - \frac{1}{\overline{\mathscr{P}}_\eta \mathscr{K}_{\mathbf{j}_*}^{Bog\,(N-2)}(z_*)\overline{\mathscr{P}}_\eta}\overline{\mathscr{P}}_\eta \mathscr{K}_{\mathbf{j}_*}^{Bog\,(N-2)}(z_*)\mathscr{P}_\eta\right]\eta \equiv \eta. \tag{3.27}$$

Furthermore, since $\mathscr{K}_{\mathbf{j}_*}^{Bog\,(N)}(z)$ is bounded-invertible for $z < z_*$ so $\mathscr{K}_{\mathbf{j}_*}^{Bog\,(N-2)}(z)$ is. Iterating this isospectrality argument, we get that $H_{\mathbf{j}_*}^{Bog} - z_*$ has ground state energy zero, i.e., $H_{\mathbf{j}_*}^{Bog}$ has ground state energy z_*, and the corresponding eigenvector is

$$\psi_{\mathbf{j}_*}^{Bog} \tag{3.28}$$

$$:= \left[Q_{\mathbf{j}_*}^{(>1)} - \frac{1}{Q_{\mathbf{j}_*}^{(0,1)}(H_{\mathbf{j}_*}^{Bog} - z_*)Q_{\mathbf{j}_*}^{(0,1)}}Q_{\mathbf{j}_*}^{(0,1)}(H_{\mathbf{j}_*}^{Bog} - z_*)Q_{\mathbf{j}_*}^{(>1)}\right] \times \tag{3.29}$$

$$\times \left\{\prod_{i=0,\,i\text{ even}}^{N-4}\left[Q_{\mathbf{j}_*}^{(>i+3)} - \frac{1}{Q_{\mathbf{j}_*}^{(i+2,i+3)}\mathscr{K}_{\mathbf{j}_*}^{Bog\,(i)}(z_*)Q_{\mathbf{j}_*}^{(i+2,i+3)}}Q_{\mathbf{j}_*}^{(i+2,i+3)}\mathscr{K}_{\mathbf{j}_*}^{Bog\,(i)}(z_*)Q_{\mathbf{j}_*}^{(>i+3)}\right]\right\}\eta.$$

In Theorem 4.6 of [32], under the standing assumptions (see the beginning of Sect. 3.1) and (3.21), it is also stated that:

(a) In the mean field limiting regime, for any space dimension $d \geq 1$, the (nondegenerate) ground state energy (of $H_{\mathbf{j}_*}^{Bog}$) is z_* and approaches $E_{\mathbf{j}_*}^{Bog}$ as $N = \rho L^d \to \infty$. In this limit, the spectral gap above z_* is not smaller than $\frac{\Delta_0}{2}$.

This is clearly not an optimal estimate of the gap, since this actually approaches

$$\min \left\{ \Delta_0 \; ; \; E_{\mathbf{j}_*}^{Bog} + \phi_{\mathbf{j}_*} \sqrt{\epsilon_{\mathbf{j}_*}^2 + 2\epsilon_{\mathbf{j}_*}} \right\} \quad , \quad \Delta_0 := \min \left\{ k_{\mathbf{j}}^2 \,|\, \mathbf{j} \in \mathbb{Z}^d \setminus \{\mathbf{0}\} \right\},$$

as $N \to \infty$; see [35]. However, and more importantly, the Feshbach–Schur flow provides the result below at *fixed* density for an arbitrarily large box.

(b) In dimension $d \geq 3$, at fixed (but large) ρ and arbitrarily large $1 < L < \infty$, the nondegenerate ground state energy (of $H_{\mathbf{j}_*}^{Bog}$) is z_* and the spectral gap (above z_*) can be estimated not smaller than

$$\min \left\{ \frac{\Delta_0}{2} \; ; \; \left(\frac{-2\sqrt{2}+3}{6} \right) \sqrt{\epsilon_{\mathbf{j}_*}} \phi_{\mathbf{j}_*} \sqrt{\epsilon_{\mathbf{j}_*}^2 + 2\epsilon_{\mathbf{j}_*}} \right\}. \tag{3.30}$$

The ground state energy z_* tends to $E_{\mathbf{j}_*}^{Bog}$ as $N = \rho L^d \to \infty$.

(c) Using the selection rules of $W_{\mathbf{j}_*}$ and $W_{\mathbf{j}_*}^*$ with respect to the Feshbach–Schur projections, it is straightforward to check that the expression in (3.29) corresponds to

$$\psi_{\mathbf{j}_*}^{Bog} = \eta \tag{3.31}$$

$$- \frac{1}{Q_{\mathbf{j}_*}^{(N-2,N-1)} \mathscr{K}_{\mathbf{j}_*}^{Bog\,(N-4)}(z_*) Q_{\mathbf{j}_*}^{(N-2,N-1)}} Q_{\mathbf{j}_*}^{(N-2,N-1)} W_{\mathbf{j}_*}^* \eta \tag{3.32}$$

$$- \sum_{j=2}^{N/2} \left\{ \prod_{r=j}^{2} \left[- \frac{1}{Q_{\mathbf{j}_*}^{(N-2r,N-2r+1)} \mathscr{K}_{\mathbf{j}_*}^{Bog\,(N-2r-2)}(z_*) Q_{\mathbf{j}_*}^{(N-2r,N-2r+1)}} W_{\mathbf{j}_*\,;\,N-2r,N-2r+2}^* \right] \right\} \times \tag{3.33}$$

$$\times \frac{1}{Q_{\mathbf{j}_*}^{(N-2,N-1)} \mathscr{K}_{\mathbf{j}_*}^{Bog\,(N-4)}(z_*) Q_{\mathbf{j}_*}^{(N-2,N-1)}} Q_{\mathbf{j}_*}^{(N-2,N-1)} W_{\mathbf{j}_*}^* \eta$$

where $\mathscr{K}_{\mathbf{j}_*}^{Bog\,(-2)}(z_*) := H_{\mathbf{j}_*}^{Bog} - z_*$.

Now, we discuss the control of the sum (3.31)–(3.33) depending on N and for arbitrarily small values of $\epsilon_{\mathbf{j}_*}$. It is not difficult to see that for $\epsilon_{\mathbf{j}} > 0$ sufficiently small and for arbitrarily large N fulfilling the standing assumption $\frac{1}{N} \leq \epsilon_{\mathbf{j}_*}^{\nu}$ and (3.21)

$$\sum_{j=2}^{N/2} \left\| \left\{ \prod_{r=j}^{2} \left[- \frac{1}{Q_{\mathbf{j}_*}^{(N-2r,N-2r+1)} \mathscr{K}_{\mathbf{j}_*}^{Bog\,(N-2r-2)}(z_*) Q_{\mathbf{j}_*}^{(N-2r,N-2r+1)}} W_{N-2r,N-2r+2}^* \right] \right\} \times \right. \tag{3.34}$$

$$\left. \times \frac{1}{Q_{\mathbf{j}_*}^{(N-2,N-1)} \mathscr{K}_{\mathbf{j}_*}^{Bog\,(N-4)}(z_*) Q_{\mathbf{j}_*}^{(N-2,N-1)}} Q_{\mathbf{j}_*}^{(N-2,N-1)} W_{\mathbf{j}_*}^* \eta \right\|$$

is bounded by a series which is convergent. Indeed, by recalling the definition in (3.6) and making use of the identity

$$Q_{\mathbf{j}_*}^{(i+2,i+3)} \mathscr{K}_{\mathbf{j}_*}^{Bog\,(i)}(z) Q_{\mathbf{j}_*}^{(i+2,i+3)} \tag{3.35}$$

$$= Q_{\mathbf{j}_*}^{(i+2,i+3)} (H_{\mathbf{j}_*}^{Bog} - z) Q_{\mathbf{j}_*}^{(i+2,i+3)} \tag{3.36}$$

$$- Q_{\mathbf{j}_*}^{(i+2,i+3)} W_{\mathbf{j}_*} R_{\mathbf{j}_*\,;\,i,i}^{Bog}(z) \sum_{l_i=0}^{\infty} \left[\Gamma_{\mathbf{j}_*\,;\,i,i}^{Bog} R_{\mathbf{j}_*\,;\,i,i}^{Bog}(z) \right]^{l_i} W_{\mathbf{j}_*}^* Q_{\mathbf{j}_*}^{(i+2,i+3)}, \tag{3.37}$$

it is evident that

$$\frac{1}{Q_{\mathbf{j}_*}^{(N-2r,N-2r+1)} \mathscr{K}_{\mathbf{j}_*}^{Bog\,(N-2r-2)}(z_*) Q_{\mathbf{j}_*}^{(N-2r,N-2r+1)}} \tag{3.38}$$

$$= \sum_{l_{N-2r}=0}^{\infty} R_{\mathbf{j}_*\,;\,N-2r,N-2r}^{Bog}(z_*) \left[\Gamma_{\mathbf{j}_*\,;\,N-2r,N-2r}^{Bog}(z_*) R_{\mathbf{j}_*\,;\,N-2r,N-2r}^{Bog}(z_*) \right]^{l_{N-2r}} \tag{3.39}$$

$$= [R_{\mathbf{j}_*\,;\,N-2r,N-2r}^{Bog}(z_*)]^{\frac{1}{2}} \sum_{l_{N-2r}=0}^{\infty} \left[[R_{N-2r,N-2r}^{Bog}(z_*)]^{\frac{1}{2}} \Gamma_{\mathbf{j}_*\,;\,N-2r,N-2r}^{Bog}(z_*) [R_{\mathbf{j}_*\,;\,N-2r,N-2r}^{Bog}(z_*)]^{\frac{1}{2}} \right]^{l_{N-2r}} \times \tag{3.40}$$

$$\times [R_{\mathbf{j}_*\,;\,N-2r,N-2r}^{Bog}(z_*)]^{\frac{1}{2}}$$

$$= [R_{\mathbf{j}_*\,;\,N-2r,N-2r}^{Bog}(z_*)]^{\frac{1}{2}} \breve{\Gamma}_{\mathbf{j}_*\,;\,N-2r,N-2r}^{Bog}(z_*) [R_{\mathbf{j}_*\,;\,N-2r,N-2r}^{Bog}(z_*)]^{\frac{1}{2}}. \tag{3.41}$$

Hence, we can write

$$\left\{ \prod_{l=j}^{2} \left[- \frac{1}{Q_{\mathbf{j}_*}^{(N-2l,N-2l+1)} \mathscr{K}_{\mathbf{j}_*}^{Bog\,(N-2l-2)}(z_*) Q_{\mathbf{j}_*}^{(N-2l,N-2l+1)}} W_{\mathbf{j}_*\,;\,N-2l,N-2l+2}^* \right] \right\} \times \tag{3.42}$$

$$\times \frac{1}{Q_{\mathbf{j}_*}^{(N-2,N-1)} \mathscr{K}_{\mathbf{j}_*}^{Bog\,(N-4)}(z_*) Q_{\mathbf{j}_*}^{(N-2,N-1)}} Q_{\mathbf{j}_*}^{(N-2,N-1)} W_{\mathbf{j}_*}^* \eta$$

$$= \left\{ \prod_{l=j}^{2} \left[- [R_{\mathbf{j}_*\,;\,N-2l,N-2l}^{Bog}(z_*)]^{\frac{1}{2}} \breve{\Gamma}_{N-2l,N-2l}^{Bog}(z_*) [R_{\mathbf{j}_*\,;\,N-2l,N-2l}^{Bog}(z_*)]^{\frac{1}{2}} W_{\mathbf{j}_*\,;\,N-2l,N-2l+2}^* \right] \right\} \times \tag{3.43}$$

$$\times [R_{\mathbf{j}_*\,;\,N-2,N-2}^{Bog}(z_*)]^{\frac{1}{2}} \breve{\Gamma}_{N-2,N-2}^{Bog}(z_*) [R_{\mathbf{j}_*\,;\,N-2,N-2}^{Bog}(z_*)]^{\frac{1}{2}} W_{\mathbf{j}_*}^* \eta$$

and estimate

$$\left\| \left\{ \prod_{l=j}^{2} \left[- \frac{1}{Q_{\mathbf{j}_*}^{(N-2l,N-2l+1)} \mathscr{K}_{\mathbf{j}_*}^{Bog\,(N-2l-2)}(z_*) Q_{\mathbf{j}_*}^{(N-2l,N-2l+1)}} W_{\mathbf{j}_*\,;\,N-2l,N-2l+2}^* \right] \right\} \times \right. \tag{3.44}$$

$$\left. \times \frac{1}{Q_{\mathbf{j}_*}^{(N-2,N-1)} \mathscr{K}_{\mathbf{j}_*}^{Bog\,(N-4)}(z_*) Q_{\mathbf{j}_*}^{(N-2,N-1)}} Q_{\mathbf{j}_*}^{(N-2,N-1)} W_{\mathbf{j}_*}^* \eta \right\|$$

$$\leq \left\| [R_{\mathbf{j}_*\,;\,N-2j,N-2j}^{Bog}(z_*)]^{\frac{1}{2}} \right\| \times$$

$$\times \left\{ \prod_{l=j}^{2} \left\| \breve{\Gamma}_{\mathbf{j}_*\,;\,N-2l,N-2l}^{Bog}(z_*) \right\| \left\| [R_{\mathbf{j}_*\,;\,N-2l,N-2l}^{Bog}(z_*)]^{\frac{1}{2}} W_{\mathbf{j}_*\,;\,N-2l,N-2l+2}^* [R_{\mathbf{j}_*\,;\,N-2l+2,N-2l+2}^{Bog}(z_*)]^{\frac{1}{2}} \right\| \right\} \times$$

$$\times \left\| \breve{\Gamma}_{\mathbf{j}_*\,;\,N-2,N-2}^{Bog}(z_*) \right\| \left\| [R_{\mathbf{j}_*\,;\,N-2,N-2}^{Bog}(z_*)]^{\frac{1}{2}} Q_{\mathbf{j}_*}^{(N-2,N-1)} W_{\mathbf{j}_*}^* \eta \right\|.$$

Then, with the help of (3.7), (3.12), and of the bound

$$\left\| [R_{\mathbf{j}_*\,;\,N-2j,N-2j}^{Bog}(z_*)]^{\frac{1}{2}} \right\| \left\| [R_{\mathbf{j}_*\,;\,N-2,N-2}^{Bog}(z_*)]^{\frac{1}{2}} Q_{\mathbf{j}_*}^{(N-2,N-1)} W_{\mathbf{j}_*}^* \eta \right\| \leq O(1) \tag{3.45}$$

we conclude that the sum

$$\sum_{j=2}^{N/2} \Big\| \Big\{ \prod_{l=j}^{2} \Big[- \frac{1}{Q_{j_*}^{(N-2l,N-2l+1)} \mathscr{K}_{j_*}^{Bog\,(N-2l-2)}(z_*) Q_{j_*}^{(N-2l,N-2l+1)}} W_{j_*}^* {}_{;\,N-2l,N-2l+2} \Big] \Big\} \times \quad (3.46)$$

$$\times \frac{1}{Q_{j_*}^{(N-2,N-1)} \mathscr{K}_{j_*}^{Bog\,(N-4)}(z_*) Q_{j_*}^{(N-2,N-1)}} Q_{j_*}^{(N-2,N-1)} W_{j_*}^* \eta \Big\| \quad (3.47)$$

is bounded by a universal constant times the series

$$\sum_{j=2}^{\infty} c_j := \sum_{j=2}^{\infty} \Big\{ \prod_{l=j}^{2} \frac{1}{\Big[1 + \sqrt{\eta a_{\epsilon_{j_*}}} - \frac{b_{\epsilon_{j_*}}/\sqrt{\eta a_{\epsilon_{j_*}}}}{2l - \epsilon_{j_*}^{\Theta}} \Big] \Big[1 + a_{\epsilon_{j_*}} - \frac{2b_{\epsilon_{j_*}}}{2l-1} - \frac{1 - c_{\epsilon_{j_*}}}{(2l-1)^2} \Big]^{\frac{1}{2}}} \Big\}, \quad (3.48)$$

which is convergent since

$$\frac{c_j}{c_{j-1}} = \frac{1}{\Big[1 + \sqrt{\eta a_{\epsilon_{j_*}}} - \frac{b_{\epsilon_{j_*}}/\sqrt{\eta a_{\epsilon_{j_*}}}}{2j - \epsilon_{j_*}^{\Theta}} \Big] \Big[1 + a_{\epsilon_{j_*}} - \frac{2b_{\epsilon_{j_*}}}{2j-1} - \frac{1 - c_{\epsilon_{j_*}}}{(2j-1)^2} \Big]^{\frac{1}{2}}} < 1 \quad (3.49)$$

for j sufficiently large.

Remark 3.7. We observe that there is no small parameter in the expansion of (3.38). In fact, the sum of the series in (3.48) is divergent in the limit $\epsilon_{j_*} \to 0$. We stress that for any $\epsilon_{j_*} > 0$, the expansion (3.31)–(3.33) of $\psi_{j_*}^{Bog}$ is well defined and controlled in terms of the parameter $\Sigma_{\epsilon_{j_*}} := \frac{1}{1 + \sqrt{\epsilon_{j_*}} + o(\sqrt{\epsilon_{j_*}})}$.

Remark 3.8. The formula in (3.31)–(3.33) gives for each j the component of the ground state vector with exactly $2j$ particles, j with momentum k_{j_*} and j with momentum $-k_{j_*}$. With the help of the further expansion in Sect. 3.4, one can verify that for low values of j the norms of these components are comparable with the norm, 1, of η, for arbitrarily large values of N.

3.4 Expansion of the Ground State Vector

Starting from expression (3.31)–(3.33), for any given $\zeta > 0$ we want to define a vector, $(\psi_{j_*}^{Bog})_\zeta$, that:

- is expressed in terms of the vector η and of a finite sum of products of the interaction terms $W_{j_*}^*$, W_{j_*}, and of the resolvent $\frac{1}{\hat{H}_{j_*}^0 - z_*}$ (see (1.11));
- in dimension $d \geq 3$ approximates $\psi_{j_*}^{Bog}$ up to a quantity in norm less than $O(\zeta)$, provided $\rho = \frac{N}{L^d}$ is sufficiently large independently of L; this implies that for fixed L the error term (of norm $O(\zeta)$) can be arbitrarily small at large but fixed $N = \rho L^d$. The expansion is *not* in $\frac{1}{N}$.

A key ingredient to be used here is the re-expansion of the operators $\Gamma^{Bog}_{\mathbf{j}_*\,;\,i,i}(z_*)$. Actually, such a re-expansion—which is based on the same estimates yielding Theorem 3.1—is also used to control the invertibility of the operator

$$\overline{\mathscr{P}_\eta}\,\mathscr{K}^{Bog\,(N-2)}_{\mathbf{j}_*}(z)\overline{\mathscr{P}_\eta}$$

in $\overline{\mathscr{P}_\eta}\mathcal{F}$ for the last step of the Feshbach–Schur flow (see Lemma 4.4 and Corollary 4.13 of [32]).

We refer to the original paper for the detailed (recursive) expression of the re-expanded operators associated with $\Gamma^{Bog}_{\mathbf{j}_*\,;\,i,i}(z_*)$. Here, we convey the idea behind it. Suppose that we want to approximate $\Gamma^{Bog}_{\mathbf{j}_*\,;\,6,6}(z)$, then we use the formula below (where we drop the index $\mathbf{j}_*$ in the symbols Γ, R, and W to shorten the notation):

$$\Gamma^{Bog}_{6,6}(z) = W_{6,4}\,(R^{Bog}_{4,4}(z))^{\frac{1}{2}} \times \tag{3.50}$$

$$\times \sum_{l_4=0}^{\infty}\Big[(R^{Bog}_{4,4}(z))^{\frac{1}{2}}\,W_{4,2}\,(R^{Bog}_{2,2}(z))^{\frac{1}{2}}\sum_{l_2=0}^{\infty}\Big[(R^{Bog}_{2,2}(z))^{\frac{1}{2}}\,W_{2,0}\,R^{Bog}_{0,0}(z)\,W^*_{0,2}\,(R^{Bog}_{2,2}(z))^{\frac{1}{2}}\Big]^{l_2}\times$$

$$\times(R^{Bog}_{2,2}(z))^{\frac{1}{2}}\,W^*_{2,4}(R^{Bog}_{4,4}(z))^{\frac{1}{2}}\Big]^{l_4}\,(R^{Bog}_{4,4}(z))^{\frac{1}{2}}\,W^*_{4,6}\,,$$

and we consider an analogous expression where the series appearing in (3.50) are truncated, up to a remainder the norm of which we estimate of order c^h where $0 < c < 1$, for some $h \in \mathbb{N}$ and $h \geq 2$. In [32], this is done consistently for each $\Gamma^{Bog}_{\mathbf{j}_*\,;\,i,i}(z)$ by defining related operators

$$[\Gamma^{Bog}_{\mathbf{j}_*\,;\,i,i}(z)]_{(r,h_-;r+2,h_-;...;i-2,h_-)} \tag{3.51}$$

such that

$$\Gamma^{Bog}_{\mathbf{j}_*\,;\,i,i}(z) = \sum_{r=2,\,r\,even}^{i-2}[\Gamma^{Bog}_{\mathbf{j}_*\,;\,i,i}(z)]_{(r,h_-;r+2,h_-;...;i-2,h_-)} + \text{corrections}\,. \tag{3.52}$$

Keeping the structure of $\Gamma^{Bog}_{\mathbf{j}_*\,;\,6,6}(z)(= \Gamma^{Bog}_{6,6}(z))$ in mind as an example, in the general case, each summand

$$[\Gamma^{Bog}_{\mathbf{j}_*\,;\,i,i}(z)]_{(r,h_-;r+2,h_-;...;i-2,h_-)}$$

in (3.52) can be thought of as the original expression $\Gamma^{Bog}_{\mathbf{j}_*\,;\,i,i}(z)$ truncated by neglecting all the series corresponding to l_j with $2 \leq j \leq r - 2$ and stopping the remaining series at $l_j = h$ for each $r \leq j \leq i - 2$. The leading terms that have been isolated and the remainder terms are controlled in Proposition 4.10 of [32] that we report below (where to shorten the notation we drop the index $\mathbf{j}_*$ appearing in the symbols Γ and R):

Proposition 3.1. *Let $\frac{1}{N} \leq \epsilon_{j_*}^{\nu}$ for some $\nu > \frac{11}{8}$ and $\epsilon_{j_*} \equiv \epsilon$ be sufficiently small. For any fixed $2 \leq h \in \mathbb{N}$ and for $N - 2 \geq i \geq 4$ and even, the splitting*

$$\Gamma_{i,i}^{Bog}(z) = \sum_{r=2,\, r\, even}^{i-2} [\Gamma_{i,i}^{Bog}(z)]_{(r,h_-;r+2,h_-;\ldots;i-2,h_-)}$$

$$+ \sum_{r=2,\, r\, even}^{i-2} [\Gamma_{i,i}^{Bog}(z)]_{(r,h_+;r+2,h_-;\ldots;i-2,h_-)} \qquad (3.53)$$

holds true for $z \leq E_{j_*}^{Bog} + (\delta - 1)\phi_{j_*}\sqrt{\epsilon_{j_*}^2 + 2\epsilon_{j_*}}$ with $\delta = 1 + \sqrt{\epsilon_{j_*}}$. Moreover, for $2 \leq r \leq i - 2$ and even, the estimates

$$\left\| (R_{i,i}^{Bog}(z))^{\frac{1}{2}} [\Gamma_{i,i}^{Bog}(z)]_{(r,h_-;r+2,h_-;\ldots;i-2,h_-)} (R_{i,i}^{Bog}(z))^{\frac{1}{2}} \right\| \qquad (3.54)$$

$$\leq \prod_{f=r+2,\, f-r\, even}^{i} \frac{K_{f,\epsilon}}{(1 - Z_{f-2,\epsilon})^2}$$

and

$$\| (R_{i,i}^{Bog}(z))^{\frac{1}{2}} [\Gamma_{i,i}^{Bog}(z)]_{(r,h_+;r+2,h_-;\ldots;i-2,h_-)} (R_{i,i}^{Bog}(z))^{\frac{1}{2}} \| \qquad (3.55)$$

$$\leq (Z_{r,\epsilon})^h \prod_{f=r+2,\, f-r\, even}^{i} \frac{K_{f,\epsilon}}{(1 - Z_{f-2,\epsilon})^2}$$

hold true, where

$$K_{i,\epsilon} := \frac{1}{4(1 + a_\epsilon - \frac{2b_\epsilon}{N-i+1} - \frac{1-c_\epsilon}{(N-i+1)^2})}\,,$$

$$Z_{i-2,\epsilon} := \frac{1}{4(1 + a_\epsilon - \frac{2b_\epsilon}{N-i+3} - \frac{1-c_\epsilon}{(N-i+3)^2})} \frac{2}{\left[1 + \sqrt{\eta a_\epsilon} - \frac{b_\epsilon/\sqrt{\eta a_\epsilon}}{N-i+4-\epsilon^\Theta}\right]} \qquad (3.56)$$

where a_ϵ, b_ϵ, and c_ϵ correspond to those defined in (3.9)–(3.11) at $\delta = 1 + \sqrt{\epsilon}$, and $\Theta := \min\{2(\nu - \frac{11}{8}) ; \frac{1}{4}\}$.

Since z_* belongs to the interval in (3.22) and the proposition above only requires $\frac{1}{N} \leq \epsilon_{j_*}^{\nu}$ for some $\nu > \frac{11}{8}$ and $\epsilon_{j_*} \equiv \epsilon$ sufficiently small, it is clear that, for a fixed box of side length L and a *large but fixed number*[6] *of particles* $N = \rho L^d$, the error term on the r-h-s of (3.55) is arbitrarily small provided h is sufficiently large.

---

[6]Due to point (b) in Theorem 4.6 of [32] reported above, for $d \geq 3$ the (sufficiently large) density ρ can be chosen independently of $L > 1$ to ensure the existence of z_*.

We can now proceed with the following operations in order to construct $(\psi_{\mathbf{j}_*}^{Bog})_\zeta$:

- Using the convergence of the series in (3.48), we truncate the sum in (3.33) at some ζ-dependent $\bar{j}$;
- For each operator of the type (3.38), we truncate the (convergent) series in (3.40) at some ζ-dependent $\bar{l}_{N-2r}$;
- In each summand obtained from expression (3.33) as a result of the previous truncations, we choose suitable ζ-dependent h and $j_\#$ to replace the operators $\Gamma_{\mathbf{j}_*\,;\,i,i}^{Bog}(z_*)$ with

$$\sum_{\substack{r=j_\#,\,r\,even}}^{i-2} [\Gamma_{\mathbf{j}_*\,;\,i,i}^{Bog}(z_*)]_{(r,h_-;r+2,h_-;...;i-2,h_-)}$$

up to a remainder of sufficiently small operator norm depending on the chosen ζ. Thereby, we replace $\Gamma_{\mathbf{j}_*\,;\,i,i}^{Bog}(z_*)$ with a finite (ζ-dependent) sum of products of the operators $W_{\mathbf{j}_*\,;\,j,j-2}$, $W_{\mathbf{j}_*\,;\,j-2,j}^*$ (where $j_\# + 2 \leq j < i$) and $R_{\mathbf{j}_*\,;\,j,j}^{Bog}(z_*)$ (where $j_\# \leq j < i$).

Remark 3.9. In the mean field limit, in each term of the (ζ-dependent) finite sum of terms that we have isolated, we can approximate z_* with $E_{\mathbf{j}_*}^{Bog}$ up to an arbitrarily small error for N sufficiently large. Therefore, for any $d \geq 1$, in the mean field limiting regime, we can provide an approximating vector (of $\psi_{\mathbf{j}_*}^{Bog}$) in terms of a finite sum of products of the interaction terms $W_{\mathbf{j}_*}^*$, $W_{\mathbf{j}_*}$, and of the resolvent $\frac{1}{\hat{H}_{\mathbf{j}_*}^0 - E_{\mathbf{j}_*}^{Bog}}$ applied to the vector η, up to any desired precision. In this case, N must be larger than some ζ-dependent N_ζ due to the replacement of z_* with $E_{\mathbf{j}_*}^{Bog}$.

References

1. Bach, V., Fröhlich, J., Sigal, I.M.: Renormalization group analysis of spectral problems in quantum field theory. Adv. Math. **137**(2), 205–298 (1998)
2. Balaban, T., Feldman, J., Knörrer, H., Trubowitz, E.: A functional integral representation for many Boson systems I: the partition function. Ann. Henri Poincare **8** (2008)
3. Balaban, T., Feldman, J., Knörrer, H., Trubowitz, E.: A functional integral representation for many Boson systems II: correlation functions. Ann. Henri Poincare **9** (2008)
4. Balaban, T., Feldman, J., Knörrer, H., Trubowitz, E.: The temporal ultraviolet limit for complex bosonic many-body models. Ann. Henri Poincare **11** (2010)
5. Balaban, T., Feldman, J., Knörrer, H., Trubowitz, E.: Complex Bosonic many-body models: overview of the small field parabolic flow. Ann. Henri Poincare **18** (2017)
6. Benedikter, N., de Oliveira, G., Schlein, B.: Quantitative derivation of the Gross-Pitaevskii equation. Commun. Pure Appl. Math. **68**, 1399–1482
7. Boccato, C., Brennecke, C., Cenatiempo, S., Schlein, B.: Complete Bose-Einstein condensation in the Gross-Pitaevskii regime. arXiv:1703.04452
8. Boccato, C., Brennecke, C., Cenatiempo, S., Schlein, B.: Bogoliubov theory in the Gross-Pitaevskii limit. arXiv:1801.01389
9. Bogoliubov, N.N.: On the theory of superfluidity. J. Phys. (U.S.S.R.) **11**, 2332 (1947)
10. Brietzke, B.: On the second order correction to the ground state energy of the dilute Bose gas. Ph.D. thesis (2017)

11. Castellani, C., Di Castro, C., Pistolesi, F., Strinati, G.C.: Infrared behavior of interacting Bosons at zero temperature. Phys. Rev. Lett. **78**(9) (1997)
12. Castellani, C., Di Castro, C., Pistolesi, F., Strinati, G.C.: Renormalization- group approach to the infrared behavior of a zero-temperature Bose system. Phys. Rev. B **69**(2) (2004)
13. Cenatiempo, S.: Low dimensional interacting Bosons. arXiv:1211.3772
14. Cenatiempo, S., Giuliani, A.: Renormalization theory of a two dimensional Bose gas: quantum critical point and quasi-condensed state. J. Stat. Phys. **157**(4), 755–829 (2014)
15. Derezinski, J., Napiorkowski, M.: Excitation spectrum of interacting Bosons in the mean-field infinite-volume limit. Ann. Henri Poincare **15**(12), 2409–2439 (2014)
16. Erdös, L., Schlein, B., Yau, H.-T.: Ground-state energy of a low-density Bose gas: a second-order upper bound. Phys. Rev. A **78**, 053627 (2008)
17. Girardeau, M.: Relationship between systems of impenetrable Bosons and fermions in one dimension. J. Math. Phys. **1**, 516–523 (1960)
18. Giuliani, A., Seiringer, R.: The ground state energy of the weakly interacting Bose gas at high density. J. Stat. Phys. **135**, 915–934 (2009)
19. Grech, P., Seiringer, R.: The excitation spectrum for weakly interacting Bosons in a trap. Commun. Math. Phys. **332**, 559–591 (2013)
20. Lewin, M., Nam, P.T., Rougerie, N.: The mean-field approximation and the non-linear Schrödinger functional for trapped Bose gases. Trans. Am. Math. Soc. **368**, 6131–6157 (2016)
21. Lewin, M., Nam, P.T., Serfaty, S., Solovej, J.P.: Bogoliubov spectrum of interacting Bose gases. Commun. Pure Appl. Math. **68**(3), 413–471
22. Lieb, E.H.: Exact analysis of an interacting Bose gas. II. The excitation spectrum. Phys. Rev. **130**, 1616–1624 (1963)
23. Lieb, E.H.: Exact analysis of an interacting Bose gas. I. The general solution and the ground state. Phys. Rev. **130**, 1605–1616 (1963)
24. Lieb, E.H., Seiringer, R.: Proof of Bose-Einstein condensation for dilute trapped gases. Phys. Rev. Lett. **88**, 170409 (2002)
25. Lieb, E.H., Solovej, J.P.: Ground state energy of the one-component charged Bose gas. Commun. Math. Phys. **217**, 127–163 (2001). Errata 225, 219–221 (2002)
26. Lieb, E.H., Solovej, J.P.: Ground state energy of the two-component charged Bose gas. Commun. Math. Phys. **252**, 485–534 (2004)
27. Lieb, E.H., Seiringer, R., Yngvason, J.: Bosons in a trap: a rigorous derivation of the Gross-Pitaevskii energy functional. Phys. Rev. A **61**, 043602 (2000)
28. Lieb, E.H., Seiringer, R., Solovej, J.P., Yngvason, J.: The Mathematics of the Bose Gas and Its Condensation. Oberwolfach Seminar Series, vol. 34. Birkhäuser, Basel (2005), expanded version available at arXiv:cond-mat/0610117
29. Nam, P.T., Seiringer, R.: Collective excitations of Bose gases in the mean-field regime. Arch. Rational Mech. Anal. **215**, 381–417 (2015)
30. Nam, P.T., Rougerie, N., Seiringer, R.: Ground states of large bosonic systems: the Gross-Pitaevskii limit revisited. Anal. PDE **9**, 459–485 (2016)
31. Nam, P.T., Napiorkowski, M., Solovej, J.P.: Diagonalization of bosonic quadratic Hamiltonians by Bogoliubov transformations. J. Funct. Anal. **270**, 4340–4368 (2016)
32. Pizzo, A.: Bose particles in a box I. A convergent expansion of the ground state of a three-modes Bogoliubov Hamiltonian. arXiv:1511.07022
33. Pizzo, A.: Bose particles in a box II. A convergent expansion of the ground state of the Bogoliubov Hamiltonian in the mean field limiting regime. arXiv:1511.07025
34. Pizzo, A.: Bose particles in a box III. A convergent expansion of the ground state of the Hamiltonian in the mean field limiting regime. arXiv:1511.07026
35. Seiringer, R.: The excitation spectrum for weakly interacting Bosons. Commun. Math. Phys. **306**, 565–578 (2011)
36. Yau, H.-T., Yin, J.: The second order upper bound for the ground energy of a Bose gas. J. Stat. Phys. **136**, 453503 (2009)

Blow-Up Profile of Rotating 2D Focusing Bose Gases

Mathieu Lewin, Phan Thành Nam and Nicolas Rougerie

Dedicated to Herbert Spohn, on the occasion of his 70th birthday.

Abstract We consider the Gross–Pitaevskii equation describing an attractive Bose gas trapped to a quasi 2D layer by means of a purely harmonic potential, and which rotates at a fixed speed of rotation Ω. First, we study the behavior of the ground state when the coupling constant approaches a_*, the critical strength of the cubic nonlinearity for the focusing nonlinear Schrödinger equation. We prove that blow-up always happens at the center of the trap, with the blow-up profile given by the Gagliardo–Nirenberg solution. In particular, the blow-up scenario is independent of Ω, to leading order. This generalizes results obtained by Guo and Seiringer (*Lett. Math. Phys.*, 2014, vol. 104, p. 141–156) in the nonrotating case. In a second part, we consider the many-particle Hamiltonian for N bosons, interacting with a potential rescaled in the mean-field manner $-a_N N^{2\beta-1} w(N^\beta x)$, with $w \geqslant 0$ a positive function such that $\int_{\mathbb{R}^2} w(x)\, dx = 1$. Assuming that $\beta < 1/2$ and that $a_N \to a_*$ sufficiently slowly, we prove that the many-body system is fully condensed on the Gross–Pitaevskii ground state in the limit $N \to \infty$.

M. Lewin (✉)
CNRS and CEREMADE, Université Paris-Dauphine, PSL Research University, 75016 Paris, France
e-mail: mathieu.lewin@math.cnrs.fr

P. Thành Nam
Department of Mathematics, Ludwig Maximilian University of Munich, Theresienstrasse 39, 80333 Munich, Germany
e-mail: nam@math.lmu.de

N. Rougerie
CNRS, LPMMC (UMR 5493), Université Grenoble-Alpes, B.P. 166, 38042 Grenoble, France
e-mail: nicolas.rougerie@grenoble.cnrs.fr

© Springer Nature Switzerland AG 2018
D. Cadamuro et al. (eds.), *Macroscopic Limits of Quantum Systems*,
Springer Proceedings in Mathematics & Statistics 270,
https://doi.org/10.1007/978-3-030-01602-9_7

1   Introduction

Because of their ability to display quantum effects at the macroscopic scale, Bose–Einstein condensates (BEC) have become an important subject of research, in particular after their first realization in the laboratory in 1995 [3, 16, 19, 36]. Condensates with *attractive* interactions are expected to behave quite differently from the better understood *repulsive* case, and they have generated many experimental, numerical or theoretical works. Some atoms like ^{7}Li indeed have a negative scattering length and were initially believed not to be able to form a condensate, until attractive BECs were finally experimentally realized in traps [6].

For attractive interactions, the Gross–Pitaevskii functional, commonly used to describe BECs, predicts a collapse of the system when Na (the number of particles times the scattering length) is too negative [5, 18, 30, 57, 70], an effect which has been observed in some experiments [24, 29]. In addition, attractive Bose–Einstein condensates are believed to respond to rotation in a rather different manner from the repulsive case. In a rotating repulsive Bose gas, a triangular lattice of vortices is formed, with the number of vortices increasing with the speed of rotation [1, 15, 17, 27]. On the contrary, it has been argued [9, 14, 52, 56, 60, 66, 67, 75] that in an attractive rotating Bose gas, vortices should be unstable and it is instead the center of mass of the system which can rotate around the axis.

In this paper, we rigorously establish two results about 2D attractive Bose–Einstein condensates in the critical regime of collapse. We consider a Bose gas trapped to a quasi 2D layer by means of a purely harmonic potential, and which rotates at a fixed speed of rotation Ω. First, we look at the Gross–Pitaevskii equation which describes the macroscopic behavior of the condensate [4]. We study its solutions in the regime where the coupling constant approaches the critical blow-up value a_*, given by the best constant in the Gagliardo–Nirenberg inequality. In this case, we prove that blow-up always happens at the center of the trap, with the blow-up profile given by the Gagliardo–Nirenberg optimizer. This shows that the rotation does not affect the general blow-up scenario, to leading order. The nonrotating case has been previously studied by Guo and Seiringer in [30]. Other similar mathematical results on the trapped nonlinear Schrödinger equation for nonrotating gases include [22, 31, 53, 62] in the stationary case and [8, 76, 77] in the time-dependent case.

In a second part, we consider the many-particle (microscopic) Hamiltonian for N such bosons, interacting with a potential rescaled in the mean-field manner

$$-a_N N^{2\beta-1} w(N^\beta x)$$

with $w \geqslant 0$ a fixed positive function such that $\int_{\mathbb{R}^2} w(x)\,dx = 1$. Assuming that $0 < \beta < 1/2$ and that $a_N \to a_*$ sufficiently slowly, we are able to show that the many-body system is fully condensed on the Gross–Pitaevskii ground state studied in the first step, in the limit $N \to \infty$. This justifies the validity of the Gross–Pitaevskii equation in this regime of collapse, with complete Bose–Einstein condensation at the point of blow-up. We do not observe fragmented condensation at this order.

The mathematical method used here follows several of our previous papers [42, 44, 46, 47]. Note that some authors have already dealt with the time-dependent equation, in the subcritical regime $a < a_*$, see [11, 35, 47, 58].

The next section contains the precise definition of our model as well as the statement of our main results. The remainder of the paper is then devoted to their proofs. In Appendix A, we mention several possible extensions of our findings, without giving the detailed mathematical proofs.

2 Models and Main Results

2.1 Collapse of the Rotating Gross–Pitaevskii Ground State

The Gross–Pitaevskii functional describing a condensed system of bosons trapped to a 2D plane and rotating along the third axis at an angular velocity Ω reads

$$\mathcal{E}_{\Omega,a}^{GP}(u) = \int_{\mathbb{R}^2} |\nabla u(x)|^2 \, dx + \int_{\mathbb{R}^2} |x|^2 |u(x)|^2 \, dx - 2\Omega\langle u, Lu \rangle - \frac{a}{2} \int_{\mathbb{R}^2} |u(x)|^4 \, dx \tag{2.1}$$

where $L = -ix \wedge \nabla = i(x_2 \partial_1 - x_1 \partial_2)$ is the angular momentum. Here, we have chosen units such that the trapping potential has a trapping frequency $\Omega_{\text{trap}} = 1$, for simplicity. The system is stable under the assumption that $|\Omega| < 1$ and $a < a_*$, where a_* is the optimal constant of the Gagliardo–Nirenberg inequality [28, 30, 53, 72, 76]

$$\left(\int_{\mathbb{R}^2} |\nabla u|^2 \right) \left(\int_{\mathbb{R}^2} |u|^2 \right) \geq \frac{a_*}{2} \int_{\mathbb{R}^2} |u|^4, \quad \forall u \in H^1(\mathbb{R}^2). \tag{2.2}$$

Equivalently,

$$a_* = \|Q\|_{L^2(\mathbb{R}^2)}^2,$$

where $Q \in H^1(\mathbb{R}^2)$ is the unique positive solution of

$$-\Delta Q + Q - Q^3 = 0 \text{ in } \mathbb{R}^2, \tag{2.3}$$

up to translations. More precisely, Q is symmetric radial decreasing and it is the unique (up to translations and dilations) optimizer for the Gagliardo–Nirenberg inequality (2.2). In the following, we therefore always assume that $0 \leqslant \Omega < 1$ and $0 < a < a_*$. The energy may also be written in the form

$$\mathcal{E}_{\Omega,a}^{GP}(u) = \int_{\mathbb{R}^2} |\nabla u(x) + i\Omega x^{\perp} u(x)|^2 \, dx + (1 - \Omega^2) \int_{\mathbb{R}^2} |x|^2 |u(x)|^2 \, dx$$

$$- \frac{a}{2} \int_{\mathbb{R}^2} |u(x)|^4 \, dx \tag{2.4}$$

where $x^{\perp} = (-x_2, x_1)$. We call

$$E_\Omega^{GP}(a) := \min_{u \in H^1(\mathbb{R}^2), \|u\|_{L^2}=1} \mathcal{E}_{\Omega,a}^{GP}(u)$$

the ground state energy and look at the limit $a \to a_*$ at fixed $0 \leqslant \Omega < 1$. The existence of ground states follows the standard direct method in the calculus of variations. Our first main result is the following.

Theorem 2.1 (Collapse of rotating Gross–Pitaevskii ground states). *Let $0 \leqslant \Omega < 1$ be any fixed rotation. Then, we have*

$$E_\Omega^{GP}(a) = E_0^{GP}(a) + o\big(E_0^{GP}(a)\big) = \sqrt{a_* - a}\left(\frac{2\lambda_*^2}{a_*} + o(1)\right), \qquad (2.5)$$

when $a \to a_*$, where

$$\lambda_* = \left(\int_{\mathbb{R}^2} |x|^2 |Q(x)|^2 dx\right)^{\frac{1}{4}}$$

and Q is the unique radial positive Gagliardo–Nirenberg solution (2.3).

In addition, for any sequence $a_N \to a_*$ and any sequence $\{u_N\}$ such that $\|u_N\|_{L^2(\mathbb{R}^2)} = 1$ and

$$\mathcal{E}_{\Omega,a_N}^{GP}(u_N) = E_\Omega^{GP}(a_N) + o(\sqrt{a_* - a_N}), \qquad (2.6)$$

(for instance u_N a minimizer of $E_\Omega(a_N)$), we have

$$\lim_{N \to \infty} \left\|u_N - e^{i\theta_N} Q_N\right\|_{L^2(\mathbb{R}^2)} = 0, \qquad (2.7)$$

for a properly chosen phase $\theta_N \in [0, 2\pi)$, where

$$Q_N(x) = (a_*)^{-1/2} \lambda_* (a_* - a_N)^{-\frac{1}{4}} Q\left(\lambda_* (a_* - a_N)^{-\frac{1}{4}} x\right) \qquad (2.8)$$

is the rescaled Gagliardo–Nirenberg optimizer which blows up at the origin at speed $(a_* - a_N)^{\frac{1}{4}}/\lambda_*$.

This theorem was proved by Guo and Seiringer in [30] in the case $\Omega = 0$, with the convergence of the ground states but not of general "approximate ground states" u_N's as in (2.7). Our theorem shows that the results found by Guo and Seiringer remain valid when the system is set in rotation, the blow-up scenario being independent of Ω to leading order. In particular, we do not see a rotation of the center of mass at this order. Our method of proof relies on the non-degeneracy of the minimizer Q, which is known to play a fundamental role in many situations [10, 73]. We expect that the non-degeneracy of Q should provide quantitative estimates for the difference between Q_N and the ground state of $E_\Omega(a_N)$, however we have not investigated this question in details.

Remark 2.2.

(i) It would be interesting to investigate the case where $\Omega = \Omega_N \to 1$ at the same time as $a_N \to a_*$. In this case, the centrifugal force almost compensates the trapping potential, and this effect could compete with the collapse scenario induced by attractive interactions.

(ii) Our proof covers more general external potentials attaining their minimum at the origin and behaving at least quadratically at zero and at infinity, like for instance the quartic-quadratic potential $V(x) = |x|^2 + k|x|^4$ with $k > 0$.   ◇

2.2 Collapse of the Many-Body System in the Gross–Pitaevskii Ground State

Next, we turn to the N-particle quantum Hamiltonian describing our trapped 2D bosons, which reads

$$H_N = \sum_{j=1}^{N} \left(-\Delta_{x_j} + |x_j|^2 - 2\Omega L_{x_j}\right) - \frac{a}{N-1} \sum_{1 \leqslant i < j \leqslant N} w_N(x_i - x_j), \qquad (2.9)$$

and acts on $\mathfrak{H}^N = \bigotimes_{\text{sym}}^N L^2(\mathbb{R}^2)$, the Hilbert space of square-integrable symmetric functions. The two-body interaction w_N approaches a Dirac delta and is chosen in the form

$$w_N(x) = N^{2\beta} w(N^\beta x) \qquad (2.10)$$

for a fixed parameter $\beta > 0$ and a fixed function w satisfying

$$w(x) = w(-x) \geqslant 0, \quad (1 + |x|)w, \ \widehat{w} \in L^1(\mathbb{R}^2), \quad \int_{\mathbb{R}^2} w = 1. \qquad (2.11)$$

Finally, $a > 0$ is a parameter which describes the strength of the interaction. We will take

$$a = a_N \to a_*$$

which is the Gagliardo–Nirenberg critical constant mentioned before.

Hamiltonians of the form (2.9) have generated a huge amount of works in the past decades, in any dimension d. The chosen coupling constant proportional to $1/(N-1)$ ensures that the kinetic and interaction energies are comparable in the limit $N \to \infty$. Due to the trapping potential, most of the particles will usually accumulate in a bounded region of space, leading to a high density of order N (in our case they will even collapse at one point).

In this paper, we are interested in the behavior of the ground state energy per particle of H_N,

$$E^Q_{\Omega,a}(N) := N^{-1} \inf_{\Psi \in \mathfrak{H}^N, \|\Psi\|=1} \langle \Psi, H_N \Psi \rangle, \tag{2.12}$$

and in the corresponding (non-necessarily unique) ground states Ψ_N, when $N \to \infty$. In the regime considered in this paper, we expect that the particles will essentially become independent (Bose–Einstein condensation), that is, in terms of wave functions:

$$\Psi_N(x_1, ..., x_N) \approx u^{\otimes N}(x_1, ..., x_N) := u(x_1)u(x_2)...u(x_N). \tag{2.13}$$

Indeed, if $w_N \equiv 0$ then the first eigenfunction Ψ_N of H_N is exactly of this form, with u the first eigenfunction of the one-body operator $-\Delta + |x|^2 - 2\Omega L$. For our interacting Hamiltonian H_N, Ψ_N will *never* be of this form, because there is no reason to believe that *all* the particles ought to be in the state u. Only of the order of N of them would suffice [45]. Nevertheless, the ansatz (2.13) provides the right energy to leading order, as well as the right density matrices, as we will prove in this paper, and as it has already been shown in many other similar situations, see [51, 64, 65] for reviews.

The energy per particle of the fully condensed trial function $u^{\otimes N}$ is given by the Hartree energy functional

$$\mathcal{E}^H_{\Omega,a,N}(u) = \frac{\langle u^{\otimes N}, H_N u^{\otimes N} \rangle}{N} = \int_{\mathbb{R}^2} \left(|\nabla u(x)|^2 + |x|^2 |u(x)|^2 \right) dx - 2\Omega \langle u, Lu \rangle$$
$$- \frac{a}{2} \iint_{\mathbb{R}^2 \times \mathbb{R}^2} w_N(x - y)|u(x)|^2 |u(y)|^2 \, dx \, dy. \tag{2.14}$$

The infimum of this functional over the set of all u's with $\|u\|_{L^2(\mathbb{R}^2)} = 1$,

$$E^H_{\Omega,a}(N) := \inf_{\|u\|_{L^2(\mathbb{R}^2)}=1} \mathcal{E}^H_{\Omega,a,N}(u) \tag{2.15}$$

is thus an upper bound to the many-body energy:

$$E^Q_{\Omega,a}(N) \leqslant E^H_{\Omega,a}(N).$$

When $N \to \infty$, since $w_N \rightharpoonup \delta_0$, the Hartree functional (2.14) *formally* boils down to the trapped nonlinear Gross–Pitaevskii functional $\mathcal{E}^{GP}_{\Omega,a}$ which we have introduced in (2.1). We can therefore expect that

$$E^H_{\Omega,a}(N) \simeq E^{GP}_{\Omega}(a),$$

and that their ground states are close. At fixed $a < a_*$, this was shown in [47], but here we need to control the limit $a_N \to a_*$ at the same time as $N \to \infty$ and the corresponding estimates will be provided later in the proof of Proposition 4.1.

In [46, 47], we have proved that if $a < a_*$ is fixed and $0 \leqslant \Omega < 1$, then the many-body ground states of H_N are condensed on the minimizer(s) of the Gross–Pitaevskii functional. In the present paper, we will consider the limit where $a_N \to a_*$ as $N \to \infty$. In that case, the Gross–Pitaevskii minimizer blows up at the center $x = 0$ of the trap, as shown in Theorem 2.1 above. We will prove that the many-particle ground state Ψ_N condensates on the exact same function Q_N, hence derive a many-body analogue to the result of Guo and Seiringer [30], at positive rotation.

As usual, the convergence of ground states is formulated using k-particles reduced density matrices, defined for any $\Psi_N \in \mathfrak{H}^N$ by a partial trace

$$\gamma_{\Psi_N}^{(k)} := \mathrm{Tr}_{k+1 \to N} |\Psi_N\rangle\langle\Psi_N|$$

or, equivalently, $\gamma_{\Psi_N}^{(k)}$ is the trace-class operator on $\mathfrak{H}^k$ with kernel

$$\gamma_{\Psi_N}^{(k)}(x_1, \ldots, x_k; y_1, \ldots, y_k) = \int_{\mathbb{R}^{2(N-k)}} \Psi_N(x_1, \ldots, x_k, Z)\overline{\Psi_N(y_1, \ldots, y_k, Z)}dZ.$$

Bose–Einstein condensation is properly expressed by the convergence in trace norm

$$\lim_{N\to\infty} \mathrm{Tr}\left|\gamma_{\Psi_N}^{(k)} - |u^{\otimes k}\rangle\langle u^{\otimes k}|\right| = 0, \quad \forall k \in \mathbb{N}.$$

Our second main result is the following

Theorem 2.3 (Collapse and condensation of the many-body ground state). *Let* $0 \leqslant \Omega < 1$, *and* $a_N = a_* - N^{-\alpha}$ *with*

$$0 < \alpha < \min\left\{\frac{4}{5}\beta, 2(1 - 2\beta)\right\}.$$

Let Ψ_N *be any ground state of* H_N. *Then, we have*

$$\lim_{N\to\infty} \mathrm{Tr}\left|\gamma_{\Psi_N}^{(k)} - |Q_N^{\otimes k}\rangle\langle Q_N^{\otimes k}|\right| = 0, \tag{2.16}$$

for all $k \in \mathbb{N}$, *where* Q_N *is the rescaled Gagliardo–Nirenberg optimizer introduced in* (2.8). *In addition, we have*

$$E_{\Omega, a_N}^Q(N) = E_\Omega^{GP}(a_N) + o\left(E_\Omega^{GP}(a_N)\right) = \sqrt{a_* - a_N}\left(\frac{2\lambda_*^2}{a_*} + o(1)\right). \tag{2.17}$$

Remark 2.4.
(i) Note that the condition $\alpha < 2(1 - 2\beta)$ implies that we consider mean-field (by opposition to dilute, see [65, Section 5.1]) interactions, i.e., their range is much larger

than the average distance between particles. The latter is set by the length scale of the GP ground state:

$$\text{distance between particles} \propto \sqrt{\frac{(a_* - a_N)^{1/2}}{N}} \propto N^{-1/2 - \alpha/4}.$$

We are in fact somewhat deep in the mean-field regime since the transition to dilute interactions would occur when

$$\text{range of the interaction} \propto N^{-\beta} \propto N^{-1/2 - \alpha/4} \propto \text{distance between particles,}$$

i.e., at $\beta \sim 1/2 + \alpha/4$.

The condition $\alpha < 4\beta/5$ is used to ensure that the Hartree and GP ground state problems are close in the limit $N \to \infty$.

(ii) By using the method of [47], we expect that our result can be extended to a dilute regime as well, i.e., to some (not too large)

$$\beta > \frac{1}{2} + \frac{\alpha}{4}.$$

Here, we assume $\beta < 1/2 - \alpha/4$ for simplicity as this ensures the stability of the many-body system immediately [47]. The approach we follow is significantly simpler than that of [47], since we use neither the moments estimates, nor the bootstrap on the energy introduced therein. The proof is less flexible however and deeply relies on the uniqueness of the limit profile for GP ground states.

(ii) When $a_N = a_*$, it is not clear to us what happens in the large N limit. While the existence of the ground state of H_N still holds true, the blow-up phenomenon becomes more complicated. The behavior of the minimizers for the Hartree functional $\mathcal{E}^{H}_{\Omega, a_*, N}(u)$ in (2.14) when $N \to \infty$ seems to be open. It seems also difficult to look at the case where Ω depends on N as well and approaches its limit of stability $\Omega_N \to 1$.   ◇

In [47] and several of our previous works [42, 43, 46, 47, 59], our approach was based on the *quantum de Finetti theorem* [34, 69], a noncommutative version of the de Finetti–Hewitt–Savage theorem for exchangeable random variables in probability theory [20, 21, 23, 33]. More precisely, we used a quantitative version of this theorem in finite-dimensional spaces, which we have proved in [44, Lemmas 3.4, 3.6] and which extends several previous results by different authors [12, 13, 25, 32, 39]. The idea of using de Finetti theorems in the context of mean-field limits is not new. For classical systems, this has been put forward by Spohn [38, 55, 68] and then extended in many directions, see, e.g., [7, 37, 64] and the references therein. For the mean-field limit of quantum systems, the older results in this spirit include [26, 61, 63, 71, 74].

Although one can follow the same strategy here, we give below a different proof of Bose–Einstein condensation, based on a Feynman–Hellman-type argument. This method is much less flexible (it relies on the uniqueness of the limit profile for GP ground states), but it allows to cover a wider range for the parameters β and α.

3 Collapse of the Rotating GP Minimizer: Proof of Theorem 2.1

In this section, we provide the proof of Theorem 2.1. It is convenient to work at the blow-up scale, and thus to rewrite everything in terms of

$$v(x) = \sqrt{\varepsilon}\, u(\sqrt{\varepsilon} x), \qquad \varepsilon = \sqrt{a_* - a}.$$

Since the angular momentum L commutes with dilations about the center of rotation, we get

$$\mathcal{E}_{\Omega,a}^{GP}(u) = \frac{1}{\varepsilon} \int_{\mathbb{R}^2} |\nabla v|^2 + \varepsilon \int_{\mathbb{R}^2} |x|^2 |v(x)|^2 \, dx - 2\Omega \langle v, Lv \rangle - \frac{a}{2\varepsilon} \int_{\mathbb{R}^2} |v(x)|^4 \, dx$$

$$= \frac{\mathcal{F}_{\Omega,\varepsilon}(v)}{\varepsilon}$$

where

$$\mathcal{F}_{\Omega,\varepsilon}(v) = \int_{\mathbb{R}^2} |\nabla v|^2 - \frac{a_*}{2} \int_{\mathbb{R}^2} |v(x)|^4 \, dx + \varepsilon^2 \int_{\mathbb{R}^2} |x|^2 |v(x)|^2 \, dx$$

$$+ \frac{\varepsilon^2}{2} \int_{\mathbb{R}^2} |v(x)|^4 \, dx - 2\varepsilon\Omega \langle v, Lv \rangle. \tag{3.1}$$

We then introduce

$$\boxed{F_\Omega(\varepsilon) = \inf_{\|v\|_{L^2}=1} \mathcal{F}_{\Omega,\varepsilon}(v) = \sqrt{a_* - a}\; E_\Omega^{GP}(a)}$$

and our goal is to prove that

$$F_\Omega(\varepsilon) = F_0(\varepsilon) + o(\varepsilon^2) = \varepsilon^2 \left(\frac{2\lambda_*^2}{a_*} + o(1) \right). \tag{3.2}$$

The behavior of $F_0(\varepsilon)$ and its associated unique ground state is studied in [30]. However, even when $\Omega = 0$ we have to prove the convergence of approximate ground states in the sense of (2.6).

Step 1. Convergence to a Gagliardo–Nirenberg optimizer. By rearrangement inequalities, $F_0(\varepsilon)$ has a radial-decreasing minimizer $\tilde{v}_\varepsilon$. Then $L\tilde{v}_\varepsilon = 0$, hence

$$F_\Omega(\varepsilon) \leqslant F_0(\varepsilon)$$

for every $0 \leqslant \Omega < 1$. It is the reverse inequality which is not obvious. From the diamagnetic inequality, we have

$$\int_{\mathbb{R}^2} |\nabla v(x) + i\varepsilon\Omega x^\perp v(x)|^2 \, dx \geqslant \int_{\mathbb{R}^2} |\nabla |v|(x)|^2 \, dx$$

and therefore we obtain

$$F_\Omega(\varepsilon) \geqslant \min_{\substack{v \in H^1(\mathbb{R}^d) \\ \|v\|_{L^2} = 1}} \left\{ \int_{\mathbb{R}^2} |\nabla v|^2 + (1 - \Omega^2)\varepsilon^2 \int_{\mathbb{R}^2} |x|^2 |v(x)|^2 \, dx \right.$$
$$\left. - \frac{a_* - \varepsilon^2}{2} \int_{\mathbb{R}^2} |v(x)|^4 \, dx \right\}$$
$$= \sqrt{1 - \Omega^2} \, F_0(\varepsilon).$$

This lower bound has the right behavior $O(\varepsilon^2)$ but not the right constant.

We can also write the energy in a different form and obtain the following lower bound:

$$\mathcal{F}_{\Omega,\varepsilon}(v) = (1 - \Omega) \int_{\mathbb{R}^2} |\nabla v(x)|^2 \, dx + \Omega \int_{\mathbb{R}^2} |\nabla v(x) + i\varepsilon x^\perp v(x)|^2 \, dx$$
$$- \frac{a_*}{2} \int_{\mathbb{R}^2} |v(x)|^4 \, dx + (1 - \Omega)\varepsilon^2 \int_{\mathbb{R}^2} |x|^2 |v(x)|^2 \, dx + \frac{\varepsilon^2}{2} \int_{\mathbb{R}^2} |v(x)|^4 \, dx$$
$$\geqslant \int_{\mathbb{R}^2} |\nabla |v(x)||^2 \, dx - \frac{a_*}{2} \int_{\mathbb{R}^2} |v(x)|^4 \, dx$$
$$+ (1 - \Omega)\varepsilon^2 \int_{\mathbb{R}^2} |x|^2 |v(x)|^2 \, dx + \frac{\varepsilon^2}{2} \int_{\mathbb{R}^2} |v(x)|^4 \, dx$$
$$\geqslant (1 - \Omega)\varepsilon^2 \int_{\mathbb{R}^2} |x|^2 |v(x)|^2 \, dx + \frac{\varepsilon^2}{2} \int_{\mathbb{R}^2} |v(x)|^4 \, dx. \tag{3.3}$$

From these bounds, we deduce that any sequence $\{v_\varepsilon\} \subset H^1(\mathbb{R}^2)$ such that $\|v\|_{L^2} = 1$ and $\mathcal{F}_{\Omega,\varepsilon}(v) = O(\varepsilon^2)$ (for instance approximate ground states) is bounded in $H^1(\mathbb{R}^2)$, and that $|x|v_\varepsilon$ is bounded in $L^2(\mathbb{R}^2)$. Such sequences are precompact in $L^p(\mathbb{R}^2)$ for all $2 \leqslant p < \infty$. Therefore, up to a subsequence, we can pass to the limit $v_\varepsilon \to v$ and obtain

$$\int_{\mathbb{R}^2} |\nabla v(x)|^2 \, dx - \frac{a_*}{2} \int_{\mathbb{R}^2} |v(x)|^4 \, dx = 0 \quad \text{with} \quad \int_{\mathbb{R}^2} |v|^2 = 1. \tag{3.4}$$

This means that v belongs to the set of the Gagliardo–Nirenberg optimizers (up to a phase)

$$\mathcal{Q} := \left\{ Q_{\lambda,X}(x) = \lambda Q_*(\lambda(x - X)), \quad \lambda > 0, \, X \in \mathbb{R}^2 \right\}. \tag{3.5}$$

Here, Q_* is the unique positive radial solution to the equation

$$-\Delta Q_* - a_* Q_*^3 = -Q_*.$$

This solution necessarily satisfies $\int_{\mathbb{R}^2} Q_*^2 = 1$ and it is just given by

$$Q_* = \|Q\|^{-1}Q = (a_*)^{-1/2}Q$$

where $-\Delta Q - Q^3 = -Q$. Note that $Q_{\lambda,X}$ solves the equation $-\Delta Q_{\lambda,X} - a_* Q_{\lambda,X}^3 = -\lambda^2 Q_{\lambda,X}$.

Note that we know from the above arguments that v_ε converges to $Q_{\lambda,X}$ strongly in $L^2(\mathbb{R}^2) \cap L^4(\mathbb{R}^2)$ and that

$$\int_{\mathbb{R}^2} |\nabla v_\varepsilon(x)|^2 \, dx - \frac{a_*}{2} \int_{\mathbb{R}^2} |v_\varepsilon(x)|^4 \, dx \to 0.$$

It follows that

$$\int_{\mathbb{R}^2} |\nabla v_\varepsilon(x)|^2 \, dx \to \frac{a_*}{2} \int_{\mathbb{R}^2} |Q_{\lambda,X}(x)|^4 \, dx = \int_{\mathbb{R}^2} |\nabla Q_{\lambda,X}(x)|^2 \, dx$$

and thus that the limit is also strong in $H^1(\mathbb{R}^2)$. For later purposes, we choose the phase of v_ε such that v_ε is the closest to its limit:

$$\|v_\varepsilon - Q_{\lambda,X}\|_{L^2} = \min_\theta \|e^{i\theta} v_\varepsilon - Q_{\lambda,X}\|_{L^2}.$$

This gives the orthogonality condition on the imaginary part of v_ε:

$$\int_{\mathbb{R}^2} Q_{\lambda,X} \operatorname{Im}(v_\varepsilon) = 0. \tag{3.6}$$

We have up to now shown that any sequence $\{v_\varepsilon\}$ such that $\mathcal{F}_\Omega(v_\varepsilon) = O(\varepsilon^2)$ converges to an element of $\mathcal{Q}$, up to a subsequence and a phase. This is optimal for sequences that have an energy of the order $O(\varepsilon^2)$. In order to determine the possible values of X and λ, we have to assume that v_ε is an approximate ground state of $F_\Omega(\varepsilon)$. In the next three steps, we actually assume v_ε is a true ground state, so that we can rely on the variational equation and get better estimates. We return to approximate ground states at the end of the proof.

Step 2. Decay of ground states. For v_ε a true ground state, the Euler–Lagrange equation takes the form

$$- \Delta v_\varepsilon - (a_* - \varepsilon^2)|v_\varepsilon|^2 v_\varepsilon + \varepsilon^2 |x|^2 v_\varepsilon - 2\Omega \varepsilon L v_\varepsilon + \mu_\varepsilon v_\varepsilon = 0, \tag{3.7}$$

with the Lagrange multiplier given by

$$\mu_\varepsilon = -\mathcal{F}_\Omega(v_\varepsilon) + \frac{a_* - \varepsilon^2}{2} \int_{\mathbb{R}^2} |v_\varepsilon|^4 \to \frac{a_*}{2} \int_{\mathbb{R}^2} Q_{\lambda,X}^4 = \lambda^2 > 0. \tag{3.8}$$

Using that $\lambda^2 > 0$ we shall obtain uniform decay estimates *à la* Agmon [2] for v_ε. First, we need to show that v_ε converges uniformly.

Lemma 3.1 (Uniform convergence). *The sequence $\{v_\varepsilon\}$ is bounded in $H^2(\mathbb{R}^2)$ and converges to $Q_{\lambda,\chi}$ strongly in $H^1(\mathbb{R}^2)$ and in $L^\infty(\mathbb{R}^2)$.*

Proof. We have

$$\left(-\Delta + \varepsilon^2|x|^2 - 2\Omega\varepsilon L + \lambda^2\right) v_\varepsilon = (\lambda^2 - \mu_\varepsilon)v_\varepsilon + (a_* - \varepsilon^2)|v_\varepsilon|^2 v_\varepsilon$$

where the right side is bounded in $L^2(\mathbb{R}^2)$. By the Cauchy–Schwarz inequality and the fact that L commutes with $-\Delta + \varepsilon^2|x|^2$, we have

$$2\varepsilon|L| \leqslant -\Delta + \varepsilon^2|x|^2$$

and

$$(2\Omega\varepsilon L)^2 \leqslant \Omega^2 \left(-\Delta + \varepsilon^2|x|^2\right)^2$$

or, equivalently,

$$\left\|2\Omega\varepsilon L \left(-\Delta + \varepsilon^2|x|^2\right)^{-1}\right\| \leqslant \Omega < 1. \tag{3.9}$$

By the resolvent formula, this proves that

$$\left\|\left(-\Delta + \varepsilon^2|x|^2\right)\left(-\Delta + \varepsilon^2|x|^2 - 2\Omega\varepsilon L\right)^{-1}\right\| \leqslant \frac{1}{1-\Omega}$$

and similarly that

$$\left\|\left(-\Delta + \varepsilon^2|x|^2 + \lambda^2\right)\left(-\Delta + \varepsilon^2|x|^2 - 2\Omega\varepsilon L + \lambda^2\right)^{-1}\right\| \leqslant \frac{1}{1-\Omega}.$$

Using the relation

$$\left\|\left(-\Delta + \varepsilon^2|x|^2 + \lambda^2\right)u\right\|^2_{L^2(\mathbb{R}^2)} = \int_{\mathbb{R}^2}|\Delta u(x)|^2\,dx + \int_{\mathbb{R}^2}(\varepsilon^2|x|^2 + \lambda^2)^2|u(x)|^2\,dx$$
$$+ 2\int_{\mathbb{R}^2}(\varepsilon^2|x|^2 + \lambda^2)|\nabla u(x)|^2\,dx - 4\varepsilon^2\int_{\mathbb{R}^2}|u(x)|^2\,dx$$

we get

$$\left(-\Delta + \varepsilon^2|x|^2 + \lambda^2\right)^2 \geqslant \left(-\Delta + \lambda^2\right)^2 - 4\varepsilon^2 \geqslant \left(-\Delta + \lambda^2/2\right)^2$$

for $\varepsilon \leqslant \sqrt{3}\lambda/4$, and thus

$$\left\|\left(-\Delta + \lambda^2/2\right)\left(-\Delta + \varepsilon^2|x|^2 - 2\Omega\varepsilon L + \lambda^2\right)^{-1}\right\| \leqslant \frac{1}{1-\Omega}.$$

Inserting in Eq. (3.7), this proves that v_ε is bounded in $H^2(\mathbb{R}^2)$. Since v_ε already converges strongly in $L^2(\mathbb{R}^2)$, it also converges strongly in $H^1(\mathbb{R}^2)$ and in $L^\infty(\mathbb{R}^2)$, by interpolation. □

Next we can prove an exponential decay estimate. We need it later only to obtain the strong convergence $|x|v_\varepsilon \to |x|Q_{\lambda,x}$ in $L^2(\mathbb{R}^2)$.

Lemma 3.2 (Exponential decay). *The function v_ε satisfies*

$$\int_{\mathbb{R}^2} e^{\lambda|x|} |v_\varepsilon(x)|^2 \, dx + \int_{\mathbb{R}^2} |\nabla e^{\frac{\lambda}{2}|x|} v_\varepsilon|^2 \leqslant C,$$

for a constant C independent of ε. In particular, $|x|v_\varepsilon \to |x|Q_{\lambda,x}$ strongly in $L^2(\mathbb{R}^2)$.

Proof. It is well known that v_ε is analytic with all its derivatives decaying fast at infinity. We seek here for an explicit bound, independent of ε. We use that

$$-\mathrm{Re}\langle v_\varepsilon, e^{\alpha|x|} \Delta v_\varepsilon \rangle = -\frac{1}{2} \int_{\mathbb{R}^2} e^{\alpha|x|} \Big(\overline{v_\varepsilon}(x) \Delta v_\varepsilon(x) + v_\varepsilon(x) \Delta \overline{v_\varepsilon}(x) \Big) dx$$

$$= -\frac{1}{2} \int_{\mathbb{R}^2} e^{\alpha|x|} \Big(\Delta|v_\varepsilon|^2(x) - 2|\nabla v_\varepsilon(x)|^2 \Big) dx$$

$$= -\frac{1}{2} \int_{\mathbb{R}^2} e^{\alpha|x|} \left(\left(\frac{\alpha}{|x|} + \alpha^2 \right) |v_\varepsilon(x)|^2 - 2|\nabla v_\varepsilon(x)|^2 \right) dx$$

$$= \int_{\mathbb{R}^2} e^{\alpha|x|} |\nabla v_\varepsilon|^2 - \frac{1}{2} \int_{\mathbb{R}^2} e^{\alpha|x|} \left(\frac{\alpha}{|x|} + \alpha^2 \right) |v_\varepsilon(x)|^2 \, dx$$

$$= \int_{\mathbb{R}^2} |\nabla e^{\frac{\alpha}{2}|x|} v_\varepsilon|^2 - \frac{\alpha^2}{4} \int_{\mathbb{R}^2} e^{\alpha|x|} |v_\varepsilon(x)|^2 \, dx.$$

Then, we integrate the Euler–Lagrange equation (3.7) against $e^{\alpha|x|} \overline{v_\varepsilon}$ and obtain

$$0 = \int_{\mathbb{R}^2} |\nabla e^{\frac{\alpha}{2}|x|} v_\varepsilon|^2 + \int_{\mathbb{R}^2} e^{\alpha|x|} \left(\varepsilon^2|x|^2 - (a_* - \varepsilon^2)|v_\varepsilon|^2 + \mu_\varepsilon - \frac{\alpha^2}{4} \right) |v_\varepsilon(x)|^2 \, dx$$

$$- 2\Omega\varepsilon \langle e^{\frac{\alpha}{2}|x|} v_\varepsilon, L e^{\frac{\alpha}{2}|x|} v_\varepsilon \rangle$$

$$\geqslant (1-\Omega) \int_{\mathbb{R}^2} |\nabla e^{\frac{\alpha}{2}|x|} v_\varepsilon(x)|^2 \, dx + (1-\Omega)\varepsilon^2 \int_{\mathbb{R}^2} e^{\alpha|x|} |x|^2 |v_\varepsilon(x)|^2 \, dx$$

$$+ \int_{\mathbb{R}^2} e^{\alpha|x|} \left(\mu_\varepsilon - (a_* - \varepsilon^2)|v_\varepsilon|^2 - \frac{\alpha^2}{4} \right) |v_\varepsilon(x)|^2 \, dx.$$

Choosing $\alpha = \lambda$ and using the uniform convergence of v_ε towards $Q_{\lambda,x}$, we can find a radius R independent of ε such that

$$\mu_\varepsilon - (a_* - \varepsilon^2)|v_\varepsilon|^2 - \frac{\alpha^2}{4} \geqslant \frac{\lambda^2}{2}, \qquad \forall |x| \geqslant R,$$

and then

$$\frac{\lambda^2}{2} \int_{\mathbb{R}^2 \setminus B_R} e^{\lambda|x|} |v_\varepsilon(x)|^2 \, dx + (1 - \Omega) \int_{\mathbb{R}^2} |\nabla(e^{\frac{1}{2}|x|} v_\varepsilon)(x)|^2 \, dx \leqslant e^{\lambda R} \left(\mu_\varepsilon + \frac{\lambda^2}{4} + a_* \|v_\varepsilon\|_{L^\infty(B_R)}^2 \right)$$

for all $\varepsilon > 0$ small enough. This proves the desired exponential decay estimate. The strong convergence of $|x| v_\varepsilon$ in $L^2(\mathbb{R}^2)$ then follows by interpolation. $\qquad \square$

Step 3. The imaginary part is (very) small. We split v_ε into real and imaginary parts

$$v_\varepsilon = q_\varepsilon + i r_\varepsilon$$

and get bounds on r_ε using energy estimates (we could similarly use the Eq. (3.7)). Recalling $x^\perp = (-x_2, x_1)$, we observe that

$$\langle v_\varepsilon, L v_\varepsilon \rangle = \int_{\mathbb{R}^2} x^\perp \cdot \text{Im}(\overline{v_\varepsilon} \nabla v_\varepsilon) = \int_{\mathbb{R}^2} x^\perp \cdot (q_\varepsilon \nabla r_\varepsilon - r_\varepsilon \nabla q_\varepsilon) = 2 \int_{\mathbb{R}^2} x^\perp \cdot q_\varepsilon \nabla r_\varepsilon \tag{3.10}$$

where we have integrated by parts and used that $\text{div} \, x^\perp = 0$. Thus,

$$|\langle v_\varepsilon, L v_\varepsilon \rangle| \leqslant C \|\nabla r_\varepsilon\|_{L^2}.$$

Here, we have used the fact that $|x| q_\varepsilon$ is bounded in $L^2(\mathbb{R}^2)$. Then the energy reads

$$\mathcal{F}_{\Omega,\varepsilon}(v_\varepsilon) \geqslant \int_{\mathbb{R}^2} |\nabla q_\varepsilon|^2 + \int_{\mathbb{R}^2} |\nabla r_\varepsilon|^2 - \frac{a_*}{2} \int_{\mathbb{R}^2} (q_\varepsilon^4 + r_\varepsilon^4 + 2 q_\varepsilon^2 r_\varepsilon^2) - C\varepsilon \|\nabla r_\varepsilon\|_{L^2}.$$

Since $q_\varepsilon \to Q_{\lambda,X}$ and $r_\varepsilon \to 0$ uniformly by Lemma 3.1, we obtain

$$\int_{\mathbb{R}^2} |q_\varepsilon^2 - Q_{\lambda,X}^2| r_\varepsilon^2 + \int_{\mathbb{R}^2} r_\varepsilon^4 = o\left(\|r_\varepsilon\|_{L^2}^2 \right).$$

Moreover, using the Gagliardo–Nirenberg inequality (2.2) for the real part q_ε, we have

$$\int_{\mathbb{R}^2} |\nabla q_\varepsilon|^2 - \frac{a_*}{2} \int_{\mathbb{R}^2} |q_\varepsilon|^4 \geqslant \left(\int_{\mathbb{R}^2} |\nabla q_\varepsilon|^2 \right) \left(1 - \int_{\mathbb{R}^2} |q_\varepsilon|^2 \right) = (\lambda^2 + o(1)) \int_{\mathbb{R}^2} |r_\varepsilon|^2.$$

Here in the second estimate, we have used the facts that $\|q_\varepsilon\|_{L^2}^2 + \|r_\varepsilon\|_{L^2}^2 = 1$ and that $q_\varepsilon \to Q_{\lambda,X}$ strongly in $H^1(\mathbb{R}^2)$ by Lemma 3.1. Thus, we can bound the energy from below as

$$\mathcal{F}_{\Omega,\varepsilon}(v_\varepsilon) \geqslant \int_{\mathbb{R}^2} |\nabla r_\varepsilon|^2 - a_* \int_{\mathbb{R}^2} Q_{\lambda,X}^2 r_\varepsilon^2 + (\lambda^2 + o(1)) \int_{\mathbb{R}^2} |r_\varepsilon|^2 - C\varepsilon \|\nabla r_\varepsilon\|_{L^2}. \tag{3.11}$$

Now, we use some non-degeneracy property of $Q_{\lambda,X}$ [10, 28, 54, 73]. Since $Q_{\lambda,X}$ is positive, it must be the first eigenfunction of the operator

$$\mathcal{L}_- := -\Delta - a_* Q_{\lambda, X}^2 + \lambda^2 \tag{3.12}$$

and the corresponding eigenvalue 0 is nondegenerate [48, Cor. 11.9]. In particular, we get

$$\langle f, \mathcal{L}_- f \rangle_{L^2(\mathbb{R}^2)} \geq \lambda_2 \|f\|_{L^2(\mathbb{R}^2)}^2 \tag{3.13}$$

for all f orthogonal to $Q_{\lambda, X}$ where $\lambda_2 > 0$ is the second eigenvalue of $\mathcal{L}_-$. Since on the other hand

$$\langle f, \mathcal{L}_- f \rangle_{L^2(\mathbb{R}^2)} \geq \|\nabla f\|_{L^2(\mathbb{R}^2)}^2 - a_* \|Q_{\lambda, X}\|_{L^\infty(\mathbb{R}^2)}^2 \|f\|_{L^2(\mathbb{R}^2)}^2 , \tag{3.14}$$

we may combine (3.13) with (3.14) (add a large constant times the second inequality to the first one) and obtain the well-known estimate [73]

$$\langle f, \mathcal{L}_- f \rangle_{L^2(\mathbb{R}^2)} \geq c \|f\|_{H^1(\mathbb{R}^2)}^2 \tag{3.15}$$

for a constant $c > 0$ and all f orthogonal to $Q_{\lambda, X}$. Inserting (3.15) in (3.11) using the fact that r_ε is orthogonal to $Q_{\lambda, X}$ as we have seen in (3.6), we obtain

$$\mathcal{F}_{\Omega, \varepsilon}(v_\varepsilon) \geq c_1 \|r_\varepsilon\|_{H^1}^2 - C\varepsilon \|\nabla r_\varepsilon\|_{L^2}$$

for a constant $c_1 > 0$. Combining with the energy upper bound $\mathcal{F}_{\Omega, \varepsilon}(v_\varepsilon) = O(\varepsilon^2)$, we conclude that

$$\|r_\varepsilon\|_{H^1(\mathbb{R}^2)} \leq C\varepsilon. \tag{3.16}$$

Step 4. Change of gauge and convergence of ground states. Since $|x| q_\varepsilon$ converges to $|x| Q_{\lambda, X}$ strongly in L^2 by Lemma 3.2, we deduce from (3.10) and (3.16) that

$$\langle v_\varepsilon, L v_\varepsilon \rangle = 2 \int_{\mathbb{R}^2} x^\perp \cdot Q_{\lambda, X} \nabla r_\varepsilon + o(\varepsilon) = -2 \int_{\mathbb{R}^2} x^\perp \cdot \nabla Q_{\lambda, X} r_\varepsilon + o(\varepsilon).$$

But $Q_{\lambda, X} = \lambda Q_*(\lambda(x - X))$ with Q_* a radial function, hence

$$(x^\perp - X^\perp) \cdot \nabla Q_{\lambda, X} = 0.$$

Inserting in the above, using (3.16) and strong L^2-convergence of $|x| q_\varepsilon$ again we obtain

$$\langle v_\varepsilon, L v_\varepsilon \rangle = 2 \int_{\mathbb{R}^2} X^\perp \cdot q_\varepsilon \nabla r_\varepsilon + o(\varepsilon).$$

Inserting this in the energy gives

$$\mathcal{F}_{\Omega,\varepsilon}(v_\varepsilon) = \int_{\mathbb{R}^2} |\nabla v_\varepsilon|^2 - \frac{a_*}{2} \int_{\mathbb{R}^2} |v_\varepsilon|^4 + 2\varepsilon\Omega \int_{\mathbb{R}^2} X^\perp \cdot \mathrm{Im}(v_\varepsilon \nabla \bar{v}_\varepsilon)$$
$$+ \varepsilon^2 \left(\frac{1}{2} \int_{\mathbb{R}^2} |v_\varepsilon|^4 + \int_{\mathbb{R}^2} |x|^2 |v_\varepsilon|^2 \right) + o(\varepsilon^2).$$

Now, define a new function f_ε by setting

$$v_\varepsilon(x) = f_\varepsilon(x)\, e^{i\varepsilon\Omega X^\perp \cdot x}$$

and observe that

$$\int_{\mathbb{R}^2} |\nabla v_\varepsilon|^2 + 2\varepsilon\Omega \int_{\mathbb{R}^2} X^\perp \cdot \mathrm{Im}(v_\varepsilon \nabla \bar{v}_\varepsilon) = \int_{\mathbb{R}^2} |\nabla f_\varepsilon|^2 - \varepsilon^2 \Omega^2 |X|^2 \int_{\mathbb{R}^2} |f_\varepsilon|^2.$$

Using the optimal Gagliardo–Nirenberg inequality again and the convergence $f_\varepsilon \to Q_{\lambda,X}$, we obtain

$$\liminf_{\varepsilon \to 0} \frac{\mathcal{F}_{\Omega,\varepsilon}(v_\varepsilon)}{\varepsilon^2} \geqslant \frac{1}{2} \int_{\mathbb{R}^2} Q_{\lambda,X}^4 + \int_{\mathbb{R}^2} \left(|x|^2 - \Omega^2 |X|^2 \right) Q_{\lambda,X}^2.$$

Recalling that $Q_{\lambda,X} = \lambda Q_*(\lambda(x - X))$ for a radial function Q_* this finally yields

$$\liminf_{\varepsilon \to 0} \frac{\mathcal{F}_{\Omega,\varepsilon}(v_\varepsilon)}{\varepsilon^2} \geqslant (1 - \Omega^2)|X|^2 + \frac{1}{\lambda^2} \int_{\mathbb{R}^2} |x|^2 |Q_*(x)|^2 \, dx + \frac{\lambda^2}{2} \int_{\mathbb{R}^2} |Q_*(x)|^4 \, dx.$$

Since $\Omega^2 < 1$, the minimum of the right side is attained for $X = 0$ and $\lambda = \lambda_*$ as in [30]. This also concludes the proof of (3.2), hence of (2.5). Also, this shows that any sequence of minimizers must, modulo rescaling, choice of a constant phase (in (3.6)) and passing to a subsequence, converge strongly in H^1 to $Q_{\lambda_*,0}$. By uniqueness of the limit, we conclude that passing to a subsequence is unnecessary, which concludes the proof of (2.7) for true ground states.

Step 5. Convergence of approximate ground states for $\Omega = 0$. Here, we assume $\Omega = 0$ and show that any sequence $\{v_\varepsilon\}$ such that

$$\mathcal{F}_{0,\varepsilon}(v_\varepsilon) = F_0(\varepsilon) + o(\varepsilon^2)$$

must converge to $Q_{\lambda_*,0}$, as we have proved before for the exact minimizers, using the Euler–Lagrange equation. Indeed, from Step 1, we already know that $v_\varepsilon \to Q_{\lambda,X}$ after extraction of a subsequence and choice of a good phase. Hence we have

$$\mathcal{F}_{0,\varepsilon}(v_\varepsilon) \geq \varepsilon^2 \left(\int_{\mathbb{R}^2} |x|^2 |v(x)|^2 \, dx + \frac{1}{2} \int_{\mathbb{R}^2} |v(x)|^4 \, dx \right)$$

$$\geq \varepsilon^2 \left(\int_{\mathbb{R}^2} |x|^2 Q_{\lambda,X}(x)^2 \, dx + \frac{1}{2} \int_{\mathbb{R}^2} Q_{\lambda,X}(x)^4 \, dx \right) + o(\varepsilon^2)$$

$$= \varepsilon^2 \left(|X|^2 + \frac{1}{\lambda^2} \int_{\mathbb{R}^2} |x|^2 Q_*(x)^2 \, dx + \frac{\lambda^2}{2} \int_{\mathbb{R}^2} Q_*(x)^4 \, dx \right) + o(\varepsilon^2).$$

Again, the minimum of the term in the parenthesis is attained uniquely for $X = 0$ and $\lambda = \lambda_*$. Since

$$\mathcal{F}_{0,\varepsilon}(v_\varepsilon) = F_0(\varepsilon) + o(\varepsilon^2) = \varepsilon^2 \left(\frac{1}{\lambda_*^2} \int_{\mathbb{R}^2} |x|^2 Q_*(x)^2 \, dx + \frac{\lambda_*^2}{2} \int_{\mathbb{R}^2} Q_*(x)^4 \, dx \right)$$

we must have $X = 0$ and $\lambda = \lambda_*$. The limit being unique, the whole sequence must converge to $Q_{\lambda_*,0}$. By usual arguments, the limit must be strong in $H^1(\mathbb{R}^2)$.

Step 6. Convergence of approximate ground states for $0 < \Omega < 1$. Next, we turn to the rotating case $\Omega < 1$. As before, we take a sequence $\{v_\varepsilon\}$ such that $\mathcal{F}_{\Omega,\varepsilon}(v_\varepsilon) = F_\Omega(\varepsilon) + o(\varepsilon^2) = F_0(\varepsilon) + o(\varepsilon^2)$. We artificially increase the rotation speed by choosing a $\eta < 1$ be such that $\Omega/\eta < 1$ and we remark that

$$F_\Omega(\varepsilon) + o(\varepsilon^2) = \mathcal{F}_{\Omega,\varepsilon}(v_\varepsilon) = \eta \mathcal{F}_{\Omega/\eta,\varepsilon}(v_\varepsilon) + (1 - \eta) \mathcal{F}_{0,\varepsilon}(v_\varepsilon)$$

$$\geq \eta F_{\Omega/\eta}(\varepsilon) + (1 - \eta) F_0(\varepsilon).$$

Since the terms $\varepsilon^{-2} F_\Omega(\varepsilon)$, $\varepsilon^{-2} F_{\Omega/\eta}(\varepsilon)$ and $\varepsilon^{-2} F_0(\varepsilon)$ all have the same limit, this proves that $\mathcal{F}_{\Omega/\eta,\varepsilon}(v_\varepsilon) = F_{\Omega/\eta}(\varepsilon) + o(\varepsilon^2)$ and $\mathcal{F}_{0,\varepsilon}(v_\varepsilon) = F_0(\varepsilon) + o(\varepsilon^2)$. But then $\{v_\varepsilon\}$ is a sequence of approximate ground states in the case $\Omega = 0$ and we can apply the previous step.

This concludes the proof of Theorem 2.1. □

4 Collapse of the Many-Body Ground State: Proof of Theorem 2.3

This section is devoted to the proof of Theorem 2.3. We start with the convergence of the ground state energy, and then settle some energy estimates for the ground state.

Step 1. Convergence of the many-body ground state energy. We provide the proof of the convergence of $E_{\Omega,a_N}^Q(N)$, using a method described in [41, Section 3]. The precise statement is the following.

Proposition 4.1 (Convergence of the many-body ground state energy). *Let $\beta \in (0, 1/2)$ and $a_* - a_N = N^{-\alpha}$ with*

$$0 < \alpha < \min\left\{\frac{4}{5}\beta, 2(1 - 2\beta)\right\}.\tag{4.1}$$

Then

$$E^Q_{\Omega,a_N}(N) = E^{GP}_\Omega(a_N) + o\big(E^{GP}_\Omega(a_N)\big) = \sqrt{a_* - a_N}\left(\frac{2\lambda^2}{a_*} + o(1)\right).\tag{4.2}$$

Proof. We start with the lower bound. From the arguments in [41, Section 3] and the fact that the Fourier transform of w_N satisfies $|\widehat{w}_N| \leqslant CN^{2\beta}$, we have

$$E^Q_{\Omega,a}(N) \geqslant \min_{\substack{\gamma = \gamma_* \geqslant 0 \\ \mathrm{Tr}\,\gamma = 1}}\left\{\mathrm{Tr}\big[(-\Delta + |x|^2 - 2\Omega L)\gamma\big] - \frac{a}{2}\int_{\mathbb{R}^d}\int_{\mathbb{R}^d} w_N(x - y)\rho_\gamma(x)\rho_\gamma(y)\,dx\,dy\right\}$$
$$- CN^{2\beta-1}.\tag{4.3}$$

Here, the one-body density is defined by writing

$$\rho_\gamma = \sum_j n_j |u_j|^2$$

with the spectral decomposition $\gamma = \sum_j n_j |u_j\rangle\langle u_j|$ ($0 \leqslant n_j \leqslant 1$ and $\sum_j n_j = 1$).

Note that the arguments in [41, Section 3] contain two ingredients. One is to get a lower bound involving the mean-field interaction following an idea from [40, 50], using auxiliary classical particles that repel each other in order to model the attractive part of the interaction. The other ingredient is to apply the Hoffmann-Ostenhof inequality[1] to the kinetic energy in order to get the Hartree energy. Here, we cannot use the second part when $\Omega \neq 0$. We thus bypass it and only bound the interaction from below. The price to pay is that we end up with the mixed state type Hartree energy on the right side of (4.3).

By the Cauchy–Schwarz inequality, we have

$$\iint |u(x)|^2|u(y)|^2 w_N(x - y)\,dx\,dy \leqslant \frac{1}{2}\iint (|u(x)|^4 + |u(y)|^4)w_N(x - y)\,dx\,dy$$
$$= \int |u(x)|^4\,dx\tag{4.4}$$

for any $u \in H^1(\mathbb{R}^2)$. Applying this to $u = \sqrt{\rho_\gamma}$, we therefore get

---

[1] Bounding the full kinetic energy from below by that of the one-body density $\sqrt{\rho_{\Psi_N}}$.

$$E^Q_{\Omega,a}(N) \geq \min_{\substack{\gamma=\gamma^*\geq 0 \\ \mathrm{Tr}\,\gamma=1}} \left\{ \mathrm{Tr}\left[(-\Delta+|x|^2-2\Omega L)\gamma\right] - \frac{a}{2}\int_{\mathbb{R}^d}\rho_\gamma(x)^2\,dx \right\} - CN^{2\beta-1}.$$

(4.5)

Here $\rho_\gamma(x) := \gamma(x,x)$ for every $\gamma = \gamma^* \geq 0$ with $\mathrm{Tr}(\gamma) = 1$. The usual Gross–Pitaevskii energy (2.1) corresponds to $\gamma = |u\rangle\langle u|$, a rank-one projection. Now, we remark that the minimum on the right of (4.5) is nothing but the original Gross–Pitaevskii minimum.

Lemma 4.2 (Mixed Gross–Pitaevskii energy). *For every $0 \leq \Omega < 1$ and every $0 \leq a < a_*$, we have*

$$\min_{\substack{\gamma=\gamma^*\geq 0 \\ \mathrm{Tr}\,\gamma=1}} \left\{ \mathrm{Tr}\left[(-\Delta+|x|^2-2\Omega L)\gamma\right] - \frac{a}{2}\int_{\mathbb{R}^d}\rho_\gamma(x)^2\,dx \right\} = E^{\mathrm{GP}}_\Omega(a).$$

(4.6)

Proof of Lemma 4.2. This follows from the concavity of the energy with respect to γ. Indeed, writing $\gamma = \sum_j n_j |u_j\rangle\langle u_j|$ with $0 \leq n_j \leq 1$ and $\sum_j n_j = 1$, we have $\rho_\gamma = \sum_j n_j |u_j|^2$, hence

$$\mathrm{Tr}\left[(-\Delta+|x|^2-2\Omega L)\gamma\right] - \frac{a}{2}\int_{\mathbb{R}^d}\rho_\gamma(x)^2\,dx \geq \sum_j n_j \,\mathcal{E}^{\mathrm{GP}}_{\Omega,a}(u_j) \geq E^{\mathrm{GP}}_\Omega(a)$$

as we wanted. □

Inserting in (4.5) and using the behavior of $E^{\mathrm{GP}}_\Omega(a_N)$ proved in Theorem 2.1, we get the simple lower bound

$$\boxed{E^Q_{\Omega,a_N}(N) \geq E^{\mathrm{GP}}_\Omega(a_N)\left(1 - CN^{2\beta-1}(a_*-a_N)^{-1/2}\right).}$$

The error term $N^{2\beta-1}(a_*-a_N)^{-1/2}$ goes to zero when $a_*-a_N = N^{-\alpha}$ and $\alpha < 2(1-2\beta)$.

Now we turn to the upper bound. By the variational principle

$$E^Q_{\Omega,a}(N) = \inf_{\Psi\in\mathfrak{H}^N,\|\Psi\|=1} \frac{\langle\Psi,H_N\Psi\rangle}{N} \leq \inf_{\|u\|_{L^2}=1} \frac{\langle u^{\otimes N},H_N u^{\otimes N}\rangle}{N} = E^{\mathrm{H}}_{\Omega,a}(N).$$

We therefore need to bound the Hartree functional from above by the Gross–Pitaevskii functional. Introducing the variable $z = N^\beta(x-y)$, we may write

$$\iint_{\mathbb{R}^2 \times \mathbb{R}^2} |u(x)|^2 N^{2\beta} w(N^\beta(x-y))|u(y)|^2 dx dy - \left(\int_{\mathbb{R}^2} w\right) \int_{\mathbb{R}^2} |u(x)|^4 dx$$

$$= \iint_{\mathbb{R}^2 \times \mathbb{R}^2} |u(x)|^2 w(z) \left(|u(x-N^{-\beta}z)|^2 - |u(x)|^2\right) dx dz$$

$$= \iint_{\mathbb{R}^2 \times \mathbb{R}^2} |u(x)|^2 w(z) \left(\int_0^1 (\nabla|u|^2)(x-tN^{-\beta}z) \cdot (N^{-\beta}z) dt\right) dx dz.$$

Using $|\nabla|u|^2| \leqslant 2|\nabla u|.|u|$ and Hölder's inequality, we find

$$\int_{\mathbb{R}^2} |u(x)|^2 |(\nabla|u|^2)(x-tN^{-\beta}z)| dx$$

$$\leqslant 2 \left(\int_{\mathbb{R}^2} |u(x)|^6 dx\right)^{1/3} \left(\int_{\mathbb{R}^2} |\nabla u(x-tN^{-\beta}z)|^2 dx\right)^{1/2} \left(\int_{\mathbb{R}^2} |u(x-tN^{-\beta}z)|^6 dx\right)^{1/6}$$

$$= 2\|\nabla u\|_{L^2(\mathbb{R}^2)} \|u\|_{L^6(\mathbb{R}^2)}^3.$$

Thus

$$|\mathcal{E}_{\Omega,a_N,N}^{\mathrm{H}}(u) - \mathcal{E}_{\Omega,a_N}^{\mathrm{GP}}(u)| = \frac{a_N}{2} \left|\iint_{\mathbb{R}^2 \times \mathbb{R}^2} |u(x)|^2 w_N(x-y) - \int_{\mathbb{R}^2} |u(x)|^4 dx\right|$$

$$= \frac{a_N}{2} \left|\iint_{\mathbb{R}^2 \times \mathbb{R}^2} |u(x)|^2 w(z) \left(\int_0^1 (\nabla|u|^2)(x-tN^{-\beta}z) \cdot (N^{-\beta}z) dt\right) dx dz\right|$$

$$\leqslant a_N N^{-\beta} \left(\int_{\mathbb{R}^2} |w(z)| |z| dz\right) \|\nabla u\|_{L^2(\mathbb{R}^2)} \|u\|_{L^6(\mathbb{R}^2)}^3$$

$$\leqslant C N^{-\beta} \|\nabla u\|_{L^2(\mathbb{R}^2)} \|u\|_{L^6(\mathbb{R}^2)}^3.$$

Now choosing the trial state Q_N as in (2.8), we find that

$$\mathcal{E}_{\Omega,a_N,N}^{\mathrm{H}}(Q_N) \leqslant \mathcal{E}_{\Omega,a_N}^{\mathrm{GP}}(Q_N) + C N^{-\beta} \|\nabla Q_N\|_{L^2} \|Q_N\|_{L^6}^3$$

$$= \sqrt{a_* - a_N} \left(\frac{2\lambda^2}{a_*} + C N^{-\beta}(a_* - a_N)^{-\frac{5}{4}}\right).$$

The error term goes to zero when $a_* - a_N \geqslant N^{-\alpha}$ with $\alpha < 4\beta/5$. We have proved the upper bound

$$\boxed{E_{\Omega,a_N}^{\mathrm{Q}}(N) \leqslant E_\Omega^{\mathrm{GP}}(a_N) \left(1 + C N^{-\beta}(a_* - a_N)^{-\frac{5}{4}}\right)}$$

which concludes the proof of Proposition 4.1. $\square$

Step 2. Convergence of reduced density matrices. This is a Feynman–Hellmann-type argument. Let $\eta > 0$ be a small parameter to be fixed later on and A be a bounded self-adjoint operator on $L^2(\mathbb{R}^2)$. Consider the perturbed Hamiltonian

$$H_{N,\eta} = \sum_{j=1}^{N} \left(-\Delta_{x_j} + |x_j|^2 - 2\Omega L_{x_j} + \eta A_j\right) - \frac{a}{N-1} \sum_{1 \leqslant i < j \leqslant N} w_N(x_i - x_j)$$

$$(4.7)$$

with ground-state energy per particle denoted E_η^Q hereafter. Introduce the associated Gross–Pitaevskii energy functional (we drop some Ω, a subscripts during this proof, for lightness of notation)

$$\mathcal{E}_\eta^{\mathrm{GP}}(u) = \int_{\mathbb{R}^2} |\nabla u(x)|^2 \, dx + \int_{\mathbb{R}^2} |x|^2 |u(x)|^2 \, dx - 2\Omega\langle u, Lu \rangle + \eta\langle u, Au \rangle - \frac{a}{2} \int_{\mathbb{R}^2} |u(x)|^4 \, dx.$$

$$(4.8)$$

In what follows, we denote by u_η a ground state for the latter and $E_\eta^{\mathrm{GP}} = \mathcal{E}_\eta^{\mathrm{GP}}(u_\eta)$ the corresponding ground-state energy.

Let Ψ_N be a ground state for $H_N = H_{N,0}$ and $\gamma_{\Psi_N}^{(1)}$ its one-body reduced density matrix. We write

$$\begin{aligned}
\eta \operatorname{Tr}\left[A\gamma_{\Psi_N}^{(1)}\right] &= N^{-1}\left(\langle \Psi_N | H_{N,\eta} | \Psi_N \rangle - \langle \Psi_N | H_{N,0} | \Psi_N \rangle\right) \\
&\geqslant E_\eta^Q - E_0^Q \\
&\geqslant E_\eta^{\mathrm{GP}} - E_0^{\mathrm{GP}} + O(N^{2\beta-1}) + O(N^{3\alpha/4-\beta}) \\
&\geqslant \mathcal{E}_\eta^{\mathrm{GP}}(u_\eta) - \mathcal{E}_0^{\mathrm{GP}}(u_\eta) + O(N^{2\beta-1}) + O(N^{3\alpha/4-\beta}) \\
&= \eta\langle u_\eta | A | u_\eta \rangle + O(N^{2\beta-1}) + O(N^{3\alpha/4-\beta}).
\end{aligned}$$

$$(4.9)$$

The first inequality is the variational principle, the second uses the estimates of the previous step. In that regard, observe that the energy lower bound applies *mutatis mutandis* to the problem perturbed by ηA. The error term in (4.3) solely comes from applying the Lévy-Leblond method to the interaction as in [41, Section 3]. It is therefore independent of the one-body term (in particular, of η and A). Lemma 4.2 generalizes to the perturbed functional, for the only property of the one-body energy used in its proof is its linearity in the density matrix. The third inequality in (4.9) is the variational principle again.

Under the assumption that

$$\alpha < \min\left(\frac{4\beta}{5}, 2(1-2\beta)\right)$$

one can pick some $\eta = \eta_N \to 0$ as $N \to \infty$, such that

$$\eta^{-1} N^{2\beta-1} + \eta^{-1} N^{3\alpha/4-\beta} \xrightarrow[N\to\infty]{} 0$$

and also

$$\eta = o(E^{GP}) = o(\sqrt{a_* - a_N}) = o(N^{-\alpha/2}).$$

Then, dividing (4.9) by η and repeating the argument with A changed to $-A$ yields

$$\langle u_\eta | A | u_\eta \rangle + o(1) \leqslant \mathrm{Tr}\left[A \gamma_{\Psi_N}^{(1)} \right] \leqslant \langle u_{-\eta} | A | u_{-\eta} \rangle + o(1). \qquad (4.10)$$

On the other hand, with the above choice of η, since

$$\mathcal{E}_0^{\mathrm{GP}}(u_\eta) = \mathcal{E}_\eta^{\mathrm{GP}}(u_\eta) + O(\eta \|A\|) \leqslant \mathcal{E}_\eta^{\mathrm{GP}}(u_0) + O(\eta \|A\|) = E_0^{\mathrm{GP}} + O(\eta \|A\|),$$

it follows that (u_η) and $(u_{-\eta})$ are sequences of quasi-minimizers for $\mathcal{E}_0^{\mathrm{GP}}$. We may apply Theorem 2.1 to them, and thus both sequences satisfy (2.7). Combining with (4.10), we get, after a dilation of space variables, trace-class weak-$\star$ convergence of $\gamma_{\Psi_N}^{(1)}$ to $|Q_N\rangle\langle Q_N|$. Since no mass is lost in the limit, the convergence must hold in trace-class norm, which gives (2.16) for $k = 1$.

To obtain (2.16) for $k > 1$, observe that, after dilation, $\gamma_{\Psi_N}^{(1)}$ converges in trace-class norm to a rank-one operator. It is well known that this implies convergence of higher order density matrices to tensor powers of the limiting operator (see, e.g., the discussion following [51, Theorem 7.1] or [42, Corollary 2.4]). $\qquad \square$

Remark 4.3. Note that we have been able to obtain the convergence of density matrices from that of the energy by a rather soft argument. What makes this possible is Lemma 4.2 and the uniqueness of (the limit of) the GP minimizer. Lemma 4.2 relies strongly on the fact that the interaction is attractive. For repulsive interactions,[2] the argument can be adapted provided the one-body part is positivity-preserving (hence, without rotation), using that the bosonic and boltzonic minimization problems then coincide [49, Theorem 3.3]. $\qquad \diamond$

Acknowledgements It is our pleasure to dedicate this paper to Herbert Spohn, on the occasion of his 70th birthday. This project has received funding from the European Research Council (ERC) under the European Union's Horizon 2020 research and innovation programme (grant agreements MDFT No 725528 and CORFRONMAT No 758620).

Appendix A. Extension to Anharmonic Potentials

The arguments given in this paper can be extended in various directions. One possibility is to consider anharmonic potentials. For completeness, we state here the corresponding result when the external potential is chosen in the form

$$V(x) = c_0 |x|^s$$

but we expect similar results when V has a unique minimizer and behaves like this in a neighborhood of this point, similarly to what was done in [30]. When $s \neq 2$ the limit $a \to a_*$ requires to have $\Omega = 0$. Although a stronger confinement $s > 2$ can control the rotating gas at infinity, it is not sufficient to control rotating effects

---

[2]There is no blow-up then, but one might want to adapt the method to obtain convergence of density matrices to the stable GP minimizers.

near the blow-up point. So we do not consider any rotation here. The many-particle Hamiltonian then takes the form

$$\tilde{H}_N = \sum_{j=1}^{N} \left(-\Delta_{x_j} + c_0 |x_j|^s\right) - \frac{a}{N-1} \sum_{1 \leqslant i < j \leqslant N} w_N(x_i - x_j). \tag{A.1}$$

The following can be proved by arguing exactly as we did for $s = 2$.

Theorem 4.4 (Collapse and condensation of the many-body ground state for anharmonic potentials). *Let $\Omega \equiv 0$, $s > 0$, $c_0 > 0$, $0 < \beta < 1/2$ and $a_N = a_* - N^{-\alpha}$ with*

$$0 < \alpha < \min \left\{ \frac{\beta(s+2)}{s+3}, \frac{(1-2\beta)(s+2)}{s} \right\}.$$

Let Ψ_N be the unique ground state of $\tilde{H}_N$. Then we have

$$\lim_{N \to \infty} \text{Tr} \left| \gamma_{\Psi_N}^{(k)} - |\tilde{Q}_N^{\otimes k}\rangle\langle \tilde{Q}_N^{\otimes k}| \right| = 0, \tag{A.2}$$

for all $k \in \mathbb{N}$, where $\tilde{Q}_N$ is the rescaled Gagliardo–Nirenberg optimizer given by

$$\tilde{Q}_N(x) = (a_*)^{-1/2} \tilde{\lambda}(a_* - a_N)^{-\frac{1}{2+s}} Q\left(\tilde{\lambda}(a_* - a_N)^{-\frac{1}{2+s}} x\right),$$

with

$$\tilde{\lambda} = \left(\frac{s}{2} c_0 \int_{\mathbb{R}^2} |x|^s |Q(x)|^2 dx\right)^{\frac{1}{2+s}}.$$

In addition, we have

$$\frac{\min \sigma(\tilde{H}_N)}{N} = (a_* - a_N)^{\frac{s}{s+2}} \left(\frac{\tilde{\lambda}^2}{a_*} \frac{s+2}{s} + o(1)\right). \tag{A.3}$$

The right side of (A.3) is of course the expansion of the Gross–Pitaevskii energy, which has already been derived in [30].

References

1. Aftalion, A.: Vortices in Bose–Einstein Condensates. Progress in Nonlinear Differential Equations and Their Applications, vol. 67. Springer, Berlin (2006)
2. Agmon, S.: Lectures on Exponential Decay of Solutions of Second-Order Elliptic Equations. Princeton University Press, Princeton (1982)
3. Anderson, M.H., Ensher, J.R., Matthews, M.R., Wieman, C.E., Cornell, E.A.: Observation of Bose-Einstein condensation in a dilute atomic vapor. Science **269**, 198–201 (1995)

4. Bao, W., Cai, Y.: Mathematical theory and numerical methods for Bose-Einstein condensation. Kinet. Relat. Models **6**, 1–135 (2013)
5. Baym, G., Pethick, C.J.: Ground-state properties of magnetically trapped Bose-condensed Rubidium gas. Phys. Rev. Lett. **76**, 6–9 (1996)
6. Bradley, C.C., Sackett, C.A., Tollett, J.J., Hulet, R.G.: Evidence of Bose-Einstein condensation in an atomic gas with attractive interactions. Phys. Rev. Lett. **75**, 1687–1690 (1995)
7. Caglioti, E., Lions, P.-L., Marchioro, C., Pulvirenti, M.: A special class of stationary flows for two-dimensional Euler equations: a statistical mechanics description. Commun. Math. Phys. **143**, 501–525 (1992)
8. Carles, R.: Critical nonlinear Schrödinger equations with and without harmonic potential. Math. Models Methods Appl. Sci. **12**, 1513–1523 (2002)
9. Carr, L.D., Clark, C.W.: Vortices in attractive Bose-Einstein condensates in two dimensions. Phys. Rev. Lett. **97**, 010403 (2006)
10. Chang, S.-M., Gustafson, S., Nakanishi, K., Tsai, T.-P.: Spectra of linearized operators for NLS solitary waves. SIAM J. Math. Anal. **39**, 1070–1111 (2008)
11. Chen, X., Holmer, J.: The rigorous derivation of the 2D cubic focusing NLS from quantum many-body evolution. Int. Math. Res. Not. IMRN, 4173–4216 (2017)
12. Chiribella, G.: On quantum estimation, quantum cloning and finite quantum de Finetti theorems. Theory of Quantum Computation, Communication, and Cryptography. Lecture Notes in Computer Science, vol. 6519. Springer, Berlin (2011)
13. Christandl, M., König, R., Mitchison, G., Renner, R.: One-and-a-half quantum de Finetti theorems. Commun. Math. Phys. **273**, 473–498 (2007)
14. Collin, A., Lundh, E., Suominen, K.-A.: Center-of-mass rotation and vortices in an attractive Bose gas. Phys. Rev. A **71**, 023613 (2005)
15. Cooper, N.R.: Rapidly rotating atomic gases. Adv. Phys. **57**, 539–616 (2008)
16. Cornell, E.A., Wieman, C.E.: Bose-Einstein condensation in a dilute gas, the first 70 years and some recent experiments. Rev. Mod. Phys. **74**, 875–893 (2002)
17. Correggi, M., Pinsker, F., Rougerie, N., Yngvason, J.: Rotating superfluids in anharmonic traps: from vortex lattices to giant vortices. Phys. Rev. A **84**, 053614 (2011)
18. Dalfovo, F., Stringari, S.: Bosons in anisotropic traps: ground state and vortices. Phys. Rev. A **53**, 2477–2485 (1996)
19. Davis, K.B., Mewes, M.O., Andrews, M.R., van Druten, N.J., Durfee, D.S., Kurn, D.M., Ketterle, W.: Bose-Einstein condensation in a gas of sodium atoms. Phys. Rev. Lett. **75**, 3969–3973 (1995)
20. de Finetti, B.: Funzione caratteristica di un fenomeno aleatorio. Atti della R. Accademia Nazionale dei Lincei, Ser. 6, Memorie, Classe di Scienze Fisiche, Matematiche e Naturali (1931)
21. de Finetti, B.: La prévision: ses lois logiques, ses sources subjectives. Ann. Inst. H. Poincaré **7**, 1–68 (1937)
22. Deng, Y., Guo, Y., Lu, L.: On the collapse and concentration of Bose-Einstein condensates with inhomogeneous attractive interactions. Calc. Var. Partial Differ. Equ. **54**, 99–118 (2015)
23. Diaconis, P., Freedman, D.: Finite exchangeable sequences. Ann. Probab. **8**, 745–764 (1980)
24. Donley, E.A., Claussen, N.R., Cornish, S.L., Roberts, J.L., Cornell, E.A., Wieman, C.E.: Dynamics of collapsing and exploding Bose-Einstein condensates. Nature **412**, 295–299 (2001)
25. Fannes, M., Vandenplas, C.: Finite size mean-field models. J. Phys. A **39**, 13843–13860 (2006)
26. Fannes, M., Spohn, H., Verbeure, A.: Equilibrium states for mean field models. J. Math. Phys. **21**, 355–358 (1980)
27. Fetter, A.: Rotating trapped Bose-Einstein condensates. Rev. Mod. Phys. **81**, 647 (2009)
28. Frank, R.L.: Ground states of semi-linear PDE. Lecture notes from the "Summerschool on Current Topics in Mathematical Physics", CIRM Marseille (2013)
29. Gerton, J.M., Strekalov, D., Prodan, I., Hulet, R.G.: Direct observation of growth and collapse of a Bose-Einstein condensate with attractive interactions. Nature **408**, 692–695 (2000)
30. Guo, Y., Seiringer, R.: On the mass concentration for Bose-Einstein condensates with attractive interactions. Lett. Math. Phys. **104**, 141–156 (2014)

31. Guo, Y., Zeng, X., Zhou, H.-S.: Energy estimates and symmetry breaking in attractive Bose-Einstein condensates with ring-shaped potentials. Ann. Inst. H. Poincaré Anal. Non Linéaire **33**, 809–828 (2016)
32. Harrow, A.: The church of the symmetric subspace (2013)
33. Hewitt, E., Savage, L.J.: Symmetric measures on Cartesian products. Trans. Am. Math. Soc. **80**, 470–501 (1955)
34. Hudson, R.L., Moody, G.R.: Locally normal symmetric states and an analogue of de Finetti's theorem. Z. Wahrscheinlichkeitstheor. und Verw. Gebiete **33**, 343–351 (1975/1976)
35. Jeblick, M., Pickl, P.: Derivation of the time dependent two dimensional focusing NLS equation (2017). arXiv:1707.06523
36. Ketterle, W.: When atoms behave as waves: Bose-Einstein condensation and the atom laser. Rev. Mod. Phys. **74**, 1131–1151 (2002)
37. Kiessling, M.K.-H.: Statistical mechanics of classical particles with logarithmic interactions. Commun. Pure Appl. Math. **46**, 27–56 (1993)
38. Kiessling, M.K.-H., Spohn, H.: A note on the eigenvalue density of random matrices. Commun. Math. Phys. **199**, 683–695 (1999)
39. König, R., Renner, R.: A de Finetti representation for finite symmetric quantum states. J. Math. Phys. **46**, 122108 (2005)
40. Lévy-Leblond, J.-M.: Nonsaturation of gravitational forces. J. Math. Phys. **10**, 806–812 (1969)
41. Lewin, M.: Mean-field limit of Bose systems: rigorous results. In: Proceedings of the International Congress of Mathematical Physics (2015)
42. Lewin, M., Nam, P.T., Rougerie, N.: Derivation of Hartree's theory for generic mean-field Bose systems. Adv. Math. **254**, 570–621 (2014)
43. Lewin, M., Nam, P.T., Rougerie, N.: Derivation of nonlinear Gibbs measures from many-body quantum mechanics. J. École Polytech. Math. **2**, 65–115 (2015)
44. Lewin, M., Nam, P.T., Rougerie, N.: Remarks on the quantum de Finetti theorem for bosonic systems. Appl. Math. Res. Express (AMRX), 48–63 (2015)
45. Lewin, M., Nam, P.T., Serfaty, S., Solovej, J.P.: Bogoliubov spectrum of interacting Bose gases. Commun. Pure Appl. Math. **68**, 413–471 (2015)
46. Lewin, M., Nam, P.T., Rougerie, N.: The mean-field approximation and the non-linear Schrödinger functional for trapped Bose gases. Trans. Am. Math. Soc. **368**, 6131–6157 (2016)
47. Lewin, M., Thành Nam, P., Rougerie, N.: A note on 2D focusing many-boson systems. Proc. Am. Math. Soc. **145**, 2441–2454 (2017)
48. Lieb, E.H., Loss, M.: Analysis. Graduate Studies in Mathematics, vol. 14, 2nd edn. American Mathematical Society, Providence (2001)
49. Lieb, E.H., Seiringer, R.: The Stability of Matter in Quantum Mechanics. Cambridge University Press, Cambridge (2010)
50. Lieb, E.H., Yau, H.-T.: The Chandrasekhar theory of stellar collapse as the limit of quantum mechanics. Commun. Math. Phys. **112**, 147–174 (1987)
51. Lieb, E.H., Seiringer, R., Solovej, J.P., Yngvason, J.: The Mathematics of the Bose Gas and Its Condensation. Oberwolfach Seminars. Birkhäuser, Basel (2005)
52. Lundh, E., Collin, A., Suominen, K.-A.: Rotational states of Bose gases with attractive interactions in anharmonic traps. Phys. Rev. Lett. **92**, 070401 (2004)
53. Maeda, M.: On the symmetry of the ground states of nonlinear Schrödinger equation with potential. Adv. Nonlinear Stud. **10**, 895–925 (2010)
54. McLeod, K.: Uniqueness of positive radial solutions of $\Delta u + f(u) = 0$ in R^n. II. Trans. Am. Math. Soc. **339**, 495–505 (1993)
55. Messer, J., Spohn, H.: Statistical mechanics of the isothermal Lane-Emden equation. J. Stat. Phys. **29**, 561–578 (1982)
56. Mottelson, B.: Yrast spectra of weakly interacting Bose-Einstein condensates. Phys. Rev. Lett. **83**, 2695–2698 (1999)
57. Mueller, E.J., Baym, G.: Finite-temperature collapse of a Bose gas with attractive interactions. Phys. Rev. A **62**, 053605 (2000)

58. Nam, P., Napiórkowski, M.: Norm approximation for many-body quantum dynamics: focusing case in low dimensions (2017). arXiv:1710.09684
59. Nam, P.T., Rougerie, N., Seiringer, R.: Ground states of large Bose systems: the Gross-Pitaevskii limit revisited (2015)
60. Pethick, C.J., Pitaevskii, L.P.: Criterion for Bose-Einstein condensation for particles in traps. Phys. Rev. A **62**, 033609 (2000)
61. Petz, D., Raggio, G.A., Verbeure, A.: Asymptotics of Varadhan-type and the Gibbs variational principle. Commun. Math. Phys. **121**, 271–282 (1989)
62. Phan, T.V.: Blow-up profile of Bose-Einstein condensate with singular potentials. J. Math. Phys. **58**, 072301, 10 (2017)
63. Raggio, G.A., Werner, R.F.: Quantum statistical mechanics of general mean field systems. Helv. Phys. Acta **62**, 980–1003 (1989)
64. Rougerie, N.: De Finetti theorems, mean-field limits and Bose-Einstein condensation (2015)
65. Rougerie, N.: Some contributions to many-body quantum mathematics. Habilitation thesis, Université de Grenoble-Alpes (2016). arXiv:1607.03833
66. Saito, H., Ueda, M.: Split-merge cycle, fragmented collapse, and vortex disintegration in rotating Bose-Einstein condensates with attractive interactions. Phys. Rev. A **69**, 013604 (2004)
67. Sakaguchi, H., Malomed, B.A.: Localized matter-wave patterns with attractive interaction in rotating potentials. Phys. Rev. A **78**, 063606 (2008)
68. Spohn, H.: On the Vlasov hierarchy. Math. Methods Appl. Sci. **3**, 445–455 (1981)
69. Størmer, E.: Symmetric states of infinite tensor products of C^*-algebras. J. Funct. Anal. **3**, 48–68 (1969)
70. Ueda, M., Leggett, A.J.: Macroscopic quantum tunneling of a Bose-Einstein condensate with attractive interaction. Phys. Rev. Lett. **80**, 1576–1579 (1998)
71. van den Berg, M., Lewis, J.T., Pulè, J.V.: The large deviation principle and some models of an interacting Boson gas. Commun. Math. Phys. **118**, 61–85 (1988)
72. Weinstein, M.I.: Nonlinear Schrödinger equations and sharp interpolation estimates. Commun. Math. Phys. **87**, 567–576 (1983)
73. Weinstein, M.I.: Modulational stability of ground states of nonlinear Schrödinger equations. SIAM J. Math. Anal. **16**, 472–491 (1985)
74. Werner, R.F.: Large deviations and mean-field quantum systems. Quantum Probability and Related Topics, QP-PQ, VII, pp. 349–381. World Scientific Publishing, River Edge (1992)
75. Wilkin, N.K., Gunn, J.M.F., Smith, R.A.: Do attractive Bosons condense? Phys. Rev. Lett. **80**, 2265 (1998)
76. Zhang, J.: Stability of attractive Bose-Einstein condensates. J. Stat. Phys. **101**, 731–746 (2000)
77. Zhang, J.: Sharp threshold for blowup and global existence in nonlinear Schrödinger equations under a harmonic potential. Commun. Partial Differ. Equ. **30**, 1429–1443 (2005)

Gross–Pitaevskii Evolution for Bose–Einstein Condensates

Benjamin Schlein

Abstract In these notes, we review the results of the paper Gross–Pitaevskii dynamics for Bose–Einstein condensates, a joint work with C. Brennecke. We consider the time evolution of initially trapped Bose–Einstein condensates, in a scaling limit where the interaction has a small scattering length. We show that the condensation is preserved by the many-body dynamics and that the evolution of the condensate wave function is governed by the nonlinear Gross–Pitaevskii equation. Compared with previous results, we prove here the convergence with optimal rates.

1 The Main Result

We are interested in the time evolution of Bose gases consisting of N particles interacting through a repulsive two-body potential with scattering length proportional to N^{-1}; this scaling limit is known as the Gross–Pitaevskii regime and it is often used to describe initially trapped Bose–Einstein condensates, like the ones that are typically produced in labs.

Particles are initially trapped by a confining external potential V_{ext}. The Hamilton operator acts on $L_s^2(\mathbb{R}^{3N})$, the subspace of $L^2(\mathbb{R}^{3N})$ consisting of functions that are symmetric with respect to permutations, and it has the form

$$H_N^{\text{trap}} = \sum_{j=1}^{N} \left[-\Delta_{x_j} + V_{\text{ext}}(x_j) \right] + \sum_{i<j}^{N} N^2 V(N(x_i - x_j)) \qquad (1.1)$$

where V is a nonnegative, spherically symmetric, short-range potential.

B. Schlein (✉)
Institute of Mathematics, University of Zurich, Winterthurerstrasse 190,
8057 Zurich, Switzerland
e-mail: Benjamin.Schlein@math.uzh.ch

© Springer Nature Switzerland AG 2018
D. Cadamuro et al. (eds.), *Macroscopic Limits of Quantum Systems*,
Springer Proceedings in Mathematics & Statistics 270,
https://doi.org/10.1007/978-3-030-01602-9_8

The interaction in (1.1) scales so that its scattering length is of the order N^{-1}. Recall here that the scattering length of a potential V is defined through the solution of the zero-energy scattering equation

$$\left[-\Delta + \frac{1}{2}V\right]f = 0$$

with the boundary condition $f(x) \to 1$, as $|x| \to \infty$. For sufficiently large $|x|$ (so that x lies outside the support of V), we must have

$$f(x) = 1 - \frac{a_0}{|x|}$$

for a constant a_0 which is known as the scattering length of V. Equivalently, the scattering length can be recovered through the identity

$$8\pi a_0 = \int V(x)f(x)dx. \tag{1.2}$$

Simple scaling then implies that

$$\left[-\Delta + \frac{N^2}{2}V(N.)\right]f(N.) = 0 \tag{1.3}$$

and therefore that a_0/N is the scattering length of the rescaled potential $N^2V(N.)$ appearing in (1.1).

At zero temperature, the Bose gas relaxes into the ground state of (1.1). From [23] it is known that the ground state energy E_N of (1.1) is given, to leading order in N, by

$$\lim_{N\to\infty} \frac{E_N}{N} = \min_{\substack{\varphi\in L^2(\mathbb{R}^3):\\ \|\varphi\|_2=1}} \mathcal{E}_{GP}(\varphi) \tag{1.4}$$

with the Gross–Pitaevskii energy functional

$$\mathcal{E}_{GP}(\varphi) = \int \left[|\nabla\varphi(x)|^2 + V_{ext}(x)|\varphi(x)|^2 + 4\pi a_0|\varphi(x)|^4\right]dx. \tag{1.5}$$

In [22], it was also shown that the ground state ψ_N of (1.1) exhibits complete Bose–Einstein condensation in the minimizer φ_0 of the Gross–Pitaevskii energy (1.5). In other words, if $\gamma_N = \text{tr}_{2,...,N} |\psi_N\rangle\langle\psi_N|$ denotes the one-particle reduced density associated with ψ_N, then, as $N \to \infty$,

$$\gamma_N \to |\varphi_0\rangle\langle\varphi_0| \tag{1.6}$$

in the trace norm topology. Equation (1.6) implies that, in the ground state of (1.1), all particles, up to a fraction vanishing as $N \to \infty$, occupy the same one-particle state φ_0. If particles are trapped in a box with fixed volume and the interaction potential is sufficiently small, the results of [22, 23] have been recently improved in [2, 3]. In [2] it was shown that in the ground state of (1.1) (and, more generally, in all low-energy states), the number of orthogonal excitations of the Bose–Einstein condensate remains finite, as $N \to \infty$ (in other words, the fraction of particles that are out of the condensate is proportional to N^{-1}, in the limit of large N). The results of [2] have then been used in [3] to compute the ground state energy of (1.1) up to errors vanishing as $N \to \infty$, showing, in particular, that the difference between E_N/N and the minimum of the Gross–Pitaevskii functional (given by $4\pi a_0$ in the translation invariant case) is of the order N^{-1}.

Notice that Bose–Einstein condensation, defined as (1.6), does not imply that the many-body wave function ψ_N is well approximated by the product $\varphi_0^{\otimes N}$. A simple computation shows that, to leading order in N,

$$\langle \varphi_0^{\otimes N}, H_N \varphi_0^{\otimes N} \rangle \simeq N \int \left[|\nabla \varphi_0(x)|^2 + V_{\text{ext}}(x)|\varphi_0(x)|^2 + \frac{\widehat{V}(0)}{2} |\varphi_0(x)|^4 \right] dx \quad (1.7)$$

which, by (1.2) (and since $f \leq 1$), is always larger than the ground state energy (1.4), with an excess energy of order N. As observed in [10, 12] and more recently in [6], correlations among particles play a fundamental role in the Gross–Pitaevskii regime; in particular, they are responsible for decreasing the energy to (1.4). It is only in the limit $N \to \infty$ that correlations, which vary on the short length scale N^{-1}, disappear and (1.6) holds true, in the sense of reduced densities.

We are interested in the time evolution of the Bose gas, resulting from a change of the external potential. To simplify notation and analysis, we consider the case in which, at time $t = 0$, the external traps are switched off and the evolution of the system is governed by the many-body Schrödinger equation

$$i\partial_t \psi_{N,t} = H_N \psi_{N,t} \quad (1.8)$$

with the translation invariant Hamilton operator

$$H_N = \sum_{j=1}^{N} -\Delta_{x_j} + \sum_{i<j}^{N} N^2 V(N(x_i - x_j)). \quad (1.9)$$

The next theorem is the main result of [4]; it describes the solution of (1.8), showing that, for every fixed $t \in \mathbb{R}$, $\psi_{N,t}$ still exhibits complete Bose–Einstein condensation and that the condensate wave function evolves according to the nonlinear time-dependent Gross–Pitaevskii equation. Moreover, it gives explicit and optimal bounds on the number of excitations of the condensate.

Theorem 1.1. *Let $V_{ext} : \mathbb{R}^3 \to \mathbb{R}$ be locally bounded, with $V_{ext}(x) \to \infty$, as $|x| \to \infty$. Let $V \in L^3(\mathbb{R}^3)$ be nonnegative, compactly supported and spherically symmetric.*

Let $\psi_N \in L^2_s(\mathbb{R}^{3N})$ be a normalized sequence, with one-particle reduced density $\gamma_N = \text{tr}_{2,\ldots,N} |\psi_N\rangle\langle\psi_N|$ (normalized so that $\text{tr}\,\gamma_N = 1$). Assume that, as $N \to \infty$,

$$
\begin{aligned}
a_N &:= 1 - \langle\varphi_0, \gamma_N\varphi_0\rangle \to 0 \\
b_N &:= \frac{1}{N}\langle\psi_N, H_N^{trap}\psi_N\rangle - \mathcal{E}_{GP}(\varphi_0) \to 0
\end{aligned}
\tag{1.10}
$$

where φ_0 is the minimizer of the Gross–Pitaevskii energy functional (1.5). Let now $\psi_{N,t} = e^{-iH_N t}\psi_N$ be the solution of the Schrödinger equation (1.8) and let $\gamma_{N,t}$ denote its one-particle reduced density. Then, there are constants $C, c > 0$ such that

$$
1 - \langle\varphi_t, \gamma_{N,t}\varphi_t\rangle \leq C(a_N + b_N + N^{-1})\exp(c\exp(c|t|)) \tag{1.11}
$$

for all $t \in \mathbb{R}$. Here, φ_t denotes the solution of the time-dependent Gross–Pitaevskii equation

$$
i\partial_t\varphi_t = -\Delta\varphi_t + 8\pi a_0|\varphi_t|^2\varphi_t \tag{1.12}
$$

with initial data $\varphi_{t=0} = \varphi_0$.

Remarks:

(1) The bound (1.11) implies that

$$
\begin{aligned}
\text{tr}\,\big|\gamma_{N,t} - |\varphi_t\rangle\langle\varphi_t|\big| &\leq 2\|\gamma_{N,t} - |\varphi_t\rangle\langle\varphi_t|\|_{HS} \\
&\leq 2^{3/2}(1 - \langle\varphi_t, \gamma_{N,t}\varphi_t\rangle)^{1/2} \\
&\leq C\sqrt{a_N + b_N + N^{-1}}\exp(c\exp(c|t|))
\end{aligned}
$$

where the first inequality is a consequence of the fact that $|\varphi_t\rangle\langle\varphi_t|$ is a rank-one projection. Theorem 1.1 is therefore a statement about the stability of Bose–Einstein condensation with respect to the many-body evolution. Assuming condensation at time $t = 0$ (through the condition $a_N \to 0$), it proves condensation at any fixed time $t \in \mathbb{R}$.

(2) Observe that, to prove condensation at time $t \in \mathbb{R}$, we do not only assume condensation at time $t = 0$. Instead, we also need the energy condition $b_N \to 0$. In particular, this condition makes sure that the initial wave function ψ_N already contains the correct short-scale correlation structure, excluding for example, by (1.7), the choice $\psi_N = \varphi^{\otimes N}$.

(3) It follows from the results of [22, 23] that ψ_N can be chosen as the ground state of the trapped Hamiltonian (1.1). If the particles are initially confined in a box with a fixed volume and periodic boundary conditions and if the interaction is small enough, it even follows from [2, 3] that (1.10) is satisfied, with the optimal bounds $a_N, b_N \simeq N^{-1}$; in this case, (1.11) implies that

$$
1 - \langle\varphi_t, \gamma_{N,t}\varphi_t\rangle \leq CN^{-1}\exp(c\exp(c|t|)).
$$

This estimate implies that the solution $\psi_{N,t}$ of (1.8) exhibits condensation with only a bounded number of orthogonal excitations, uniformly in N.

(4) Theorem 1.1 is stated for the time evolution generated by the translation invariant Hamilton operator (1.9). The theorem and its proof can be easily generalized to describe the time evolution generated by Hamilton operators containing an external potential. Physically, this would correspond to experiments where the external traps are modified, at time $t = 0$, rather than switched off.

(5) Theorem 1.1 is designed to describe the evolution of Bose gases prepared in the ground state of a trapped Hamiltonian of the form (1.1). It is also possible to state the result differently, with no reference to an external potential. Given a normalized sequence $\psi_N \in L^2_s(\mathbb{R}^{3N})$ with

$$
\begin{aligned}
\widetilde{a}_N &= \operatorname{tr} \, |\gamma_N - |\varphi\rangle\langle\varphi|| \to 0 \\
\widetilde{b}_N &= \frac{1}{N}\langle\psi_N, H_N\psi_N\rangle - \int \left[|\nabla\varphi|^2 + 4\pi a_0|\varphi|^4\right] dx \to 0
\end{aligned}
\tag{1.13}
$$

for an arbitrary $\varphi \in L^2(\mathbb{R}^3)$, we find that the solution $\psi_{N,t} = e^{-iH_N t}\psi_N$ of (1.8) with initial data ψ_N is such that

$$
1 - \langle\varphi_t, \gamma_{N,t}\varphi_t\rangle \le C\left[\widetilde{a}_N + \widetilde{b}_N + N^{-1}\right]\exp(c\exp(c|t|))
\tag{1.14}
$$

for all $t \in \mathbb{R}$. Here, φ_t is the solution of the nonlinear Gross–Pitaevskii equation (1.12) with initial data $\varphi_{t=0} = \varphi$. Compared with Theorem 1.1, the only difference is that, if the initial condensate wave function $\varphi \in L^2(\mathbb{R}^3)$ is not the minimizer of (1.5) (for an appropriate external potential), then we need to assume condensation in a slightly stronger sense (because we always have $1 - \langle\varphi, \gamma_N\varphi\rangle \le \operatorname{tr}|\gamma_N - |\varphi\rangle\langle\varphi||$).

(6) While the bound (1.11) is optimal in its N-dependence, it is not expected to be optimal in its dependence on t. It is probably possible to improve the time dependence on the r.h.s. of (1.11) assuming sufficient decay of the solution of the nonlinear Gross–Pitaevskii equation (1.12) (this observation goes back to [19], in the mean-field setting). Another approach leading to better time dependence has been developed in [13, 14] for less singular interactions.

(7) Theorem 1.1 is not the first result establishing convergence of the many-body evolution toward the Gross–Pitaevskii dynamics. The first proof of convergence has been obtained in a series of papers [8, 9, 11, 12] that are based on the study of the BBGKY hierarchy introduced for mean-field systems in [26]. An important step of this derivation (the proof of the uniqueness of the infinite Gross–Pitaevskii hierarchy) has been recently simplified in [7], using also ideas from [18]. This approach does not provide control on the rate of the convergence. An alternative derivation of the Gross–Pitaevskii equation has been proposed in [24]; in this case, convergence is established with rate $N^{-\eta}$, for some unspecified $\eta > 0$. Control of the rate of convergence toward the Gross–Pitaevskii equation has been obtained in [1], for approximately coherent initial data on the bosonic Fock space,

extending techniques developed for mean-field interactions in [15, 16, 25]. In Theorem 1.1, we combine the analysis of [1], with techniques developed in [20] for the study of the time evolution of N-particle Bose–Einstein condensates in the mean- field regime; this allows us to get optimal rates of convergence, for the physically relevant class of N-particle initial data. Finally, let us remark that, recently, a derivation of the Gross–Pitaevskii equation for potential with some small negative part (in particular, so small that the scattering length remains positive) has been achieved in [17] (with no control on the rate of the convergence). For interactions with negative scattering length, the nonlinear evolution (1.12) may exhibit blow up in finite time, leading to the collapse of the condensate; in this case, there is, so far, no rigorous result in the Gross–Pitaevskii scaling.

2 Some Ideas from the Proof

The starting point of our analysis is the observation, due to [21], that, for a given $\varphi \in L^2(\mathbb{R}^3)$, every N-particle wave function $\psi_N \in L^2_s(\mathbb{R}^{3N})$ can be uniquely decomposed as

$$\psi_N = \alpha_0 \varphi^{\otimes N} + \alpha_1 \otimes_s \varphi^{\otimes(N-1)} + \cdots + \alpha_N \tag{2.1}$$

where $\otimes_s$ denotes the symmetrized tensor product and where, for $j = 1, \ldots, N$, $\alpha_j \in L^2_{\perp\varphi}(\mathbb{R}^3)^{\otimes_s j}$ is a j-particle wave function, orthogonal to φ in each one of its variables. Since, moreover,

$$\|\psi_N\|^2 = \sum_{j=0}^{N} \|\alpha_j\|_2^2$$

we can define a unitary map

$$U_\varphi : L^2_s(\mathbb{R}^{3N}) \to \mathcal{F}^{\leq N}_{\perp\varphi} = \bigoplus_{j=0}^{N} L^2_{\perp\varphi}(\mathbb{R}^3)^{\otimes_s j} \tag{2.2}$$

by setting $U_\varphi \psi_N = \{\alpha_0, \alpha_1, \ldots, \alpha_N\}$. In applications, φ is going to be the condensate wave function, while the sequence $\{\alpha_0, \ldots, \alpha_N\}$ will describe orthogonal excitations of the condensate.

Using the unitary map (2.2) we can construct a fluctuation dynamics. To prove Theorem 1.1, we want to compare the many-body evolution governed by (1.8) with the Gross–Pitaevskii equation (1.12). For technical reasons, it is convenient to compare first (1.8) with the modified, N-dependent, Gross–Pitaevskii equation given by

$$i\partial_t \varphi_{N,t} = -\Delta\varphi_{N,t} + \left[N^3 V(N.) f(N.) * |\varphi_{N,t}|^2\right]\varphi_{N,t} \tag{2.3}$$

again with the initial data $\varphi_{t=0} = \varphi_0$. At the end, using (1.2), it is possible to estimate the difference between the solution of (2.3) and the solution of the limiting Gross–Pitaevskii equation (1.12); we find that

$$\|\varphi_t - \varphi_{N,t}\|_2 \leq C N^{-1} \exp(c|t|). \tag{2.4}$$

To compare the many-body evolution (1.8) and the modified Gross–Pitaevskii dynamics, we can proceed as follows. From the assumption (1.10), the initial wave function ψ_N exhibits condensation in the one-particle state φ_0, minimizing (1.5). Defining the initial excitation vector $\theta_N \in \mathcal{F}_{\perp\varphi_0}^{\leq N}$ by

$$\theta_N = U_{\varphi_0}\psi_N$$

we find $\psi_N = U_{\varphi_0}^*\theta_N$. Similarly, at time $t \in \mathbb{R}$, we expect $\psi_{N,t}$ to exhibit condensation in $\varphi_{N,t}$. Hence, we write $\psi_{N,t} = U_{\varphi_{N,t}}^*\theta_{N,t}$, with

$$\theta_{N,t} = U_{\varphi_{N,t}}\psi_{N,t} = U_{\varphi_{N,t}}e^{-iH_Nt}\psi_N = U_{\varphi_{N,t}}e^{-iH_Nt}U_{\varphi_0}^*\theta_N = S_N(t;0)\theta_N \tag{2.5}$$

where we defined the fluctuation dynamics

$$S_N(t;s) = U_{\varphi_{N,t}}e^{-iH_N(t-s)}U_{\varphi_{N,s}}^*.$$

A simple computation shows that

$$
\begin{aligned}
1 - \langle \varphi_{N,t}\gamma_{N,t}\varphi_{N,t}\rangle &= 1 - \frac{1}{N}\langle \psi_{N,t}, a^*(\varphi_{N,t})a(\varphi_{N,t})\psi_{N,t}\rangle \\
&= 1 - \frac{1}{N}\langle \theta_{N,t}, U_{\varphi_{N,t}}a^*(\varphi_{N,t})a(\varphi_{N,t})U_{\varphi_{N,t}}^*\theta_{N,t}\rangle \\
&= \frac{1}{N}\langle \theta_{N,t}, \mathcal{N}\theta_{N,t}\rangle = \frac{1}{N}\langle \theta_N, S_N^*(t;0)\mathcal{N}S_N(t;0)\theta_N\rangle
\end{aligned}
\tag{2.6}
$$

where, for $f \in L^2(\mathbb{R}^3)$, $a^*(f), a(f)$ are the usual creation and annihilation operators satisfying canonical commutation relations and where we used the identity

$$U_{\varphi_{N,t}}a^*(\varphi_{N,t})a(\varphi_{N,t})U_{\varphi_{N,t}}^* = N - \mathcal{N}.$$

Equation (2.6) shows that Bose–Einstein condensation and the bound (1.11) in Theorem 1.1 would follow from an estimate on the number of particles in the excitation vector $\theta_{N,t}$. At time $t = 0$, such a bound holds because of the assumption $a_N \to 0$ in (1.10); hence, to prove Theorem 1.1, it would be enough to control the growth of the number of particles with respect to the fluctuation dynamics S_N.

While this approach works well in the mathematically simpler mean-field regime, see [20], it does not work in the Gross–Pitaevskii limit that we consider here. From $\psi_{N,t} = U_{\varphi_{N,t}}^*\theta_{N,t}$ and since $U_{\varphi_{N,t}}^*$ only generates product wave functions, correlations among particles that are so important in the Gross–Pitaevskii regime are included, by definition, in the excitation vector $\theta_{N,t}$. While this does not mean that the expectation of the number of particles in the state $\theta_{N,t}$ is large (by (2.6), this would contradict Bose–Einstein condensation and Theorem 1.1), it does imply that the vector $\theta_{N,t}$, as defined in (2.5), carries a large energy. This observation prevents us from using the energy to control the growth of $\mathcal{N}$ with respect to the fluctuation dynamics $\mathcal{S}_N$, which seems to make the problem very difficult.

To circumvent this issue, we are going to define a new excitation vector $\xi_{N,t}$, extracting from $\theta_{N,t}$ the correlations that make its energy large. To reach this goal, we would like to follow [1] and use Bogoliubov transformations to implement the desired correlation structure. For a kernel $k \in L^2(\mathbb{R}^3 \times \mathbb{R}^3)$, we define the corresponding Bogoliubov transformation

$$\widetilde{B}(k) = \exp\left(\frac{1}{2}\int dxdy \, \left[k(x;y)a_x^*a_y^* - \bar{k}(x;y)a_xa_y\right]\right) \qquad (2.7)$$

where a_x^*, a_x denote the usual operator valued distributions, formally creating and, respectively, annihilating, a particle at position $x \in \mathbb{R}^3$.

A nice feature of Bogoliubov transformations is the fact that their action on creation and annihilation operators is explicit. For any $f \in L^2(\mathbb{R}^3)$, we find that

$$\widetilde{B}(k)^*a(f)\widetilde{B}(k) = a(\cosh_k(f)) + a^*(\sinh_k(\bar{f})) \qquad (2.8)$$

where the operators $\cosh_k$ and $\sinh_k$ are defined through the convergent series

$$\cosh_k = \sum_{n\geq 0} \frac{(k\bar{k})^n}{(2n)!}, \qquad \sinh_k = \sum_{n\geq 0} \frac{(k\bar{k})^n k}{(2n+1)!}.$$

With a slight abuse of notation, we denote here by k the Hilbert–Schmidt operator with integral kernel given by k.

Unfortunately, while Bogoliubov transformations work well for approximately coherent Fock space initial data (where the number of particles is allowed to fluctuate), they are more difficult to use when dealing with N-particle initial data (for an exception, where Bogoliubov transformations are used to study the evolution of N-particle states, see [5]). The problem is that, even if the kernel k is chosen to be orthogonal to $\varphi_{N,t}$ in both its variables, the operator $\widetilde{B}(k)$ does not leave the space $\mathcal{F}_{\perp\varphi_{N,t}}^{\leq N}$ invariant, because it does not respect the truncation on the number of particles.

To solve this issue, we introduce modified field operators on $\mathcal{F}_{\perp\varphi_{N,t}}^{\leq N}$; for $f \in L^2_{\perp\varphi_{N,t}}(\mathbb{R}^3)$, we define

$$b^*(f) = a^*(f)\sqrt{\frac{N-\mathcal{N}}{N}}, \qquad b(f) = \sqrt{\frac{N-\mathcal{N}}{N}}a(f).$$

From

$$U_{\varphi_{N,t}}^* b^*(f) U_{\varphi_{N,t}} = a^*(f)\frac{a(\varphi_{N,t})}{\sqrt{N}}, \qquad U_{\varphi_{N,t}}^* b(f) U_{\varphi_{N,t}} = \frac{a^*(\varphi_{N,t})}{\sqrt{N}}a(f)$$

we understand the action of the modified creation and annihilation operators; $b^*(f)$ creates a particle with wave function $f \perp \varphi_{N,t}$ and, at the same time, it annihilates a particle from the condensate. Similarly, $b(f)$ annihilate a particle with wave function f and it creates a particle in the condensate. In other words, the operators $b^*(f)$ and $b(f)$ create and, respectively, annihilate excitations, preserving, however, the total number of particles. This is the reason why, in contrast with the operators $a^*(f)$ and $a(f)$, they leave the space $\mathcal{F}_{\perp\varphi_{N,t}}^{\leq N}$ invariant. It is also useful to observe that, when acting on states exhibiting Bose–Einstein condensation in the one-particle state $\varphi_{N,t}$, we have $a^*(\varphi_{N,t}), a(\varphi_{N,t}) \simeq \sqrt{N}$ and thus $b^*(f) \simeq a^*(f)$ and $b(f) \simeq a(f)$. In this sense, the modified creation and annihilation operators $b^*(f), b(f)$ are the correct replacement for $a^*(f), a(f)$ on the excitation space $\mathcal{F}_{\perp\varphi_{N,t}}^{\leq N}$.

Using the modified field operators, we can define generalized Bogoliubov transformations. Given a kernel $k \in L^2(\mathbb{R}^3 \times \mathbb{R}^3)$, orthogonal to $\varphi_{N,t}$ in both its arguments, we define

$$B(k) = \exp\left(\frac{1}{2}\int dxdy \, [k(x; y)b_x^* b_y^* - \text{h.c.}]\right)$$

where b_x^*, b_x are the operator valued distributions associated with the modified fields b^*, b. Then, $B(k)$ is a unitary operator, mapping $\mathcal{F}_{\perp\varphi_{N,t}}^{\leq N}$ back into itself. The price that we have to pay for working with generalized Bogoliubov transformation of the form $B(k)$, rather than with the standard Bogoliubov transformations having the form (2.7), is the absence of explicit formula describing the action on generalized creation and annihilation operators. It turns out, however, that the simple relation (2.8) can be replaced, for an arbitrary $f \in L_{\perp\varphi_{N,t}}^2(\mathbb{R}^3)$, by the identity

$$B(k)^* b(f) B(k) = b(\cosh_k(f)) + b^*(\sinh_k(\bar{f})) + d_k(f)$$

where the remainder operator $d_k(f)$ is small on states exhibiting condensation in $\varphi_{N,t}$, in the sense that, if $\|k\|_2$ is sufficiently small, there exists $C > 0$ such that

$$\|d_k(f)\xi\|, \|d_k^*(f)\psi\| \leq \frac{C}{N}\|f\|_2\|(\mathcal{N}+1)^{3/2}\xi\| \qquad (2.9)$$

for all $\xi \in \mathcal{F}_{\perp\varphi_{N,t}}^{\leq N}$. More complicated bounds are also available for products of remainder operators.

Now, we use the generalized Bogoliubov transformation to implement correlations and to obtain a better approximation for the many-body evolution. We fix $\ell > 0$ (independently of N) and we consider the solution of the Neumann ground state problem

$$\left[-\Delta + \frac{N^2}{2} V(N.) \right] f_{N,\ell} = \lambda_{N,\ell} f_{N,\ell}$$

on the ball $|x| \le \ell$. We impose the normalization $f_{N,\ell}(x) = 1$ for $|x| = \ell$ and we extend then $f_{N,\ell}(x) = 1$ for all $|x| \ge \ell$. Notice that, since ℓ is of order one and the potential has range of the order N^{-1}, $f_{N,\ell}(x)$ is close, for $|x| \le \ell$, to the solution of the zero-energy scattering equation (1.3); in other words,

$$f_{N,\ell}(x) \simeq 1 - \frac{a_0}{|x|} \tag{2.10}$$

for $|x| \le \ell$ outside the range of the potential $N^2 V(N.)$ (for our purposes, it is important to cut off correlations at finite distances $\ell > 0$; this is why we cannot work directly with the solution of (1.3)).

We define $w_{N,\ell} = 1 - f_{N,\ell}$ and, for any $t \in \mathbb{R}$, we introduce the kernel

$$\eta_t(x; y) = -N w_{N,\ell}(x - y) \varphi_{N,t}(x) \varphi_{N,t}(y).$$

According to the approximation (2.10), we have

$$\eta_t(x; y) \simeq -\frac{a_0}{|x - y|} \varphi_{N,t}(x) \varphi_{N,t}(y) \tag{2.11}$$

for $x, y \in \mathbb{R}^3$ with $CN^{-1} \le |x - y| \le \ell$ (for an appropriate constant C, depending on the radius of the support of V). We can also project η_t orthogonally to $\varphi_{N,t}$, defining

$$k_t = (1 - |\varphi_{N,t}\rangle\langle\varphi_{N,t}|) \otimes (1 - |\varphi_{N,t}\rangle\langle\varphi_{N,t}|) \eta_t.$$

It follows from (2.11) that $k_t \in L^2(\mathbb{R}^3 \times \mathbb{R}^3)$ with $\|k_t\|_2 \le C$, uniformly in N (choosing $\ell > 0$ sufficiently small, we can make sure that $\|k_t\|_2$ is so small that (2.9) holds true). On the other hand, (2.11) also implies that the H^1-norm of k_t diverges, as $N \to \infty$; we find that $\|k_t\|_{H^1} \simeq N^{1/2}$. It follows that conjugation with the generalized Bogoliubov transformation $B(k_t)$ only increase the number of particles by a quantity of order one, independent of N, while it increases the energy by order N. In other words, the Bogoliubov transformation $B(k_t)$ generates finitely many particles, carrying, however, a macroscopically large energy of order N.

Given a sequence of initial wave functions ψ_N satisfying the assumptions (1.10), we define now a modified excitation vector

$$\xi_N = B(k_0) U_{\varphi_0} \psi_N \in \mathcal{F}_{\perp\varphi_0}^{\le N}$$

so that $\psi_N = U_{\varphi_0}^* B(k_0)^* \xi_N$. Similarly, we write the solution of the Schrödinger equation (1.8) as $\psi_{N,t} = U_{\varphi_{N,t}}^* B(k_t)^* \xi_{N,t}$ with the modified excitation vector

$$\xi_{N,t} = B(k_t)U_{\varphi_{N,t}}\psi_{N,t}$$
$$= B(k_t)U_{\varphi_{N,t}}e^{-iH_Nt}\psi_N$$
$$= B(k_t)U_{\varphi_{N,t}}e^{-iH_Nt}U_{\varphi_0}^*B(k_0)^*\xi_N$$
$$= \mathcal{W}_N(t;0)\xi_N .$$

Here, we introduced the modified fluctuation dynamics

$$\mathcal{W}(t;s) = B(k_t)^*U_{\varphi_{N,t}}e^{-iH_N(t-s)}U_{\varphi_{N,s}}^*B(k_s) : \mathcal{F}_{\perp\varphi_{N,s}}^{\leq N} \rightarrow \mathcal{F}_{\perp\varphi_{N,t}}^{\leq N} .$$

Arguing as in (2.6), and using the fact that the Bogoliubov transformation $B(k_t)$ does not change the number of particles substantially (meaning that $B(k_t)\mathcal{N}B(k_t)^* \leq C(\mathcal{N}+1)$ for a constant $C > 0$ independent of N), we obtain

$$1 - \langle\varphi_{N,t}, \gamma_{N,t}\varphi_{N,t}\rangle \leq \frac{C}{N}\langle\xi_{N,t}, \mathcal{N}\xi_{N,t}\rangle = \frac{C}{N}\langle\xi_N, \mathcal{W}_N^*(t;0)\mathcal{N}\mathcal{W}_N(t;0)\xi_N\rangle .$$

$$(2.12)$$

Hence, Theorem 1.1 follows, if we can control the growth of the number of particles with respect to the modified fluctuation dynamics $\mathcal{W}_N$. Compared with (2.6), however, (2.12) puts us in a better position, because we have more information on the initial modified fluctuation vector ξ_N than we previously had on θ_N. In fact, while the assumption $a_N \to 0$ gives a bound for the number of particles in ξ_N as it did for θ_N, the assumption $b_N \to 0$ also gives us a bound on the energy of ξ_N. This means that now, to control the growth of the number of particles with respect to the dynamics $\mathcal{W}_N$, we can also use the energy, and that is exactly what we do.

More precisely, we consider the generator

$$\mathcal{G}_{N,t} = \left[i\partial_t B(k_t)^*\right]B(k_t) + B(k_t)^*\left(\left[i\partial_t U_{\varphi_{N,t}}\right]U_{\varphi_{N,t}}^* + U_{\varphi_{N,t}}H_N U_{\varphi_{N,t}}^*\right)B(k_t)$$

of the fluctuation dynamics $\mathcal{W}_N$, which is defined so that

$$i\partial_t \mathcal{W}_N(t;s) = \mathcal{G}_{N,t}\mathcal{W}_N(t;s) .$$

The main tool to control the growth of the number of particles are the following estimates for the generator $\mathcal{G}_{N,t}$. For an appropriate (and explicit) choice of $C_{N,t}$ depending on N and t, there are constants $C, c > 0$ such that, as operator inequalities on $\mathcal{F}_{\perp\varphi_{N,t}}^{\leq N}$, we have

$$\frac{1}{2} - Ce^{c|t|}(\mathcal{N}+1) \leq \mathcal{G}_{N,t} - C_{N,t} \leq 2\mathcal{H}_N + Ce^{c|t|}(\mathcal{N}+1)$$
$$\pm\left[i\mathcal{N}, \mathcal{G}_{N,t}\right] \leq \mathcal{H}_N + Ce^{c|t|}(\mathcal{N}+1) \qquad (2.13)$$
$$\pm\partial_t(\mathcal{G}_{N,t} - C_{N,t}) \leq \mathcal{H}_N + Ce^{c|t|}(\mathcal{N}+1)$$

with the Hamilton operator

$$\mathcal{H}_N = \int dx \, \nabla_x a_x^* \nabla_x a_x + \frac{1}{2} \int dx dy \, N^2 V(N(x-y)) a_x^* a_y^* a_y a_x \, .$$

The proof of (2.13) is the main step in the analysis of [4]. It is based on a precise computation of the generator $\mathcal{G}_{N,t}$ (bounds of the form (2.9) allow us to control all error terms that are not explicit) and it crucially makes use of the choice of the kernel k_t.

Using the bounds (2.13), it is easy to conclude that, for an appropriate choice of $C_0, c > 0$, there exist $C > 0$ such that

$$\left| \frac{d}{dt} \langle \mathcal{W}(t;0)\xi_N, \left[(\mathcal{G}_{N,t} - C_{N,t}) + C_0 e^{c|t|}(\mathcal{N}+1) \right] \mathcal{W}(t;0)\xi_N \rangle \right|$$
$$\leq C \langle \mathcal{W}(t;0)\xi_N, \left[(\mathcal{G}_{N,t} - C_{N,t}) + C_0 e^{c|t|}(\mathcal{N}+1) \right] \mathcal{W}(t;0)\xi_N \rangle \, .$$

With Gronwall's inequality, we conclude that

$$\langle \mathcal{W}(t;0)\xi_N, \mathcal{N}\mathcal{W}(t;0)\xi_N \rangle \leq C \langle \xi_N, \left[(\mathcal{G}_N - C_{N,0}) + (\mathcal{N}+1) \right] \xi_N \rangle \, . \tag{2.14}$$

It is important to stress the fact that, to close Gronwall's argument, it is not enough to consider the number of particles operator and that, instead, we have to include also the energy term $\mathcal{G}_{N,t} - C_{N,t}$ (because, as it appears from (2.13), it is impossible for us to bound the commutator $[i\mathcal{N}, \mathcal{G}_{N,t}]$ just by $\mathcal{N}$).

To obtain the desired estimate for the growth of the number of particles operator with respect to the fluctuation dynamics $\mathcal{W}_N$, we still have to control the right-hand side of (2.14). This is where the assumptions (1.10) on the initial data are used. Since the Bogoliubov transformation $B(k_0)$ does not change the number of particles substantially, we find

$$\begin{aligned}
\langle \xi_N, \mathcal{N}\xi_N \rangle &= \langle B(k_0) U_{\varphi_0} \psi_N, \mathcal{N} B(k_0) U_{\varphi_0} \psi_N \rangle \\
&\leq C \left[\langle U_{\varphi_0} \psi_N, \mathcal{N} U_{\varphi_0} \psi_N \rangle + 1 \right] \\
&= C \left[N - \langle \psi_N, a^*(\varphi_0) a(\varphi_0) \psi_N \rangle + 1 \right] \\
&= CN(1 - \langle \varphi_0, \gamma_N \varphi_0 \rangle) + C = CN a_N + C \, .
\end{aligned} \tag{2.15}$$

On the other hand, the expectation of the energy term $\mathcal{G}_N - C_{N,0}$ can be bounded using the assumption $b_N \to 0$ in (1.10). Combined again with $a_N \to 0$ and with the fact that φ_0 is the minimizer of the Gross–Pitaevskii energy functional (1.5), we find

$$\langle \xi_N, (\mathcal{G}_{N,0} - C_{N,0})\xi_N \rangle \leq C \left[N a_N + N b_N + 1 \right] \, . \tag{2.16}$$

Inserting (2.15), (2.16) into (2.14), we conclude that

$$\langle \mathcal{W}(t;0)\xi_N, \mathcal{N}\mathcal{W}(t;0)\xi_N \rangle \leq C \left[N a_N + N b_N + 1 \right] \, .$$

Arguing similarly as in (2.15), this implies that

$$1 - \langle \varphi_{N,t}, \gamma_{N,t} \varphi_{N,t} \rangle \leq C \left[a_N + b_N + N^{-1} \right].$$

Finally, using the estimate (2.4), we can replace $\varphi_{N,t}$ with the solution of the limiting Gross–Pitaevskii equation. We obtain

$$1 - \langle \varphi_t, \gamma_{N,t} \varphi_t \rangle \leq C \left[a_N + b_N + N^{-1} \right]$$

which concludes the proof of the main theorem. As for Remark 5 after Theorem 1.1, if we do not assume the initial data of the Gross–Pitaevskii equation to minimize the functional (1.5), we have to replace (2.16) with the weaker estimate

$$\langle \xi_N, (\mathcal{G}_{N,0} - C_{N,0}) \xi_N \rangle \leq C \left[N \, \widetilde{a}_N + N \, \widetilde{b}_N + 1 \right]$$

with $\widetilde{a}_N, \widetilde{b}_N$ as defined in (1.13); this leads to (1.14). For further details, we refer to [4].

Acknowledgements The author gratefully acknowledges support from the NCCR SwissMAP and from the Swiss National Foundation of Science through the SNF Grant "Dynamical and energetic properties of Bose–Einstein condensates".

References

1. Benedikter, N., de Oliveira, G., Schlein, B.: Quantitative derivation of the Gross-Pitaevskii equation. Commun. Pure Appl. Math. (2014)
2. Boccato, C., Brennecke, C., Cenatiempo, S., Schlein, B.: Complete Bose-Einstein condensation in the Gross-Pitaevskii regime. To appear in Commun. Math. Phys. arXiv:1703.04452
3. Boccato, C., Brennecke, C., Cenatiempo, S., Schlein, B.: Bogoliubov theory in the Gross-Pitaevskii limit. arXiv:1801.01389
4. Brennecke, C., Schlein, B.: Gross-Pitaevskii dynamics for Bose-Einstein condensates. arXiv:1702.05625
5. Brennecke, C., Nam, P.T., Napiórkowski, M., Schlein, B.: Fluctuations of N-particle quantum dynamics around the nonlinear Schrödinger equation. arXiv:1710.09743
6. Chen, X., Holmer, J.: Correlation structures, many-body scattering processes and the derivation of the Gross-Pitaevskii hierarchy. arXiv:1409.1425
7. Chen, T., Hainzl, C., Pavlović, N., Seiringer, R.: Unconditional uniqueness for the cubic Gross-Pitaevskii hierarchy via quantum de Finetti. arXiv:1307.3168
8. Erdős, L., Schlein, B., Yau, H.-T.: Derivation of the cubic nonlinear Schrödinger equation from quantum dynamics of many-body systems. Invent. Math. **167**, 515–614 (2006)
9. Erdős, L., Schlein, B., Yau, H.-T.: Rigorous derivation of the Gross-Pitaevskii equation. Phys. Rev. Lett. **98**(4), 040404 (2007)
10. Erdős, L., Michelangeli, A., Schlein, B.: Dynamical formation of correlations in a Bose-Einstein condensate. Commun. Math. Phys. **289**(3), 1171–1210 (2009)
11. Erdős, L., Schlein, B., Yau, H.-T.: Rigorous derivation of the Gross-Pitaevskii equation with a large interaction potential. J. Am. Math. Soc. **22**, 1099–1156 (2009)
12. Erdős, L., Schlein, B., Yau, H.-T.: Derivation of the Gross-Pitaevskii equation for the dynamics of Bose-Einstein condensate. Ann. Math. (2) **172**(1), 291–370 (2010)

13. Grillakis, M., Machedon, M.: Pair excitations and the mean field approximation of interacting Bosons, I. Commun. Math. Phys. **324**, 601–636 (2013)
14. Grillakis, M., Machedon, M.: Pair excitations and the mean field approximation of interacting Bosons, II. Commun. PDE **42**, 24–67 (2017)
15. Grillakis, M., Machedon, M., Margetis, D.: Second-order corrections to mean-field evolution of weakly interacting Bosons. I. Commun. Math. Phys. **294**(1), 273–301 (2010)
16. Hepp, K.: The classical limit for quantum mechanical correlation functions. Commun. Math. Phys. **35**, 265–277 (1974)
17. Jeblick, M., Pickl, P.: Derivation of the time dependent Gross-Pitaevskii equation for a class of non purely positive potentials. arXiv:1801.04799
18. Klainerman, S., Machedon, M.: On the uniqueness of solutions to the Gross-Pitaevskii hierarchy. Commun. Math. Phys. **279**(1), 169–185 (2008)
19. Knowles, A., Pickl, P.: Mean-field dynamics: singular potentials and rate of convergence. Commun. Math. Phys. **298**(1), 101–138 (2010)
20. Lewin, M., Nam, P.T., Schlein, B.: Fluctuations around Hartree states in the mean-field regime. arXiv:1307.0665
21. Lewin, M., Nam, P.T., Serfaty, S., Solovej, J.P.: Bogoliubov spectrum of interacting Bose gases. arXiv:1211.2778
22. Lieb, E.H., Seiringer, R.: Proof of Bose-Einstein condensation for dilute trapped gases. Phys. Rev. Lett. **88**, 170409 (2002)
23. Lieb, E.H., Seiringer, R., Yngvason, J.: Bosons in a trap: A rigorous derivation of the Gross-Pitaevskii energy functional. Phys. Rev. A **61**, 043602 (2000)
24. Pickl, P.: Derivation of the time dependent Gross-Pitaevskii equation with external fields. arXiv:1001.4894
25. Rodnianski, I., Schlein, B.: Quantum fluctuations and rate of convergence towards mean-field dynamics. Commun. Math. Phys. **291**(1), 31–61 (2009)
26. Spohn, H.: Kinetic equations from Hamiltonian dynamics: Markovian limits. Rev. Mod. Phys. **52**(3), 569–615 (1980)

Mean-Field Limits of Particles in Interaction with Quantized Radiation Fields

Nikolai Leopold and Peter Pickl

Dedicated to Herbert Spohn on the occasion of his 70th birthday.

Abstract We report on a novel strategy to derive mean-field limits of quantum mechanical systems in which a large number of particles weakly couple to a second-quantized radiation field. The technique combines the method of counting and the coherent state approach to study the growth of the correlations among the particles and in the radiation field. As an instructional example, we derive the Schrödinger–Klein–Gordon system of equations from the Nelson model with ultraviolet cutoff and possibly massless scalar field. In particular, we prove the convergence of the reduced density matrices (of the nonrelativistic particles and the field bosons) associated with the exact time evolution to the projectors onto the solutions of the Schrödinger–Klein–Gordon equations in trace norm. Furthermore, we derive explicit bounds on the rate of convergence of the one-particle reduced density matrix of the nonrelativistic particles in Sobolev norm.

Keywords Mean-field limit · Nelson model · Schrödinger–Klein–Gordon system

MSC Class 35Q40 · 81Q05 · 82C10

N. Leopold (✉)
IST Austria (Institute of Science and Technology Austria), Am Campus 1,
3400 Klosterneuburg, Austria
e-mail: nikolai.leopold@ist.ac.at

P. Pickl
Duke Kunshan University, Duke Avenue 8, Kunshan 215316, China
e-mail: peter.pickl@dukekunshan.edu.cn; pickl@math.lmu.de

P. Pickl
Ludwig-Maximilians-Universität München, Theresienstraße 39,
80333 München, Germany

© Springer Nature Switzerland AG 2018
D. Cadamuro et al. (eds.), *Macroscopic Limits of Quantum Systems*,
Springer Proceedings in Mathematics & Statistics 270,
https://doi.org/10.1007/978-3-030-01602-9_9

1 Introduction

Quantum systems with many degrees of freedom are difficult to analyze. The situation is especially complicated in the presence of quantized radiation fields which are described by Fock spaces with infinitely many degrees of freedom. The dynamics of such systems can, however, be studied in special scaling limits by means of simpler effective evolution equations. These involve fewer degrees of freedom and are less exact but easier to investigate. Effective evolution equations for particles that interact with quantized radiation fields have rigorously been derived for example in [1–8]. The general setting in these works is given by the tensor product of two Hilbert spaces

$$\mathcal{H}^{(N)} = \mathcal{H}_p^{(N)} \otimes \mathcal{F}. \tag{1}$$

The space $\mathcal{H}_p^{(N)}$ describes N nonrelativistic particles and $\mathcal{F}$ (usually a bosonic Fock space) models the quantized radiation field in terms of field bosons. The dynamics of the system is governed by the Schrödinger equation with a Hamiltonian of the form

$$H^N := H_0^N + H_f + \sum_{j=1}^{N} H_{int,j}. \tag{2}$$

Here, H_0^N and H_f (solely acting on $\mathcal{H}_p^{(N)}$ and $\mathcal{F}$ respectively) denote the free Hamiltonians of the particles and the radiation field. The term $H_{int,j}$ establishes an interaction between the jth particle and the radiation field. A typical question of interest is whether the quantized radiation field can be approximated by a classical field and the evolution of the whole system described by a system of simple effective equations. Usually, one considers initial data $\Psi_{N,0} = \Phi_{N,0} \otimes W(\gamma^{1/2}\alpha_0)\Omega$ with no correlations between the particles and the field bosons, sometimes referred to as Pekar product state [3]. The state $W(\gamma^{1/2}\alpha_0)\Omega \in \mathcal{F}$ denotes field bosons in the coherent state α_0 with a mean particle number $\gamma\,||\alpha_0||^2$, see (16). Here, γ is a model-dependent scaling parameter, for instance, the number of particles [1, 2, 7] or the strong coupling parameter in the Polaron model [3, 4, 6]. From physics literature, it is commonly known that coherent states with a high occupation number of field bosons can approximately be described by a classical radiation field [9, Chapter III.C.4]. This allows us to describe the system in the limit $\gamma \to \infty$ (in a suitable sense, see Sect. 3) effectively by the state of the particles $\Phi_{N,0}$ and a classical radiation field with mode function α_0. The arising question is if at later time t, one can still approximate the system by the pair $(\Phi_{N,t}, \alpha_t)$ which evolves according to a set of simple effective equations with initial datum $(\Phi_{N,0}, \alpha_0)$:

$$
\begin{array}{ccc}
\Psi_{N,0} & \xrightarrow{\gamma\to\infty} & (\Phi_{N,0}, \alpha_0) \\
{\scriptstyle\text{Many-body dynamics}}\Big\downarrow & & \Big\downarrow{\scriptstyle\text{Effective dynamics}} \\
\Psi_{N,t} & \xrightarrow{\gamma\to\infty} & (\Phi_{N,t}, \alpha_t).
\end{array}
\tag{3}
$$

This holds, if the field bosons, that are created during the time evolution, are either in a coherent state or subleading with respect to γ. The effect of the particles on the radiation field is typically negligible, if one considers a fixed number of particles, a coupling constant that tends to zero in a suitable sense and a coherent state, whose mean particle number scales with the parameter γ [5]. Otherwise, the state of the particles must have a special structure to ensure that the contributing field bosons are coherent [9, Complement B_{III}]. This is expected, if one considers slow and heavy particles [8] or a condensate of particles that weakly couple to the radiation field. In this work, we are interested in the latter situation. More explicitly, we study the dynamics of initial states $\Psi_{N,0} = \varphi_0^{\otimes N} \otimes W(N^{1/2}\alpha_0)\Omega$ with one-particle wave function φ_0 in the limit $N = \gamma \to \infty$ where the fields in the interaction Hamiltonian $H_{int,j}$ are multiplied by $N^{-1/2}$ (see Sect. 2). We refer to this limit as mean-field limit, because it implies that the source term of the radiation field is replaced by its mean value in the effective description. So far, such kind of limits has been studied either by the coherent state approach [2, 5, 10] or by means of Wigner measures [1].[1] While the method of Wigner measures allows us to derive limiting equations for an extensive class of initial states, it does in contrast to the coherent state approach not provide quantitative bounds on the rate of convergence. In the following, we present a strategy which is designed for systems with fixed particle number. Such systems usually arise in the nonrelativistic limit when the creation and annihilation of the nonrelativistic particles is suppressed.[2] The technique is a combination of the method of counting [11] and the coherent state approach [12]. It provides explicit error bounds on the rate of convergence and arises in a natural way if one extends the method of counting with ideas inspired by [2, 13]. As an instructional example, we derive the Schrödinger–Klein–Gordon system of equations from the Nelson model with ultraviolet cutoff. Our strategy seems applicable to more complicated models and we hope it will be helpful in the derivation of other mean-field limits. In [7], it was already used to derive the Maxwell–Schrödinger system of equations from the bosonic Pauli–Fierz Hamiltonian.

2　Setting of the Problem

We consider a system of N identical charged bosons interacting with a quantized scalar field, described by a wave function $\Psi_{N,t} \in \mathcal{H}^{(N)}$. The Hilbert space is given by

$$\mathcal{H}^{(N)} := L^2\left(\mathbb{R}^{3N}\right) \otimes \mathcal{F}, \tag{4}$$

---

[1]These approaches usually embed the N particle states of $\mathcal{H}_p^{(N)}$ in a bosonic Fock space for the particles $\mathcal{F}_p$ and consider the Hilbert space $\mathcal{F}_p \otimes \mathcal{F}$.

[2]For the sake of clarity, we want to stress that only the number of the nonrelativistic particles is fixed while field bosons are created and destroyed during the time evolution.

where the scalar field is represented by elements of the Fock space $\mathcal{F} := \bigoplus_{n \geq 0} L^2(\mathbb{R}^3)^{\otimes_s^n}$. The subscript s indicates symmetry under the interchange of field bosons. An element $\Psi_N \in \mathcal{H}^{(N)}$ is a sequence $\{\Psi_N^{(n)}\}_{n \in \mathbb{N}_0}$ in $L^2(\mathbb{R}^{3N+3n})$ with[3]

$$||\Psi_N||^2 = \sum_{n=0}^{\infty} \int d^{3N}x \, d^{3n}k \, |\Psi_N^{(n)}(x_1, \ldots, x_N, k_1, \ldots, k_n)|^2 < \infty. \tag{5}$$

On $\mathcal{H}^{(N)}$, we define the (pointwise) annihilation and creation operators by[4]

$$(a(k)\Psi_N)^{(n)}(X_N, k_1, \ldots, k_n) = (n+1)^{1/2}\Psi_N^{(n+1)}(X_N, k, k_1, \ldots, k_n),$$

$$\left(a^*(k)\Psi_N\right)^{(n)}(X_N, k_1, \ldots, k_n) = n^{-1/2}\sum_{j=1}^{n}\delta(k-k_j)\Psi_N^{(n)}(X_N, k_1, \ldots, \hat{k}_j, \ldots, k_n).$$
$$\tag{6}$$

They are operator valued distributions and satisfy the commutation relations

$$[a(k), a^*(l)] = \delta(k-l), \quad [a(k), a(l)] = [a^*(k), a^*(l)] = 0. \tag{7}$$

The time evolution of the wave function $\Psi_{N,t}$ is governed by the Schrödinger equation

$$i\partial_t \Psi_{N,t} = H_N \Psi_{N,t}. \tag{8}$$

Here,

$$H_N = \sum_{j=1}^{N}\left(-\Delta_j + \frac{\widehat{\Phi}_\kappa(x_j)}{\sqrt{N}}\right) + H_f \tag{9}$$

denotes the Nelson Hamiltonian and

$$\widehat{\Phi}_\kappa(x) = \int d^3k \, \frac{\tilde{\kappa}(k)}{\sqrt{2\omega(k)}}\left(e^{ikx}a(k) + e^{-ikx}a^*(k)\right). \tag{10}$$

The field bosons evolve according to the dispersion relation $\omega(k) = (|k|^2 + m^2)^{1/2}$ with mass $m \geq 0$ and

$$\tilde{\kappa}(k) = (2\pi)^{-3/2}\,\mathbb{1}_{|k| \leq \Lambda}(k), \quad \text{with } \mathbb{1}_{|k| \leq \Lambda}(k) = \begin{cases} 1 & \text{if } |k| \leq \Lambda, \\ 0 & \text{otherwise,} \end{cases} \tag{11}$$

---

[3]Note that $\Psi_N^{(n)}$ is symmetric in the variables $k_1, \ldots k_n$. For notational convenience, we will use the shorthand notation $\Psi_N^{(n)}(X_N, K_n) = \Psi_N^{(n)}(x_1, \ldots, x_N, k_1, \ldots k_n)$.
[4]Here, $\hat{k}_j$ means that k_j is left out in the argument of the function.

cuts off the high frequency modes of the radiation field. On the domain

$$
\mathcal{D}(H_f) = \left\{ \Psi_N \in \mathcal{H}^{(N)} : \sum_{n=1}^{\infty} \int d^{3N}x \, d^{3n}k \mid \sum_{j=1}^{n} \omega(k_j) \Psi_N^{(n)}(X_N, K_n)|^2 < \infty \right\}
$$

(12)

the free Hamiltonian of the scalar field is defined by

$$
\left(H_f \Psi_N \right)^{(n)} = \sum_{j=1}^{n} \omega(k_j) \Psi_N^{(n)}.
$$

(13)

By means of the creation and annihilation operators, it can be written as

$$
H_f = \int d^3k \, \omega(k) a^*(k) a(k).
$$

(14)

The Nelson model was originally introduced to describe the interaction of non-relativistic nucleons with a meson field. By standard estimates of the field operator and Kato's theorem, it is easily shown that H_N is a self-adjoint operator with $\mathcal{D}(H_N) = \mathcal{D}(\sum_{j=1}^{N} -\Delta_j + H_f)$ [14, 15]. The scaling in front of the interaction is chosen in a way that we can derive an interesting effective system in which the kinetic energy and the interaction between the nonrelativistic particle and the radiation field are of the same order. For simplicity, we are first interested in the evolution of initial states of the product form

$$
\varphi_0^{\otimes N} \otimes W(\sqrt{N}\alpha_0)\Omega.
$$

(15)

Here, Ω denotes the vacuum in $\mathcal{F}$ and $W(f)$ is the Weyl operator

$$
W(f) := \exp\left(\int d^3k \, f(k) a^*(k) - \overline{f(k)} a(k) \right),
$$

(16)

where $f \in L^2(\mathbb{R}^3)$. This choice of initial states corresponds to situations in which no correlations among the particles and the field bosons are present. The interaction between the particles and the field bosons leads to the development of correlations and the time evolved state will no longer be of product form. However, for large N and times of order one, it can be approximated, in a sense more specified below, by a state of the form $\varphi_t^{\otimes N} \otimes W(\sqrt{N}\alpha_t)\Omega$, where (φ_t, α_t) solves the Schrödinger–Klein–Gordon equations[5]

---

[5]We use $\mathcal{F}T[\Phi](k, t) = (2\pi)^{-3/2} \int_{\mathbb{R}^3} d^3x \, e^{-ikx} \Phi(x, t)$ to denote the Fourier transform of $\Phi(x, t)$ and $(\kappa * \Phi)(x, t) = \int d^3k \, e^{ikx} \tilde{\kappa}(k) \mathcal{F}T[\Phi](k, t)$.

$$\begin{cases} i\partial_t \varphi_t(x) & = H^{\text{eff}} \varphi_t(x) = [-\Delta + (\kappa * \Phi)(x, t)] \varphi_t(x), \\ i\partial_t \alpha_t(k) & = \omega(k)\alpha_t(k) + (2\pi)^{3/2} \frac{\tilde{\kappa}(k)}{\sqrt{2\omega(k)}} \mathcal{FT}\left[|\varphi_t|^2\right](k), \\ \Phi(x, t) & = \int d^3k \, (2\pi)^{-3/2} \frac{1}{\sqrt{2\omega(k)}} \left(e^{ikx}\alpha_t(k) + e^{-ikx}\overline{\alpha_t(k)} \right), \end{cases} \quad (17)$$

with $(\varphi_0, \alpha_0) \in L^2(\mathbb{R}^3) \oplus L^2(\mathbb{R}^3)$. This system of equations determines the evolution of a single quantum particle in interaction with a classical scalar field. In the literature, it is better known in its formally equivalent form

$$\begin{cases} i\partial_t \varphi_t(x) & = [-\Delta + (\kappa * \Phi)(x, t)] \varphi_t(x), \\ \left[\partial_t^2 - \Delta + m^2\right] \Phi(x, t) & = -\left(\kappa * |\varphi_t|^2\right)(x). \end{cases} \quad (18)$$

3 Main Result

The physical situation we are interested in is the dynamical description of a Bose–Einstein condensate of charges. We start initially with a product state (15) and show that the condensate persists during the time evolution, i.e., correlations are small also at later times. Let

$$\mathcal{N} := \int d^3k \, a^*(k)a(k) \quad (19)$$

be the number (of field bosons) operator with domain

$$\mathcal{D}(\mathcal{N}) = \left\{ \Psi_N \in \mathcal{H}^{(N)} : \sum_{n=1}^{\infty} n^2 \int d^{3N}x \, d^{3n}k \, |\Psi_N^{(n)}(X_N, K_n)|^2 < \infty \right\} \quad (20)$$

and $\Psi_{N,t} \in \left(L_s^2\left(\mathbb{R}^{3N}\right) \otimes \mathcal{F}\right) \cap \mathcal{H}^{(N)} \cap \mathcal{D}(\mathcal{N})$ such that $\left\|\Psi_{N,t}\right\|_{\mathcal{H}^{(N)}} = 1$. On the Hilbert space $L^2(\mathbb{R}^3)$, we define the "one-particle reduced density matrix of the charges" by

$$\gamma_{N,t}^{(1,0)} := \text{Tr}_{2,\dots,N} \otimes \text{Tr}_{\mathcal{F}} |\Psi_{N,t}\rangle\langle\Psi_{N,t}|, \quad (21)$$

where $\text{Tr}_{2,\dots,N}$ denotes the partial trace over the coordinates $x_2, \dots, x_N$ and $\text{Tr}_{\mathcal{F}}$ the trace over Fock space. Then, the charged particles of the many-body state $\Psi_{N,t}$ are said to exhibit complete asymptotic Bose–Einstein condensation at time t, if there exists $\varphi_t \in L^2(\mathbb{R}^3)$ with $\|\varphi_t\| = 1$, such that

$$\text{Tr}_{L^2(\mathbb{R}^3)}|\gamma_{N,t}^{(1,0)} - |\varphi_t\rangle\langle\varphi_t|| \to 0, \quad (22)$$

as $N \to \infty$. Such φ_t is called the condensate wave function. For other indicators of condensation and their relation, we refer to [16]. Moreover, we introduce the "one-particle reduced density matrix of the field bosons" with kernel

$$\gamma_{N,t}^{(0,1)}(k, k') := N^{-1}\langle \Psi_{N,t}, a^*(k')a(k)\Psi_{N,t}\rangle_{\mathcal{H}^{(N)}}. \tag{23}$$

$\gamma_{N,t}^{(0,1)}$ is a positive trace class operator with $\mathrm{Tr}_{L^2(\mathbb{R}^3)}(\gamma_{N,t}^{(0,1)}) = N^{-1}\langle \Psi_{N,t}, \mathcal{N}\Psi_{N,t}\rangle_{\mathcal{H}^{(N)}}$. It is worth noting that (23) differs from the usual definition $\langle \Psi_{N,t}, \mathcal{N}\Psi_{N,t}\rangle^{-1}$ $\langle \Psi_{N,t}, a^*(k')a(k)\Psi_{N,t}\rangle$ by a constant. Our choice ensures that we only measure deviations from the classical mode function that are at least of order N. This is reasonable because Fock space vectors with a mean particle number smaller than of order N only have a subleading effect on the dynamics of the charged particles. We say the field bosons exhibit "asymptotic Bose–Einstein condensation", if there exists a state $\alpha_t \in L^2(\mathbb{R}^3)$, such that

$$\mathrm{Tr}_{L^2(\mathbb{R}^3)}|\gamma_{N,t}^{(0,1)} - |\alpha_t\rangle\langle\alpha_t|| \to 0, \tag{24}$$

as $N \to \infty$.

In order to derive our main result, the solutions of the Schrödinger–Klein–Gordon equations have to satisfy the following assumptions.

Definition 3.1. Let $m \in \mathbb{N}$, $H^m(\mathbb{R}^3)$ denote the Sobolev space of order m and $L_m^2(\mathbb{R}^3)$ a weighted L^2-space with norm $||\alpha||_{L_m^2(\mathbb{R}^3)} = ||(1 + |\cdot|^2)^{m/2}\alpha||_{L^2(\mathbb{R}^3)}$. We define two sets of solutions of the Schrödinger–Klein–Gordon equations:

$$(\varphi_t, \alpha_t) \in \mathcal{G}_1 \Leftrightarrow \begin{cases} (\varphi_t, \alpha_t) \text{ is a strongly differentiable } H^2(\mathbb{R}^3) \oplus L_1^2(\mathbb{R}^3)\text{-valued function} \\ \text{on } [0, \infty) \text{that satisfies (17) and } ||\varphi_0||_{L^2(\mathbb{R}^3)} = 1. \end{cases} \tag{25}$$

$$(\varphi_t, \alpha_t) \in \mathcal{G}_2 \Leftrightarrow \begin{cases} (\varphi_t, \alpha_t) \text{ is a strongly differentiable} H^4(\mathbb{R}^3) \oplus L_2^2(\mathbb{R}^3)\text{-valued function} \\ \text{on } [0, \infty) \text{that satisfies (17) and } ||\varphi_0||_{L^2(\mathbb{R}^3)} = 1. \end{cases} \tag{26}$$

Then, $||\varphi_t||_{L^2(\mathbb{R}^3)} = 1 \ \forall t \in [0, \infty)$ follows because the L^2-norm of φ_t is conserved in the Schrödinger–Klein–Gordon system.

These assumptions are expected to follow from appropriately chosen initial data.

Proposition 3.2. (unproven). *Let $(\varphi_0, \alpha_0) \in H^{2n}(\mathbb{R}^3) \oplus L_n^2(\mathbb{R}^3)$ for $1 \leq n \leq 2$. Then, there is a strongly differentiable $(H^{2n}(\mathbb{R}^3) \oplus L_n^2(\mathbb{R}^3))$-valued function $(\varphi(t), \alpha(t))$ on $[0, \infty)$ that satisfies (17).*

For $m = 1$ and in the absence of an ultraviolet cutoff, it is a consequence of [17, Theorem 1.4] that the Schrödinger–Klein–Gordon system (17) is globally well posed in $H^s \oplus L_\sigma^2$ for any $s \geq 0$ and $s - 1/2 \leq \sigma \leq s + 1/2$. We expect this result to hold also in the presence of the cutoff function (11). Moreover, we suppose that Proposition

(3.2) can be proven by a standard fixed-point argument. Especially due to the cutoff in the radiation field, it seems possible to make use of [18, Theorem X.74].
Our main theorem is the following.

Theorem 3.3. *Let* $(\varphi_t, \alpha_t) \in \mathcal{G}_1$ *and* $\Psi_{N,0} \in \left(L_s^2(\mathbb{R}^{3N}) \otimes \mathcal{F}\right) \cap \mathcal{D}(\mathcal{N}) \cap \mathcal{D}(\mathcal{N}H_N)$ *with* $\|\Psi_{N,0}\| = 1$ *such that* [6]

$$a_N = Tr_{L^2(\mathbb{R}^3)} |\gamma_{N,0}^{(1,0)} - |\varphi_0\rangle\langle\varphi_0|| \to 0 \text{ and} \tag{27}$$

$$b_N = N^{-1} \langle W^{-1}(\sqrt{N}\alpha_0)\Psi_{N,0}, \mathcal{N}W^{-1}(\sqrt{N}\alpha_0)\Psi_{N,0}\rangle_{\mathcal{H}^{(N)}} \to 0 \tag{28}$$

as $N \to \infty$. *Let* $\Psi_{N,t}$ *be the unique solution of* (8) *with initial data* $\Psi_{N,0}$. *Then, there exists a generic constant* C *independent of* N, Λ *and* t *such that*

$$Tr_{L^2(\mathbb{R}^3)} |\gamma_{N,t}^{(1,0)} - |\varphi_t\rangle\langle\varphi_t|| \le \sqrt{a_N + b_N + N^{-1}} e^{\Lambda^2 Ct}, \tag{29}$$

$$Tr_{L^2(\mathbb{R}^3)} |\gamma_{N,t}^{(0,1)} - |\alpha_t\rangle\langle\alpha_t|| \le \sqrt{a_N + b_N + N^{-1}} e^{\Lambda^2 Ct} C\left(1 + \|\alpha_t\|\right) \tag{30}$$

for any $t \in [0, \infty)$. [7] *In particular, for* $\Psi_{N,0} = \varphi_0^{\otimes N} \otimes W(\sqrt{N}\alpha_0)\Omega$, *one obtains*

$$Tr_{L^2(\mathbb{R}^3)} |\gamma_{N,t}^{(1,0)} - |\varphi_t\rangle\langle\varphi_t|| \le N^{-1/2} e^{C\Lambda^2 t}, \tag{31}$$

$$Tr_{L^2(\mathbb{R}^3)} |\gamma_{N,t}^{(0,1)} - |\alpha_t\rangle\langle\alpha_t|| \le N^{-1/2} e^{\Lambda^2 Ct} C\left(1 + \|\alpha_t\|\right). \tag{32}$$

Moreover, let $(\varphi_t, \alpha_t) \in \mathcal{G}_2$ *and* $\Psi_{N,0} \in \left(L_s^2(\mathbb{R}^{3N}) \otimes \mathcal{F}\right) \cap \mathcal{D}(\mathcal{N}) \cap \mathcal{D}(\mathcal{N}H_N) \cap \mathcal{D}\left(H_N^2\right)$ *such that*

$$c_N = \left\|\nabla_1 \left(1 - |\varphi_0\rangle\langle\varphi_0| \otimes \mathbb{1}_{L^2(\mathbb{R}^{3(N-1)})} \otimes \mathbb{1}_{\mathcal{F}}\right) \Psi_{N,0}\right\|_{\mathcal{H}^{(N)}}^2 \to 0 \tag{33}$$

as $N \to \infty$. *Then, there exists a positive monotone increasing function* $C(s)$ *of the norms* $\|\alpha_s\|_{L^2(\mathbb{R}^3)}$ *and* $\|\varphi_s\|_{H^1(\mathbb{R}^3)}$ *such that*

$$Tr_{L^2(\mathbb{R}^3)} |\sqrt{1-\Delta} \left(\gamma_{N,t}^{(1,0)} - |\varphi_t\rangle\langle\varphi_t|\right) \sqrt{1-\Delta}| \le \sqrt{a_N + b_N + c_N + N^{-1}} C(t) e^{\Lambda^4 \int_0^t C(s)ds}. \tag{34}$$

For $\Psi_{N,0} = \varphi_0^{\otimes N} \otimes W(\sqrt{N}\alpha_0)\Omega$, *one obtains*

$$Tr_{L^2(\mathbb{R}^3)} |\sqrt{1-\Delta} \left(\gamma_{N,t}^{(1,0)} - |\varphi_t\rangle\langle\varphi_t|\right) \sqrt{1-\Delta}| \le N^{-1/2} C(t) e^{\Lambda^4 \int_0^t C(s)ds}. \tag{35}$$

Remark 3.4. The convergence of the reduced density matrices in trace norm with rate N^{-1} was already shown in [2] for special classes of initial states (coherent and

---

[6]Here, $W^{-1}(\sqrt{N}\alpha_0) = W(-\sqrt{N}\alpha_0)$ is the inverse of the unitary Weyl operator $W(\sqrt{N}\alpha_0)$, see Sect. 9.

[7]To ease the presentation, we have chosen for given t the scaling parameter N large enough such that $0 \le \beta(t) \le 1$ and $0 \le \beta_2(t) \le 1$ (see Sects. 8.2 and 8.3).

product states).[8] Theorem 3.3 extends this result to more general states but only with error estimates of order $N^{-1/2}$. Moreover, we present the first explicit bounds on the rate of convergence of the one-particle reduced density matrix of the charges in Sobolev norm. It seems possible to prove the results of Theorem 3.3 with convergence rate N^{-1}, if one regards (similar to [19]) fluctuations around the mean-field dynamics.

4 Comparison with the Literature

In [5], Ginibre, Nironi, and Velo studied the partial limit of the renormalized Nelson model without ultraviolet cutoff. They considered a finite number of charged bosons, a coupling constant that tends to zero and a coherent state of field bosons whose particle number goes to infinity. The number of field bosons that are created during the time evolution is negligible in this case and it is possible to approximate the quantized scalar field by an external potential which evolves according to the Klein–Gordon equation without source term. Falconi [2] derived the Schrödinger–Klein–Gordon system of equations in the setting of the present paper by means of the coherent state approach. A comparison between his result and Theorem 3.3 is given in Remark 3.4. Making use of a Wigner measure approach, Ammari and Falconi [1] were able to establish the classical limit (without quantitative bounds on the rate of convergence) of the renormalized Nelson model without cutoff. Teufel [8] considered the adiabatic limit of the Nelson model and showed that the interaction mediated by the quantized radiation field is well approximated by a direct Coulomb interaction. In [7], we used the strategy of the present paper to derive the Maxwell–Schrödinger equations from the bosonic Pauli–Fierz Hamiltonian. Here, additional technical difficulties arise from the minimal coupling term in the Pauli–Fierz Hamiltonian.

5 Notations

The Fourier transform of a function f is denoted by $\tilde{f}$ or $\mathcal{FT}[f]$. $H^s(\mathbb{R}^3)$ stands for the Sobolev space with norm $||f||_{H^s(\mathbb{R}^3)} = \left|\left|(1 + |\cdot|^2)^{s/2}\mathcal{FT}[f]\right|\right|_{L^2(\mathbb{R}^3)}$ and $L^2_m(\mathbb{R}^3)$ is the weighted L^2 space with $||f||_{L^2_m(\mathbb{R}^3)} = \left|\left|(1 + |\cdot|^2)^{m/2}f\right|\right|_{L^2(\mathbb{R}^3)}$. Moreover, we use $||A||_{HS} = \sqrt{Tr A^* A}$ to denote the Hilbert–Schmidt norm. With a slight abuse of notation, we write Φ and F to indicate the scalar and auxiliary field but also their respective Fourier transforms. If we use $\Phi(t)$ or $F(t)$, we always refer to the coordinate representation of the fields. Furthermore, we apply the shorthand notation $\Phi_\kappa(x, t) := (\kappa * \Phi)(x, t)$.

---

[8]For the precise definition of the initial data, we refer to [2, (I.3) and Theorem 3].

6 The Strategy

We are interested in the evolution of product states of the form (15) under the dynamics (8). The scalar field in the Nelson Hamiltonian establishes an interaction between the charges and the field modes with wave vectors smaller than Λ.[9] This changes the state of the charges, leads to the creation and annihilation of field bosons, and causes initially factorized states to build correlations between the charges, the field bosons as well as among charges and field bosons. To study the growth of these correlations, we combine the method of counting with the coherent state approach. The key idea is to prove condensation not in terms of reduced density matrices but to consider a different indicator of condensation. To control the correlations between the charges, we introduce a functional β^a, which counts the relative number of particles that are not in the state of the condensate wave function φ_t.

Definition 6.1. For any $N \in \mathbb{N}$, $\varphi_t \in L^2(\mathbb{R}^3)$ with $\|\varphi_t\| = 1$ and $1 \leq j \leq N$, we define the time-dependent projectors $p_j^{\varphi_t} : L^2(\mathbb{R}^{3N}) \to L^2(\mathbb{R}^{3N})$ and $q_j^{\varphi_t} : L^2(\mathbb{R}^{3N}) \to L^2(\mathbb{R}^{3N})$ by

$$p_j^{\varphi_t} f(x_1, \ldots, x_N) := \varphi_t(x_j) \int d^3x_j \, \overline{\varphi_t(x_j)} f(x_1, \ldots, x_N) \quad \text{for all } f \in L^2(\mathbb{R}^{3N}) \tag{36}$$

and $q_j^{\varphi_t} := 1 - p_j^{\varphi_t}$.[10] Let $\Psi_{N,t} \in \mathcal{H}^{(N)}$. Then $\beta^a : \mathcal{H}^{(N)} \times L^2(\mathbb{R}^3) \to [0, \infty)$ is given by

$$\beta^a\left(\Psi_{N,t}, \varphi_t\right) := \left\langle \Psi_{N,t}, q_1^{\varphi_t} \otimes \mathbb{1}_{\mathcal{F}} \, \Psi_{N,t}\right\rangle. \tag{37}$$

Remark 6.2. The functional β^a is used in the method of counting. The counting method was introduced in [11] and applied in [19–34] to derive the Hartree and Gross–Pitaevskii equation.

The situation is slightly different in the radiation sector because the number of field bosons is not preserved during the time evolution. From physics literature [9, Chapter III.C.4] it is known that the field bosons behave classically if the radiation field is in a coherent state with a high occupation number. This is a state not only with little correlations but also with a Poisson distributed number of field bosons. In order to investigate if the state of the radiation field is coherent, we define a functional, referred to as β^b, which measures the fluctuations of the field modes around the classical mode function α_t at each time.

Definition 6.3. Let $\alpha_t \in L^2(\mathbb{R}^3)$ and $\Psi_{N,t} \in \mathcal{H}^{(N)} \cap \mathcal{D}(\mathcal{N})$. Then $\beta^b : \mathcal{H}^{(N)} \cap \mathcal{D}(\mathcal{N}) \times L^2(\mathbb{R}^3) \to [0, \infty)$ is given by

---

[9]One should note that the high frequency modes of the radiation field do not interact with the nonrelativistic particles and evolve according to the free dynamics.

[10]Occasionally, we use the bra–ket notation $p_j^{\varphi_t} = |\varphi_t(x_j)\rangle\langle\varphi_t(x_j)| = |\varphi_t\rangle\langle\varphi_t|_j$.

$$\beta^b \left(\Psi_{N,t}, \alpha_t \right) := \int d^3k \left\langle \left(\frac{a(k)}{\sqrt{N}} - \alpha_t(k) \right) \Psi_{N,t}, \left(\frac{a(k)}{\sqrt{N}} - \alpha_t(k) \right) \Psi_{N,t} \right\rangle. \quad (38)$$

Remark 6.4. Let $\alpha_0 \in L^2(\mathbb{R}^3)$, $\Psi_{N,0} = W(\sqrt{N}\alpha_0)\Psi$ for some $\Psi \in \mathcal{H}^{(N)} \cap \mathcal{D}(\mathcal{N})$ and $\mathcal{U}_N(t; 0) = W^*(\sqrt{N}\alpha_t)e^{-iH_N t}W(\sqrt{N}\alpha_0)$ be the fluctuation dynamics of the coherent state approach (compare for example with [35, (3.24)]) . Then, the functional β^b can be written as[11]

$$\beta^b \left(\Psi_{N,t}, \alpha_t \right) = N^{-1} \langle \mathcal{U}_N(t; 0)\Psi, \mathcal{N}\mathcal{U}_N(t; 0)\Psi \rangle. \quad (39)$$

Thus, β^b is the same measure that is used in the coherent state approach to study the field boson fluctuations around the effective evolution. The coherent state approach, which is inspired by ideas from [36, 37], was introduced in [12] and applied in [2, 38–45] to derive the Hartree, Gross–Pitaevskii and Schrödinger–Klein–Gordon equations. A comprehensive introduction to the method is given in [35].

Remark 6.5. While β^a appears to be the natural quantity to consider for condensates with fixed particle number, β^b is perfectly suited to control whether the radiation field is in a coherent state.

Finally, the counting functional is defined by

Definition 6.6. Let $N \in \mathbb{N}$, $\varphi_t \in L^2(\mathbb{R}^3)$ with $\|\varphi_t\| = 1$, $\alpha_t \in L^2(\mathbb{R}^3)$ and $\Psi_{N,t} \in \mathcal{H}^{(N)} \cap \mathcal{D}(\mathcal{N})$. Then $\beta : \mathcal{H}^{(N)} \cap \mathcal{D}(\mathcal{N}) \times L^2(\mathbb{R}^3) \times L^2(\mathbb{R}^3) \rightarrow [0, \infty)$ is defined by[12]

$$\beta \left(\Psi_{N,t}, \varphi_t, \alpha_t \right) := \beta^a \left(\Psi_{N,t}, \varphi_t \right) + \beta^b \left(\Psi_{N,t}, \alpha_t \right). \quad (40)$$

In summary, the functional has the following properties:

(i) β^a measures if the nonrelativistic particles exhibit condensation.
(ii) β^b examines whether the radiation field is in a coherent state.
(iii) $\beta \left(\Psi_{N,t}, \varphi_t, \alpha_t \right) \rightarrow 0$ as $N \rightarrow \infty$ implies condensation in terms of reduced density matrices (Lemma 7.1).
(iv) $\beta \left(\Psi_{N,t}, \varphi_t, \alpha_t \right) = 0$ if $\Psi_{N,t} = \varphi_t^{\otimes N} \otimes W(\sqrt{N}\alpha_t)\Omega$ (see Lemma 9.2).

In order to show that the product structure (15) is preserved asymptotically during the time evolution, we apply the following strategy:

1. We choose initial states φ_0, α_0 and $\Psi_{N,0}$ such that $\beta \left(\Psi_{N,0}, \varphi_0, \alpha_0 \right) \leq a_N + b_N \rightarrow 0$ as $N \rightarrow \infty$.
2. For each $t \in [0, \infty)$, we estimate $|\frac{d}{dt}\beta \left(\Psi_{N,t}, \varphi_t, \alpha_t \right)| \leq C\Lambda^2 \left(\beta(\Psi_{N,t}, \varphi_t, \alpha_t) + N^{-1} \right)$ for some $C \in [0, \infty)$. Then, Gronwall's lemma establishes the bound $\beta \left(\Psi_{N,t}, \varphi_t, \alpha_t \right) \leq e^{C\Lambda^2 t} \left(\beta \left(\Psi_{N,0}, \varphi_0, \alpha_0 \right) + N^{-1} \right)$.

---

[11]This is a simple consequence of $W(\sqrt{N}\alpha_t)$ being unitary and $W^*(\sqrt{N}\alpha_t)a(k) = a(k)W^*(\sqrt{N}\alpha_t) + \sqrt{N}W^*(\sqrt{N}\alpha_t)\alpha_t(k)$, see (129).

[12]We sometimes apply the shorthand notation $\beta(t) = \beta(\Psi_{N,t}, \varphi_t, \alpha_t)$.

3. By means of property (iii), we conclude condensation in terms of reduced density matrices.

To show the convergence of $\gamma_{N,t}^{(1,0)}$ to the projector onto the condensate wave function in Sobolev norm, we include $\beta^c(\Psi_{N,t}, \varphi_t) := \left\| \nabla_1 q_1^{\varphi_t} \Psi_{N,t} \right\|^2$ in the definition of the functional. This allows us to control the kinetic energy of the nonrelativistic particles which are not in the condensate.

Definition 6.7. Let $N \in \mathbb{N}$, $\varphi_t \in H^2(\mathbb{R}^3)$ with $\|\varphi_t\| = 1$, $\alpha_t \in L^2(\mathbb{R}^3)$ and $\Psi_{N,t} \in \mathcal{D}(H_N) \cap \mathcal{D}(\mathcal{N})$. Then $\beta_2 : \mathcal{D}(H_N) \cap \mathcal{D}(\mathcal{N}) \times H^2(\mathbb{R}^3) \times L^2(\mathbb{R}^3) \to [0, \infty)$ is defined by

$$\beta_2\left(\Psi_{N,t}, \varphi_t, \alpha_t\right) := \beta\left(\Psi_{N,t}, \varphi_t, \alpha_t\right) + \beta^c\left(\Psi_{N,t}, \varphi_t\right)$$
$$= \beta\left(\Psi_{N,t}, \varphi_t, \alpha_t\right) + \left\| \nabla_1 q_1^{\varphi_t} \Psi_{N,t} \right\|^2 . \tag{41}$$

We would like to remark, that the ultraviolet cutoff (11) is essential for the proof because:

1. The finiteness of $\|\eta\|_2$ (see (67)) is needed to establish a connection between the difference of the radiation fields and the functional β^b by means of the auxiliary fields (64).
2. The cutoff Λ imposes regularity on the radiation fields which will be used to estimate the time derivative of $\left\| \nabla_1 q_1 \Psi_{N,t} \right\|^2$. In spirit, this is opposite to the usual treatment of the polaron [46], where regularity of the electron state is used to obtain a sufficient decay in the field modes with large wave vectors.

7   Relation to Reduced Density Matrices

In this section, we relate the functional β to the trace norm distance of the one-particle reduced density matrices.

Lemma 7.1. Let $N \in \mathbb{N}$, $\varphi_t \in L^2(\mathbb{R}^3)$ with $\|\varphi_t\| = 1$, $\alpha_t \in L^2(\mathbb{R}^3)$ and $\Psi_{N,t} \in \mathcal{H}^{(N)} \cap \mathcal{D}(\mathcal{N})$. Then,

$$\beta^a(\Psi_{N,t}, \varphi_t) \leq Tr_{L^2(\mathbb{R}^3)} |\gamma_{N,t}^{(1,0)} - |\varphi_t\rangle\langle\varphi_t|| \leq \sqrt{8\beta^a(\Psi_{N,t}, \varphi_t)}, \tag{42}$$

$$Tr_{L^2(\mathbb{R}^3)} |\gamma_{N,t}^{(0,1)} - |\alpha_t\rangle\langle\alpha_t|| \leq 3\beta^b(\Psi_{N,t}, \alpha_t) + 6\|\alpha_t\|_{L^2(\mathbb{R}^3)} \sqrt{\beta^b(\Psi_{N,t}, \alpha_t)}. \tag{43}$$

For $\varphi_t \in H^2(\mathbb{R}^3)$ with $\|\varphi_t\| = 1$ and $\Psi_{N,t} \in \mathcal{H}^{(N)} \cap \mathcal{D}(H_N)$, we have

$$Tr_{L^2(\mathbb{R}^3)} |\sqrt{1 - \Delta} \left(\gamma_{N,t}^{(1,0)} - |\varphi_t\rangle\langle\varphi_t|\right) \sqrt{1 - \Delta}| \leq \left(1 + \|\varphi_t\|_{H^1(\mathbb{R}^3)}^2\right) \times$$
$$\times \left(\beta^a(\Psi_{N,t}, \varphi_t) + \beta^c(\Psi_{N,t}, \varphi_t)\right) + 2\|\varphi_t\|_{H^1(\mathbb{R}^3)} \sqrt{\beta^a(\Psi_{N,t}, \varphi_t) + \beta^c(\Psi_{N,t}, \varphi_t)}. \tag{44}$$

Proof. The lower bound of (42) is proven by

$$\beta^a(t) = 1 - \langle \Psi_{N,t}, p_1^{\varphi_t} \Psi_{N,t} \rangle = 1 - \langle \varphi_t, \gamma_{N,t}^{(1,0)} \varphi_t \rangle = \text{Tr}_{L^2(\mathbb{R}^3)}(|\varphi_t\rangle\langle\varphi_t| - |\varphi_t\rangle\langle\varphi_t|\gamma_{N,t}^{(1,0)})$$

$$\leq \|p_1\|_{\text{op}} \text{Tr}_{L^2(\mathbb{R}^3)} |\gamma_{N,t}^{(1,0)} - |\varphi_t\rangle\langle\varphi_t|| = \text{Tr}_{L^2(\mathbb{R}^3)} |\gamma_{N,t}^{(1,0)} - |\varphi_t\rangle\langle\varphi_t||. \quad (45)$$

To obtain the upper bound, we use that

$$\text{Tr}|\gamma - p| \leq 2\|\gamma - p\|_{HS} + \text{Tr}(\gamma - p) \quad (46)$$

is valid for any one-dimensional projector p and nonnegative density matrix γ. The original argument of the proof was first observed by Robert Seiringer, see [12]. We present a version that is found in [20]: Let $(\lambda_n)_{n\in\mathbb{N}}$ be the sequence of eigenvalues of the trace class operator $A := \gamma - p$. Since p is a rank one projection, A has at most one negative eigenvalue. If there is no negative eigenvalue, $\text{Tr}|A| = \text{Tr}(A)$ and (46) holds. If there is one negative eigenvalue λ_1, we have $\text{Tr}|A| = |\lambda_1| + \sum_{n\geq2} \lambda_n = 2|\lambda_1| + \text{Tr}(A)$. Inequality (46) then follows from $|\lambda_1| \leq \|A\|_{\text{op}} \leq \|A\|_{HS}$. This shows

$$\text{Tr}_{L^2(\mathbb{R}^3)} |\gamma_{N,t}^{(1,0)} - |\varphi_t\rangle\langle\varphi_t|| \leq 2 \left\| \gamma_{N,t}^{(1,0)} - |\varphi_t\rangle\langle\varphi_t| \right\|_{HS} \quad (47)$$

because $\text{Tr}_{L^2(\mathbb{R}^3)}(\gamma_{N,t}^{(1,0)} - |\varphi_t\rangle\langle\varphi_t|) = 0$. The upper bound of (42) is obtained by

$$\text{Tr}_{L^2(\mathbb{R}^3)}(\gamma_{N,t}^{(1,0)} - |\varphi_t\rangle\langle\varphi_t|)^2 = 1 - 2\text{Tr}_{L^2(\mathbb{R}^3)}(|\varphi_t\rangle\langle\varphi_t|\gamma_{N,t}^{(1,0)}) + \text{Tr}_{L^2(\mathbb{R}^3)}((\gamma_{N,t}^{(1,0)})^2)$$

$$\leq 2(1 - \text{Tr}_{L^2(\mathbb{R}^3)}(|\varphi_t\rangle\langle\varphi_t|\gamma_{N,t}^{(1,0)})) = 2\beta^a(t). \quad (48)$$

To prove (43), it is useful to write the kernel of $\gamma_{N,t}^{(0,1)} - |\alpha_t\rangle\langle\alpha_t|$ as

$$(\gamma_{N,t}^{(0,1)} - |\alpha_t\rangle\langle\alpha_t|)(k, l) = N^{-1}\langle \Psi_N, a^*(l)a(k)\Psi_N \rangle - \overline{\alpha_t(l)}\alpha_t(k)$$

$$= \langle (N^{-1/2}a(l) - \alpha_t(l)) \Psi_N, (N^{-1/2}a(k) - \alpha_t(k)) \Psi_N \rangle$$

$$+ \alpha_t(k)\langle (N^{-1/2}a(l) - \alpha_t(l)) \Psi_N, \Psi_N \rangle$$

$$+ \overline{\alpha_t(l)}\langle \Psi_N, (N^{-1/2}a(k) - \alpha_t(k)) \Psi_N \rangle. \quad (49)$$

By means of Schwarz's inequality, we have

$$|(\gamma_{N,t}^{(0,1)} - |\alpha_t\rangle\langle\alpha_t|)(k, l)|^2 \leq \left\| (N^{-1/2}a(k) - \alpha_t(k)) \Psi_N \right\|^2 \left\| (N^{-1/2}a(l) - \alpha_t(l)) \Psi_N \right\|^2$$

$$+ |\alpha_t(l)|^2 \left\| (N^{-1/2}a(k) - \alpha_t(k)) \Psi_N \right\|^2$$

$$+ |\alpha_t(k)|^2 \left\| (N^{-1/2}a(l) - \alpha_t(l)) \Psi_N \right\|^2 \quad (50)$$

and

$$\left\|\gamma_{N,t}^{(0,1)} - |\alpha_t\rangle\langle\alpha_t|\right\|_{HS}^2 = \int d^3k \int d^3l \,|(\gamma_{N,t}^{(0,1)} - |\alpha_t\rangle\langle\alpha_t|)(k,l)|^2$$

$$\leq (\beta^b(t))^2 + 2\,||\alpha_t||_{L^2(\mathbb{R}^3)}^2\,\beta^b(t). \tag{51}$$

Similarly, one obtains

$$Tr_{L^2(\mathbb{R}^3)}(\gamma_{N,t}^{(0,1)} - |\alpha_t\rangle\langle\alpha_t|) \leq \int d^3k \,|(\gamma_{N,t}^{(0,1)} - |\alpha_t\rangle\langle\alpha_t|)(k,k)|$$

$$\leq \int d^3k \,\left\|(N^{-1/2}a(k) - \alpha_t(k))\,\Psi_N\right\|_{\mathcal{H}^{(N)}}^2$$

$$+ 2\int d^3k \,|\alpha_t(k)|\,\left\|(N^{-1/2}a(k) - \alpha_t(k))\,\Psi_N\right\|_{\mathcal{H}^{(N)}}. \tag{52}$$

Applying Schwarz's inequality in the last line leads to

$$Tr_{L^2(\mathbb{R}^3)}(\gamma_{N,t}^{(0,1)} - |\alpha_t\rangle\langle\alpha_t|) \leq \beta^b(t) + 2\,||\alpha_t||_{L^2(\mathbb{R}^3)}\sqrt{\beta^b(t)}. \tag{53}$$

Inequality (43) follows from the monotonicity of the square root and (46). The estimate (44) originates from [19]. One starts with the relation

$$Tr_{L^2(\mathbb{R}^3)}|\sqrt{1-\Delta}(\gamma_{N,t}^{(1,0)} - |\varphi_t\rangle\langle\varphi_t|)\sqrt{1-\Delta}|$$

$$= \sup_{||A_1||\leq 1} |Tr_{L^2(\mathbb{R}^3)}(A_1\sqrt{1-\Delta}(\gamma_{N,t}^{(1,0)} - |\varphi_t\rangle\langle\varphi_t|)\sqrt{1-\Delta})|, \tag{54}$$

where the supremum is applied to all compact operators A_1 on $L^2(\mathbb{R}^3)$ with norm smaller or equal to one. Then, one continues with

$$Tr_{L^2(\mathbb{R}^3)}(A_1\sqrt{1-\Delta_1}(\gamma_{N,t}^{(1,0)} - |\varphi_t\rangle\langle\varphi_t|)\sqrt{1-\Delta_1})$$

$$= \langle\Psi_N, p_1^{\varphi_t}\sqrt{1-\Delta_1}A_1\sqrt{1-\Delta_1}p_1^{\varphi_t}\Psi_N\rangle - \langle\varphi_t, \sqrt{1-\Delta_1}A_1\sqrt{1-\Delta_1}\varphi_t\rangle \tag{55}$$

$$+ \langle\Psi_N, q_1^{\varphi_t}\sqrt{1-\Delta_1}A_1\sqrt{1-\Delta_1}p_1^{\varphi_t}\Psi_N\rangle \tag{56}$$

$$+ \langle\Psi_N, p_1^{\varphi_t}\sqrt{1-\Delta_1}A_1\sqrt{1-\Delta_1}q_1^{\varphi_t}\Psi_N\rangle \tag{57}$$

$$+ \langle\Psi_N, q_1^{\varphi_t}\sqrt{1-\Delta_1}A_1\sqrt{1-\Delta_1}q_1^{\varphi_t}\Psi_N\rangle. \tag{58}$$

By means of

$$\left\|\sqrt{1-\Delta_1}q_1^{\varphi_t}\Psi_N\right\|^2 = \left\|q_1^{\varphi_t}\Psi_N\right\|^2 + \left\|\nabla_1 q_1^{\varphi_t}\Psi_N\right\|^2 \leq \beta^a(t) + \beta^c(t) \tag{59}$$

and

$$\left\|\sqrt{1-\Delta_1}p_1^{\varphi_t}\right\|^2_{op} \le \langle \varphi_t, (1-\Delta_1)\,\varphi_t \rangle = ||\varphi_t||^2_{H^1(\mathbb{R}^3)} \tag{60}$$

we estimate

$$|(55)| \le |\langle \varphi_t, \sqrt{1-\Delta}A_1\sqrt{1-\Delta}\varphi_t \rangle| |\langle \Psi_N, p_1^{\varphi_t}\Psi_N \rangle - 1| \le ||A_1||_{op}\,||\varphi_t||^2_{H^1(\mathbb{R}^3)}\,\beta^a(t),$$

$$|(57)| \le 2\,||A_1||_{op}\,||\varphi_t||_{H^1(\mathbb{R}^3)}\,\sqrt{\beta^a(t)+\beta^c(t)},$$

$$|(58)| \le ||A_1||_{op}\left(\beta^a(t)+\beta^c(t)\right). \tag{61}$$

This leads to

$$\mathrm{Tr}_{L^2(\mathbb{R}^3)}|\sqrt{1-\Delta}\left(\gamma_{N,t}^{(1,0)} - |\varphi_t\rangle\langle\varphi_t|\right)\sqrt{1-\Delta}| \le \left(1 + ||\varphi_t||^2_{H^1(\mathbb{R}^3)}\right)\left(\beta^a(t)+\beta^c(t)\right)$$
$$+2\,||\varphi_t||_{H^1(\mathbb{R}^3)}\,\sqrt{\beta^a(t)+\beta^c(t)}. \tag{62}$$

$$\square$$

8 Estimates on the Time Derivative

8.1 Preliminary Estimates

In the following, we control the change of β in time by separately estimating the time derivative of β^a and β^b. On the one hand, a change in β^a is caused by the fraction of particles which are not in the condensate state φ_t. On the other hand, there will be a change due to the fact that the particles of the many-body system couple to the quantized radiation field, whereas the condensate wave function is in interaction with the classical field. To control the difference between the quantized and classical field by the functional β^b, we divide (in a similar manner to [47]) the radiation fields into two parts.

$$\widehat{\Phi}^+_\kappa(x) := \int d^3k\,\frac{\tilde{\kappa}(k)}{\sqrt{2\omega(k)}}e^{ikx}a(k), \quad \widehat{\Phi}^-_\kappa(x) := \int d^3k\,\frac{\tilde{\kappa}(k)}{\sqrt{2\omega(k)}}e^{-ikx}a^*(k),$$

$$\Phi^+_\kappa(x,t) := \int d^3k\,\frac{\tilde{\kappa}(k)}{\sqrt{2\omega(k)}}e^{ikx}\alpha_t(k), \quad \Phi^-_\kappa(x,t) := \int d^3k\,\frac{\tilde{\kappa}(k)}{\sqrt{2\omega(k)}}e^{-ikx}\overline{\alpha_t(k)}. \tag{63}$$

For technical reasons, it is then helpful to introduce the following (less singular) auxiliary fields

$$\hat{F}_\kappa^+(x) := \int d^3k\, \tilde{\kappa}(k) e^{ikx} a(k), \qquad \hat{F}_\kappa^-(x) := \int d^3k\, \tilde{\kappa}(k) e^{-ikx} a^*(k),$$

$$F_\kappa^+(x,t) := \int d^3k\, \tilde{\kappa}(k) e^{ikx} \alpha_t(k), \quad F_\kappa^-(x,t) := \int d^3k\, \tilde{\kappa}(k) e^{-ikx} \overline{\alpha_t(k)}. \quad (64)$$

By means of the cutoff function

$$\tilde{\eta}(k) := \frac{\tilde{\kappa}(k)}{\sqrt{2\omega(k)}} = \frac{(2\pi)^{-3/2}}{\sqrt{2\omega(k)}} \mathbb{1}_{|k|\le\Lambda}(k) \qquad (65)$$

we are able to express the scalar fields in terms of the auxiliary fields.

Lemma 8.1. *Let η be the Fourier transform of (65), then*

$$\widehat{\Phi}_\kappa^+(x) = \left(\eta * \hat{F}_\kappa^+\right)(x), \qquad \widehat{\Phi}_\kappa^-(x) = \left(\eta * \hat{F}_\kappa^-\right)(x),$$

$$\Phi_\kappa^+(x,t) = \left(\eta * F_\kappa^+\right)(x,t), \quad \Phi_\kappa^-(x,t) = \left(\eta * F_\kappa^-\right)(x,t). \quad (66)$$

Proof. The proof is a simple application of convolutions theorem. □

In the following, we will integrate the form-factor η of the radiation field and estimate the difference in the auxiliary fields. This requires that the L^2-norms of the cutoff functions

$$\|\kappa\|_2^2 = \Lambda^3/(6\pi^2) \quad \text{and} \quad \|\eta\|_2^2 \le \Lambda^2/(4\pi^2) \qquad (67)$$

are finite. Subsequently, we use Plancherel's theorem and estimate the difference in the auxiliary fields by

$$\int d^3y \left\|\left(N^{-1/2}\hat{F}_\kappa^+(y) - F_\kappa^+(y,t)\right)\Psi_{N,t}\right\|^2 = \int d^3k \left\|\left(N^{-1/2}\hat{F}_\kappa^+(k) - F_\kappa^+(k,t)\right)\Psi_{N,t}\right\|^2$$

$$= \int_{|k|\le\Lambda} d^3k \left\langle\left(N^{-1/2}a(k) - \alpha_t(k)\right)\Psi_{N,t}, \left(N^{-1/2}a(k) - \alpha_t(k)\right)\Psi_{N,t}\right\rangle \le \beta^b\left(\Psi_{N,t}, \alpha_t\right). \quad (68)$$

Pulling the pieces together, we get

Lemma 8.2. *Let $\alpha_t \in L^2(\mathbb{R}^3)$ and $\Psi_{N,t} \in \mathcal{H}^{(N)} \cap \mathcal{D}(\mathcal{N})$. Then, there exists a generic constant C independent of N, Λ and t such that*

$$\left\|\left(N^{-1/2}\widehat{\Phi}_\kappa(x_1) - \Phi_\kappa(x_1,t)\right)\Psi_{N,t}\right\|^2 \le C\Lambda^2\left(\beta^b\left(\Psi_{N,t},\alpha_t\right) + N^{-1}\right), \quad (69)$$

$$\left\|\left(N^{-1/2}\widehat{\Phi}_\kappa^-(x_1) - \Phi_\kappa^-(x_1,t)\right)\Psi_{N,t}\right\|^2 \le C\Lambda^2\left(\beta^b\left(\Psi_{N,t},\alpha_t\right) + N^{-1}\right), \quad (70)$$

$$\left\|\left(N^{-1/2}\widehat{\Phi}_\kappa^+(x_1) - \Phi_\kappa^+(x_1,t)\right)p_1\Psi_{N,t}\right\|^2 \le C\Lambda^2\beta^b\left(\Psi_{N,t},\alpha_t\right). \quad (71)$$

Proof. From the canonical commutation relations (7), we obtain

$$\left[\left(N^{-1/2}\widehat{\Phi}_\kappa^+(x) - \Phi_\kappa^+(x,t)\right), \left(N^{-1/2}\widehat{\Phi}_\kappa^-(x) - \Phi_\kappa^-(x,t)\right)\right] = N^{-1}\|\eta\|_2^2 \quad (72)$$

and estimate

$$\left|\left|\left(N^{-1/2}\widehat{\Phi}_\kappa(x_1) - \Phi_\kappa(x_1, t)\right)\Psi_N\right|\right|^2$$
$$\leq 2\left|\left|\left(N^{-1/2}\widehat{\Phi}_\kappa^+(x_1) - \Phi_\kappa^+(x_1, t)\right)\Psi_N\right|\right|^2 + 2\left|\left|\left(N^{-1/2}\widehat{\Phi}_\kappa^-(x_1) - \Phi_\kappa^-(x_1, t)\right)\Psi_N\right|\right|^2$$
$$\leq 4\left|\left|\left(N^{-1/2}\widehat{\Phi}_\kappa^+(x_1) - \Phi_\kappa^+(x_1, t)\right)\Psi_N\right|\right|^2 + 2N^{-1}||\eta||_2^2. \tag{73}$$

By means of Lemma 8.1, we have

$$\left|\left|\left(N^{-1/2}\widehat{\Phi}_\kappa^+(x_1) - \Phi_\kappa^+(x_1, t)\right)\Psi_N\right|\right|^2$$
$$= \langle \int d^3y\, \eta(x_1 - y)\left(N^{-1/2}\hat{F}_\kappa^+(y) - F_\kappa^+(y, t)\right)\Psi_N, \int d^3z\, \eta(x_1 - z)\left(N^{-1/2}\hat{F}_\kappa^+(z) - F_\kappa^+(z, t)\right)\Psi_N\rangle$$
$$\leq \int d^3y \int d^3z\, |\langle \overline{\eta(x_1 - z)}\left(N^{-1/2}\hat{F}_\kappa^+(y) - F_\kappa^+(y, t)\right)\Psi_N, \overline{\eta(x_1 - y)}\left(N^{-1/2}\hat{F}_\kappa^+(z) - F_\kappa^+(z, t)\right)\Psi_N\rangle|. \tag{74}$$

Cauchy–Schwarz inequality and the estimate $ab \leq 1/2\left(a^2 + b^2\right)$ give rise to

$$||(N^{-1/2}\widehat{\Phi}_\kappa^+(x_1) - \Phi_\kappa^+(x_1, t))\Psi_N||^2$$
$$\leq \int d^3y \int d^3z\, \left|\left|\overline{\eta(x_1 - z)}\left(N^{-1/2}\hat{F}_\kappa^+(y) - F_\kappa^+(y, t)\right)\Psi_N\right|\right| \left|\left|\overline{\eta(x_1 - y)}\left(N^{-1/2}\hat{F}_\kappa^+(z) - F_\kappa^+(z, t)\right)\Psi_N\right|\right|$$
$$\leq \int d^3y \int d^3z\, \left|\left|\overline{\eta(x_1 - z)}\left(N^{-1/2}\hat{F}_\kappa^+(y) - F_\kappa^+(y, t)\right)\Psi_N\right|\right|^2$$
$$= \int d^3y\, \langle\left(N^{-1/2}\hat{F}_\kappa^+(y) - F_\kappa^+(y, t)\right)\Psi_N, \int d^3z\, |\eta(x_1 - z)|^2\left(N^{-1/2}\hat{F}_\kappa^+(y) - F_\kappa^+(y, t)\right)\Psi_N\rangle$$
$$= ||\eta||_2^2 \int d^3y\, \left|\left|\left(N^{-1/2}\hat{F}_\kappa^+(y) - F_\kappa^+(y, t)\right)\Psi_N\right|\right|^2 \leq ||\eta||_2^2\, \beta^b\left(\Psi_{N,t}, \alpha_t\right). \tag{75}$$

In total, we get

$$\left|\left|\left(N^{-1/2}\widehat{\Phi}_\kappa(x_1) - \Phi_\kappa(x_1, t)\right)\Psi_{N,t}\right|\right|^2 \leq ||\eta||_2^2\left(4\beta^b\left(\Psi_{N,t}, \alpha_t\right) + 2N^{-1}\right)$$
$$\leq C\Lambda^2\left(\beta^b\left(\Psi_{N,t}, \alpha_t\right) + N^{-1}\right). \tag{76}$$

The second and third inequality are shown analogously. In the derivation of (71), it is helpful to recall that $\left[p_1, \hat{F}_\kappa^+(y)\right] = \left[p_1, F_\kappa^+(y)\right] = 0$. $\qquad\square$

8.2 Estimate on the Time Derivative of β

Subsequently, we control the change of $\beta\left(\Psi_{N,t}, \varphi_t, \alpha_t\right)$ in time.

Lemma 8.3. Let $(\varphi_t, \alpha_t) \in \mathcal{G}_1$ and $\Psi_{N,t}$ be the unique solution of (8) with initial data $\Psi_{N,0} \in \left(L_s^2(\mathbb{R}^{3N}) \otimes \mathcal{F}\right) \cap \mathcal{D}(\mathcal{N}) \cap \mathcal{D}(\mathcal{N}H_N)$ such that $||\Psi_{N,0}|| = 1$. Then

$$\frac{d}{dt}\beta^a(t) = -2\text{Im}\langle\Psi_{N,t}, \left(N^{-1/2}\widehat{\Phi}_\kappa(x_1) - \Phi_\kappa(x_1,t)\right)q_1^{\varphi_t}\Psi_{N,t}\rangle,$$

$$\frac{d}{dt}\beta^b(t) = 2\text{Im}\langle\Psi_{N,t}, \left(\int d^3k\,\tilde{\eta}(k)(2\pi)^{3/2}\overline{\mathcal{FT}[|\varphi_t|^2](k)}\left(N^{-1/2}a(k) - \alpha_t(k)\right)\right)\Psi_{N,t}\rangle$$

$$- 2\text{Im}\langle\Psi_{N,t}, \left(\int d^3k\,\tilde{\eta}(k)e^{ikx_1}\left(N^{-1/2}a(k) - \alpha_t(k)\right)\right)\Psi_{N,t}\rangle. \qquad (77)$$

Proof. The structure of the proof is best understood by the following formal calculation. A rigorous derivation which requires to show the invariance of the domain $\mathcal{D}(\mathcal{N}) \cap \mathcal{D}(\mathcal{N}H_N)$ during the time evolution is presented in [48, Appendix 2.11]. The functional $\beta^a(t)$ is time-dependent, because $\Psi_{N,t}$ and φ_t evolve according to (8) and (17) respectively. The derivative of the projector q^{φ_t} is given by

$$\frac{d}{dt}q_1^{\varphi_t} = -i\left[H_1^{\text{eff}}, q_1^{\varphi_t}\right], \qquad (78)$$

where $H_1^{\text{eff}} = -\Delta_1 + \Phi_\kappa(x_1, t)$ is the effective Hamiltonian acting on the first variable. This leads to

$$\frac{d}{dt}\beta^a(t) = \frac{d}{dt}\langle\Psi_{N,t}, q_1^{\varphi_t}\Psi_{N,t}\rangle = i\langle\Psi_{N,t}, \left[\left(H_N - H_1^{\text{eff}}\right), q_1^{\varphi_t}\right]\Psi_{N,t}\rangle$$

$$= i\langle\Psi_{N,t}, \left[\left(N^{-1/2}\widehat{\Phi}_\kappa(x_1) - \Phi_\kappa(x_1,t)\right), q_1^{\varphi_t}\right]\Psi_{N,t}\rangle$$

$$= -2\text{Im}\langle\Psi_{N,t}, \left(N^{-1/2}\widehat{\Phi}_\kappa(x_1) - \Phi_\kappa(x_1,t)\right)q_1^{\varphi_t}\Psi_{N,t}\rangle. \qquad (79)$$

We calculate the commutator

$$i\left[H_N, \left(N^{-1/2}a(k) - \alpha_t(k)\right)\right] = -i\omega(k)N^{-1/2}a(k) - iN^{-1}\sum_{j=1}^{N}\tilde{\eta}(k)e^{-ikx_j} \qquad (80)$$

by means of the canonical commutation relations (7) and continue with

$$\frac{d}{dt}\beta^b(t) = \int d^3k\,\frac{d}{dt}\langle\left(N^{-1/2}a(k) - \alpha_t(k)\right)\Psi_{N,t}, \left(N^{-1/2}a(k) - \alpha_t(k)\right)\Psi_{N,t}\rangle$$

$$= \int d^3k\,\langle i\left[H_N, \left(N^{-1/2}a(k) - \alpha_t(k)\right)\right]\Psi_{N,t}, \left(N^{-1/2}a(k) - \alpha_t(k)\right)\Psi_{N,t}\rangle$$

$$+ \int d^3k\,\langle\left(N^{-1/2}a(k) - \alpha_t(k)\right)\Psi_{N,t}, i\left[H_N, \left(N^{-1/2}a(k) - \alpha_t(k)\right)\right]\Psi_{N,t}\rangle$$

$$- \int d^3k\,\langle(\partial_t\alpha_t)(k)\Psi_{N,t}, \left(N^{-1/2}a(k) - \alpha_t(k)\right)\Psi_{N,t}\rangle$$

$$- \int d^3k\,\langle\left(N^{-1/2}a(k) - \alpha_t(k)\right)\Psi_{N,t}, (\partial_t\alpha_t)(k)\Psi_{N,t}\rangle$$

$$= 2\int d^3k\,\text{Re}\langle i\left[H_N, \left(N^{-1/2}a(k) - \alpha_t(k)\right)\right]\Psi_{N,t}, \left(N^{-1/2}a(k) - \alpha_t(k)\right)\Psi_{N,t}\rangle$$

$$-2\int d^3k\,\mathrm{Re}\big\langle(\partial_t\alpha_t)(k)\Psi_{N,t},\big(N^{-1/2}a(k)-\alpha_t(k)\big)\Psi_{N,t}\big\rangle$$

$$= 2\int d^3k\,\mathrm{Re}\{i\omega(k)\langle\big(N^{-1/2}a(k)-\alpha_t(k)\big)\Psi_{N,t},\big(N^{-1/2}a(k)-\alpha_t(k)\big)\Psi_{N,t}\rangle\}$$

$$+2\int d^3k\,\mathrm{Re}\{i\langle N^{-1}\sum_{j=1}^{N}\tilde{\eta}(k)e^{-ikx_j}\Psi_{N,t},\big(N^{-1/2}a(k)-\alpha_t(k)\big)\Psi_{N,t}\rangle\}$$

$$-2\int d^3k\,\mathrm{Re}\{i\langle(2\pi)^{3/2}\tilde{\eta}(k)\mathcal{FT}[|\varphi_t|^2](k)\Psi_{N,t},\big(N^{-1/2}a(k)-\alpha_t(k)\big)\Psi_{N,t}\rangle\}. \tag{81}$$

So if we use the symmetry of the wave function and $\mathrm{Re}\{iz\}=-\mathrm{Im}\{z\}$, we get

$$\frac{d}{dt}\beta^b(t)=-2\int d^3k\,\mathrm{Im}\{\omega(k)\langle\big(N^{-1/2}a(k)-\alpha_t(k)\big)\Psi_{N,t},\big(N^{-1/2}a(k)-\alpha_t(k)\big)\Psi_{N,t}\rangle\}$$

$$-2\int d^3k\,\mathrm{Im}\{\langle\tilde{\eta}(k)e^{-ikx_1}\Psi_{N,t},\big(N^{-1/2}a(k)-\alpha_t(k)\big)\Psi_{N,t}\rangle\}$$

$$+2\int d^3k\,\mathrm{Im}\{\langle(2\pi)^{3/2}\tilde{\eta}(k)\mathcal{FT}[|\varphi_t|^2](k)\Psi_{N,t},\big(N^{-1/2}a(k)-\alpha_t(k)\big)\Psi_{N,t}\rangle\}$$

$$= 2\mathrm{Im}\{\langle\Psi_{N,t},\Big(\int d^3k\,(2\pi)^{3/2}\tilde{\eta}(k)\overline{\mathcal{FT}[|\varphi_t|^2](k)}\big(N^{-1/2}a(k)-\alpha_t(k)\big)\Big)\Psi_{N,t}\rangle\}$$

$$-2\mathrm{Im}\{\langle\Psi_{N,t},\Big(\int d^3k\,\tilde{\eta}(k)e^{ikx_1}\big(N^{-1/2}a(k)-\alpha_t(k)\big)\Big)\Psi_{N,t}\rangle\}. \tag{82}$$

$\square$

Lemma 8.4. *Let* $(\varphi_t,\alpha_t)\in\mathcal{G}_1$ *and* $\Psi_{N,t}$ *be the unique solution of* (8) *with initial data* $\Psi_{N,0}\in\big(L_s^2(\mathbb{R}^{3N})\otimes\mathcal{F}\big)\cap\mathcal{D}(\mathcal{N})\cap\mathcal{D}(\mathcal{N}H_N)$ *such that* $\|\Psi_{N,0}\|=1$. *Then for any* $t\in[0,\infty)$, *there exists a generic constant* C *independent of* N, Λ *and* t *such that*

$$|\frac{d}{dt}\beta\big(\Psi_{N,t},\varphi_t,\alpha_t\big)|\leq C\Lambda^2\big(\beta\big(\Psi_{N,t},\varphi_t,\alpha_t\big)+N^{-1}\big), \tag{83}$$

$$\beta\big(\Psi_{N,t},\varphi_t,\alpha_t\big)\leq e^{C\Lambda^2 t}\big(\beta\big(\Psi_{N,0},\varphi_0,\alpha_0\big)+N^{-1}\big). \tag{84}$$

Proof. Schwarz's inequality and $ab\leq 1/2(a^2+b^2)$ let us estimate the first line of Lemma 8.3 by

$$|\frac{d}{dt}\beta^a(t)|\leq 2|\langle\Psi_{N,t},\big(N^{-1/2}\widehat{\Phi}_\kappa(x_1)-\Phi_\kappa(x_1,t)\big)q_1^{\varphi_t}\Psi_{N,t}\rangle|$$

$$\leq\big\|\big(N^{-1/2}\widehat{\Phi}_\kappa(x_1)-\Phi_\kappa(x_1,t)\big)\Psi_{N,t}\big\|^2+\big\|q_1^{\varphi_t}\Psi_{N,t}\big\|^2. \tag{85}$$

By Lemma 8.2, we obtain

$$|\frac{d}{dt}\beta^a(t)|\leq C\Lambda^2\big(\beta(t)+N^{-1}\big). \tag{86}$$

In order to estimate $\frac{d}{dt}\beta^b(t)$, we notice that

$$\int d^3k\,\tilde{\eta}(k)e^{ikx_1}\left(N^{-1/2}a(k)-\alpha(k,t)\right) = \int d^3y\,\eta(x_1-y)\left(N^{-1/2}\hat{F}_\kappa^+(y)-F_\kappa^+(y,t)\right)$$

$$= \left(N^{-1/2}\hat{\Phi}_\kappa^+(x_1)-\Phi_\kappa^+(x_1,t)\right) \tag{87}$$

and

$$\int d^3k\,\tilde{\eta}(k)(2\pi)^{3/2}\overline{\mathcal{FT}[|\varphi_t|^2](k)}\left(N^{-1/2}a(k)-\alpha_t(k)\right)$$

$$= \int d^3y\left(\eta * |\varphi_t|^2\right)(y,t)\left(N^{-1/2}\hat{F}_\kappa^+(y)-F_\kappa^+(y,t)\right) \tag{88}$$

follow from the convolution theorem. This gives

$$\frac{d}{dt}\beta^b(t) = -2\mathrm{Im}\int d^3y\left\langle\Psi_{N,t},\eta(x_1-y)\left(N^{-1/2}\hat{F}_\kappa^+(y)-F_\kappa^+(y,t)\right)\Psi_{N,t}\right\rangle$$

$$+2\mathrm{Im}\int d^3y\left\langle\Psi_{N,t},\left(\eta * |\varphi_t|^2\right)(y,t)\left(N^{-1/2}\hat{F}_\kappa^+(y)-F_\kappa^+(y,t)\right)\Psi_{N,t}\right\rangle. \tag{89}$$

We see that not only present field boson fluctuations around the coherent state lead to a growth in $\beta^b(t)$ but an additional change appears, because the second-quantized radiation field couples to the mean particle density of the many-body system while the source of the classical field is given by the density of the condensate wave function. In order to estimate the difference between the densities by the functional $\beta^a(t)$, we insert the identity $1 = p_1^{\varphi_t}+q_1^{\varphi_t}$.

$$\frac{d}{dt}\beta^b(t) = -2\mathrm{Im}\int d^3y\left\langle\Psi_{N,t},\,p_1^{\varphi_t}\eta(x_1-y)p_1^{\varphi_t}\left(N^{-1/2}\hat{F}_\kappa^+(y)-F_\kappa^+(y,t)\right)\Psi_{N,t}\right\rangle$$

$$+2\mathrm{Im}\int d^3y\left\langle\Psi_{N,t},\left(\eta * |\varphi_t|^2\right)(y,t)\left(N^{-1/2}\hat{F}_\kappa^+(y)-F_\kappa^+(y,t)\right)\Psi_{N,t}\right\rangle$$

$$-2\mathrm{Im}\int d^3y\left\langle\Psi_{N,t},\,q_1^{\varphi_t}\eta(x_1-y)p_1^{\varphi_t}\left(N^{-1/2}\hat{F}_\kappa^+(y)-F_\kappa^+(y,t)\right)\Psi_{N,t}\right\rangle$$

$$-2\mathrm{Im}\int d^3y\left\langle\Psi_{N,t},\,\eta(x_1-y)q_1^{\varphi_t}\left(N^{-1/2}\hat{F}_\kappa^+(y)-F_\kappa^+(y,t)\right)\Psi_{N,t}\right\rangle. \tag{90}$$

The first two lines are the most important. They become small, because the mean particle density of the many-body system is approximately given by the density of the condensate wave function. From $\eta(-x) = \eta(x)$, we conclude

$$p_1^{\varphi_t}\eta(x_1-y)p_1^{\varphi_t} = p_1^{\varphi_t}\int d^3z\,\eta(z-y)|\varphi_t|^2(z,t) = p_1^{\varphi_t}\left(\eta * |\varphi_t|^2\right)(y,t) \tag{91}$$

and continue with

$$\frac{d}{dt}\beta^b(t) = -2\mathrm{Im}\int d^3y \left\langle \Psi_{N,t}, (p_1^{\varphi_t} - 1)\left(\eta * |\varphi_t|^2\right)(y,t)\left(N^{-1/2}\hat{F}_\kappa^+(y) - F_\kappa^+(y,t)\right)\Psi_{N,t}\right\rangle$$

$$-2\mathrm{Im}\left\langle \Psi_{N,t}, q_1\int d^3y\, \eta(x_1 - y)\left(N^{-1/2}\hat{F}_\kappa^+(y) - F_\kappa^+(y,t)\right)p_1^{\varphi_t}\Psi_{N,t}\right\rangle$$

$$-2\mathrm{Im}\left\langle \Psi_{N,t}, \int d^3y\, \eta(x_1 - y)\left(N^{-1/2}\hat{F}_\kappa^+(y) - F_\kappa^+(y,t)\right)q_1^{\varphi_t}\Psi_{N,t}\right\rangle$$

$$= 2\mathrm{Im}\int d^3y \left\langle \Psi_{N,t}, q_1^{\varphi_t}\left(\eta * |\varphi_t|^2\right)(y,t)\left(N^{-1/2}\hat{F}_\kappa^+(y) - F_\kappa^+(y,t)\right)\Psi_{N,t}\right\rangle \quad (92)$$

$$-2\mathrm{Im}\left\langle \Psi_{N,t}, q_1^{\varphi_t}\left(N^{-1/2}\widehat{\Phi}_\kappa^+(x_1) - \Phi_\kappa^+(x_1,t)\right)p_1^{\varphi_t}\Psi_{N,t}\right\rangle \quad (93)$$

$$-2\mathrm{Im}\left\langle \Psi_{N,t}, \left(N^{-1/2}\widehat{\Phi}_\kappa^+(x_1) - \Phi_\kappa^+(x_1,t)\right)q_1^{\varphi_t}\Psi_{N,t}\right\rangle. \quad (94)$$

In the following, we estimate each line separately.

$$|(91)| \leq 2\left|\int d^3y \left\langle \left(\eta * |\varphi_t|^2\right)(y,t)q_1^{\varphi_t}\Psi_{N,t}, \left(N^{-1/2}\hat{F}_\kappa^+(y) - F_\kappa^+(y,t)\right)\Psi_{N,t}\right\rangle\right|$$

$$\leq \int d^3y \left\langle q_1^{\varphi_t}\Psi_{N,t}, |\left(\eta * |\varphi_t|^2\right)(y,t)|^2 q_1^{\varphi_t}\Psi_{N,t}\right\rangle$$

$$+ \int d^3y \left\|\left(N^{-1/2}\hat{F}_\kappa^+(y) - F_\kappa^+(y,t)\right)\Psi_{N,t}\right\|^2$$

$$\leq \left\|\eta * |\varphi_t|^2\right\|_2^2 \left\langle \Psi_{N,t}, q_1^{\varphi_t}\Psi_{N,t}\right\rangle + \beta^b(t) \leq C\Lambda^2\beta(t). \quad (95)$$

Here, we have used that

$$\left\|\eta * |\varphi_t|^2\right\|_2 \leq \|\eta\|_2 \left\||\varphi_t|^2\right\|_1 = \|\eta\|_2 \|\varphi_t\|_2^2 \leq C\Lambda \quad (96)$$

holds due to Young's inequality and (67). Lemma 8.2 leads to

$$|(92)| \leq 2\left|\left\langle q_1^{\varphi_t}\Psi_{N,t}, \left(N^{-1/2}\widehat{\Phi}_\kappa^+(x_1) - \Phi_\kappa^+(x_1,t)\right)p_1^{\varphi_t}\Psi_{N,t}\right\rangle\right|$$

$$\leq \left\|\left(N^{-1/2}\widehat{\Phi}_\kappa^+(x_1) - \Phi_\kappa^+(x_1,t)\right)p_1^{\varphi_t}\Psi_{N,t}\right\|^2 + \left\|q_1^{\varphi_t}\Psi_{N,t}\right\|^2 \leq C\Lambda^2\beta(t) \quad (97)$$

and

$$|(93)| \leq 2\left|\left\langle \left(N^{-1/2}\widehat{\Phi}_\kappa^-(x_1) - \Phi_\kappa^-(x_1,t)\right)\Psi_{N,t}, q_1^{\varphi_t}\Psi_{N,t}\right\rangle\right|$$

$$\leq \left\|\left(N^{-1/2}\widehat{\Phi}_\kappa^-(x_1) - \Phi_\kappa^-(x_1,t)\right)\Psi_{N,t}\right\|^2 + \left\|q_1^{\varphi_t}\Psi_{N,t}\right\|^2$$

$$\leq C\Lambda^2\left(\beta(t) + N^{-1}\right). \quad (98)$$

In total, we have

$$\left|\frac{d}{dt}\beta^b(t)\right| \leq C\Lambda^2\left(\beta(t) + N^{-1}\right). \quad (99)$$

Now, we can put the terms together to get

$$\frac{d}{dt}\beta(t) \le |\frac{d}{dt}\beta^a(t)| + |\frac{d}{dt}\beta^b(t)| \le C\Lambda^2\left(\beta(t) + N^{-1}\right). \tag{100}$$

Applying Gronwall's lemma proves

$$\beta(t) \le e^{C\Lambda^2 t}\left(\beta(0) + N^{-1}\right). \tag{101}$$

$$\square$$

8.3 Control of the Kinetic Energy

In order to prove the convergence of the one-particle reduced density matrix of the charges in Sobolev norm, it is necessary to control the kinetic energy of the particles which are not in the condensate (see Sect. 7). To this end, we include $\beta^c(\Psi_{N,t}, \varphi_t) :=$ $\left\|\nabla_1 q_1^{\varphi_t}\Psi_{N,t}\right\|^2$ in the definition of the functional and perform a Gronwall estimate for the redefined functional $\beta_2(\Psi_{N,t}, \varphi_t, \alpha_t)$.

Lemma 8.5. *Let $(\varphi_t, \alpha_t) \in \mathcal{G}_2$ and $\Psi_{N,t}$ be the unique solution of* (8) *with initial data $\Psi_{N,0} \in \left(L_s^2(\mathbb{R}^{3N}) \otimes \mathcal{F}\right) \cap \mathcal{D}(\mathcal{N}) \cap \mathcal{D}(\mathcal{N}H_N) \cap \mathcal{D}\left(H_N^2\right)$ such that $\left\|\Psi_{N,0}\right\| = 1$. Then*

$$\frac{d}{dt}\beta^c(\Psi_{N,t}, \varphi_t) = \quad 2\mathrm{Im}\langle p_1^{\varphi_t}\left(N^{-1/2}\widehat{\Phi}_\kappa(x_1) - \Phi_\kappa(x_1, t)\right)\Psi_{N,t}, (-\Delta_1)q_1^{\varphi_t}\Psi_{N,t}\rangle \tag{102}$$

$$- 2\mathrm{Im}\langle\left(N^{-1/2}\widehat{\Phi}_\kappa(x_1) - \Phi_\kappa(x_1, t)\right)p_1^{\varphi_t}\Psi_{N,t}, (-\Delta_1)q_1^{\varphi_t}\Psi_{N,t}\rangle \tag{103}$$

$$- 2\mathrm{Im}\langle N^{-1/2}\widehat{\Phi}_\kappa(x_1)q_1^{\varphi_t}\Psi_{N,t}, (-\Delta_1)q_1^{\varphi_t}\Psi_{N,t}\rangle. \tag{104}$$

Proof. We infer $\Psi_{N,t} \in \left(L_s^2(\mathbb{R}^{3N}) \otimes \mathcal{F}\right) \cap \mathcal{D}(\mathcal{N}) \cap \mathcal{D}(\mathcal{N}H_N) \cap \mathcal{D}\left(H_N^2\right)$ for all $t \in [0, \infty)$ from $\Psi_{N,0} \in \left(L_s^2(\mathbb{R}^{3N}) \otimes \mathcal{F}\right) \cap \mathcal{D}(\mathcal{N}) \cap \mathcal{D}(\mathcal{N}H_N) \cap \mathcal{D}\left(H_N^2\right)$ by Stone's Theorem and the invariance of $\mathcal{D}(\mathcal{N}) \cap \mathcal{D}(\mathcal{N}H_N)$ during the time evolution (see [48, Appendix 2.11]). This ensures that the following expressions are well defined. The derivative of $\beta^c(t)$ is determined by

$$\begin{aligned}
\frac{d}{dt}\beta^c(t) = \quad &i\langle q_1^{\varphi_t}H_N\Psi_{N,t}, (-\Delta_1)q_1^{\varphi_t}\Psi_{N,t}\rangle \quad - i\langle q_1^{\varphi_t}\Psi_{N,t}, (-\Delta_1)q_1^{\varphi_t}H_N\Psi_{N,t}\rangle \\
&+ i\langle[H_1^{\mathrm{eff}}, q_1^{\varphi_t}]\Psi_{N,t}, (-\Delta_1)q_1^{\varphi_t}\Psi_{N,t}\rangle - i\langle q_1^{\varphi_t}\Psi_{N,t}, (-\Delta_1)[H_1^{\mathrm{eff}}, q_1^{\varphi_t}]\Psi_{N,t}\rangle \\
= \quad &i\langle q_1^{\varphi_t}H_N\Psi_{N,t}, (-\Delta_1)q_1^{\varphi_t}\Psi_{N,t}\rangle \quad - i\langle(-\Delta_1)q_1^{\varphi_t}\Psi_{N,t}, q_1^{\varphi_t}H_N\Psi_{N,t}\rangle \\
&+ i\langle[H_1^{\mathrm{eff}}, q_1^{\varphi_t}]\Psi_{N,t}, (-\Delta_1)q_1^{\varphi_t}\Psi_{N,t}\rangle - i\langle(-\Delta_1)q_1^{\varphi_t}\Psi_{N,t}, [H_1^{\mathrm{eff}}, q_1^{\varphi_t}]\Psi_{N,t}\rangle \\
= \quad &-2\mathrm{Im}\langle q_1^{\varphi_t}H_N\Psi_{N,t}, (-\Delta_1)q_1^{\varphi_t}\Psi_{N,t}\rangle - 2\mathrm{Im}\langle[H_1^{\mathrm{eff}}, q_1^{\varphi_t}]\Psi_{N,t}, (-\Delta_1)q_1^{\varphi_t}\Psi_{N,t}\rangle. \tag{105}
\end{aligned}$$

Since $\langle q_1^{\varphi_t} \left(-\Delta_i + N^{-1/2}\widehat{\Phi}_\kappa(x_i)\right) \Psi_{N,t}, (-\Delta_1)q_1^{\varphi_t}\Psi_{N,t}\rangle$ and $\langle q_1^{\varphi_t} H_f \Psi_{N,t}, (-\Delta_1)$ $q_1^{\varphi_t}\Psi_{N,t}\rangle$ are real numbers for $i \in \{2, 3, \ldots, N\}$, this becomes

$$
\begin{aligned}
\frac{d}{dt}\beta^c(t) = & -2\mathrm{Im}\langle q_1^{\varphi_t}\left(-\Delta_1 + N^{-1/2}\widehat{\Phi}_\kappa(x_1)\right)\Psi_{N,t}, (-\Delta_1)q_1^{\varphi_t}\Psi_{N,t}\rangle \\
& + 2\mathrm{Im}\langle q_1^{\varphi_t} H^{\mathrm{eff}}\Psi_{N,t}, (-\Delta_1)q_1^{\varphi_t}\Psi_{N,t}\rangle \\
& - 2\mathrm{Im}\langle H^{\mathrm{eff}}q_1^{\varphi_t}\Psi_{N,t}, (-\Delta_1)q_1^{\varphi_t}\Psi_{N,t}\rangle \\
= & -2\mathrm{Im}\langle q_1^{\varphi_t}\left(N^{-1/2}\widehat{\Phi}_\kappa(x_1) - \Phi_\kappa(x_1, t)\right)\Psi_{N,t}, (-\Delta_1)q_1^{\varphi_t}\Psi_{N,t}\rangle \\
& - 2\mathrm{Im}\langle \Phi_\kappa(x_1, t)q_1^{\varphi_t}\Psi_{N,t}, (-\Delta_1)q_1^{\varphi_t}\Psi_{N,t}\rangle \\
& - 2\mathrm{Im}\left\|(-\Delta_1)q_1^{\varphi_t}\Psi_{N,t}\right\|^2 \\
= & -2\mathrm{Im}\langle q_1^{\varphi_t}\left(N^{-1/2}\widehat{\Phi}_\kappa(x_1) - \Phi_\kappa(x_1, t)\right)\Psi_{N,t}, (-\Delta_1)q_1^{\varphi_t}\Psi_{N,t}\rangle \\
& - 2\mathrm{Im}\langle \Phi_\kappa(x_1, t)q_1^{\varphi_t}\Psi_{N,t}, (-\Delta_1)q_1^{\varphi_t}\Psi_{N,t}\rangle. \quad (106)
\end{aligned}
$$

The identity $q_1^{\varphi_t}\mathcal{O} = \mathcal{O}p_1^{\varphi_t} + \mathcal{O}q_1^{\varphi_t} - p_1^{\varphi_t}\mathcal{O}$ (for any operator $\mathcal{O}$) and

$$
\begin{aligned}
-\langle \Phi_\kappa(x_1, t)q_1^{\varphi_t}\Psi_{N,t}, (-\Delta_1)q_1^{\varphi_t}\Psi_{N,t}\rangle = & \langle\left(N^{-1/2}\widehat{\Phi}_\kappa(x_1) - \Phi_\kappa(x_1, t)\right) q_1^{\varphi_t}\Psi_{N,t}, (-\Delta_1)q_1^{\varphi_t}\Psi_{N,t}\rangle \\
& - \langle N^{-1/2}\widehat{\Phi}_\kappa(x_1)q_1^{\varphi_t}\Psi_{N,t}, (-\Delta_1)q_1^{\varphi_t}\Psi_{N,t}\rangle \quad (107)
\end{aligned}
$$

then lead to Lemma 8.5. $\qquad\square$

Lemma 8.6. *Let $(\varphi_t, \alpha_t) \in \mathcal{G}_2$ and $\Psi_{N,t}$ be the unique solution of (8) with initial data $\Psi_{N,0} \in \left(L_s^2(\mathbb{R}^{3N}) \otimes \mathcal{F}\right) \cap \mathcal{D}(\mathcal{N}) \cap \mathcal{D}(\mathcal{N}H_N) \cap \mathcal{D}\left(H_N^2\right)$ such that $\left\|\Psi_{N,0}\right\| = 1$. Then, there exists a positive monotone increasing function $C(s)$ of the norms $\left\|\alpha_s\right\|_{L^2(\mathbb{R}^3)}$ and $\left\|\varphi_s\right\|_{H^1(\mathbb{R}^3)}$ such that*

$$
\begin{aligned}
&\left|\frac{d}{dt}\beta_2\left(\Psi_{N,t}, \varphi_t, \alpha_t\right)\right| \leq \Lambda^4 C(t)\left(\beta_2\left(\Psi_{N,t}, \varphi_t, \alpha_t\right) + N^{-1}\right), \\
&\beta_2\left(\Psi_{N,t}, \varphi_t, \alpha_t\right) \leq e^{\Lambda^4 \int_0^t C(s)ds}\left(\beta_2\left(\Psi_{N,0}, \varphi_0, \alpha_0\right) + N^{-1}\right) \quad (108)
\end{aligned}
$$

hold for any $t \in [0, \infty)$.

Proof. In order to estimate $\frac{d}{dt}\beta^c(t)$ by $\beta(t)$ and $\left\|\nabla_1 q_1^{\varphi_t}\Psi_{N,t}\right\|$, we will integrate by parts and apply Schwarz's inequality. The gradient will then occasionally act on the radiation fields, which will give rise to the vector fields

$$
\begin{aligned}
(\nabla\widehat{\Phi}_\kappa)(x) &= \int d^3k\, \tilde{\eta}(k)ki\left(e^{ikx}a(k) - e^{-ikx}a^*(k)\right), \\
(\nabla\Phi_\kappa)(x, t) &= \int d^3k\, \tilde{\eta}(k)ki\left(e^{ikx}\alpha_t(k) - e^{-ikx}\overline{\alpha_t(k)}\right). \quad (109)
\end{aligned}
$$

We define the vector field $\tilde{\Theta}(k) := \tilde{\eta}(k)k$ and its Fourier transform Θ with $\sum_{i=1}^3$ $\left\|\Theta^i\right\|_2^2 \leq \Lambda^4/(16\pi^2)$. This allows us to obtain the relations

$$(\nabla\widehat{\Phi}_\kappa^+)(x) = i\left(\Theta * \hat{F}_\kappa^+\right)(x), \quad (\nabla\Phi_\kappa^+)(x,t) = i\left(\Theta * F_\kappa^+\right)(x) \qquad (110)$$

between the vector fields and the auxiliary fields (64). In analogy to Lemma 8.2, one proves the estimates

$$\left\|\left(N^{-1/2}(\nabla\widehat{\Phi}_\kappa)(x_1) - (\nabla\Phi_\kappa)(x_1,t)\right)p_1\Psi_{N,t}\right\|^2 \le C\Lambda^4\left(\beta^b(t) + N^{-1}\right),$$

$$\left\|\left(N^{-1/2}(\nabla\widehat{\Phi}_\kappa)(x_1) - (\nabla\Phi_\kappa)(x_1,t)\right)q_1\Psi_{N,t}\right\|^2 \le C\Lambda^4\left(\beta^b(t) + N^{-1}\right),$$

$$\left\|\left(N^{-1/2}\widehat{\Phi}_\kappa(x_1) - \Phi_\kappa(x_1,t)\right)\nabla_1 p_1\Psi_{N,t}\right\|^2 \le C\Lambda^2\|\nabla\varphi_t\|_2^2\left(\beta^b(t) + N^{-1}\right).$$
$$(111)$$

The first term of $\frac{d}{dt}\beta^c(t)$ is estimated by

$$\begin{aligned}
|(101)| &\le 2|\langle p_1^{\varphi_t}\left(N^{-1/2}\widehat{\Phi}_\kappa(x_1) - \Phi_\kappa(x_1,t)\right)\Psi_{N,t}, (-\Delta_1)q_1^{\varphi_t}\Psi_{N,t}\rangle| \\
&= 2|\langle\nabla_1 p_1^{\varphi_t}\left(N^{-1/2}\widehat{\Phi}_\kappa(x_1) - \Phi_\kappa(x_1,t)\right)\Psi_{N,t}, \nabla_1 q_1^{\varphi_t}\Psi_{N,t}\rangle| \\
&\le \left\|\nabla_1 p_1\left(N^{-1/2}\widehat{\Phi}_\kappa(x_1) - \Phi_\kappa(x_1,t)\right)\Psi_N\right\|^2 + \|\nabla_1 q_1\Psi_N\|^2 \\
&\le \|\nabla\varphi_t\|^2\left\|\left(N^{-1/2}\widehat{\Phi}_\kappa(x_1) - \Phi_\kappa(x_1,t)\right)\Psi_N\right\|^2 + \|\nabla_1 q_1\Psi_N\|^2. \quad (112)
\end{aligned}$$

Lemma 8.2 gives rise to

$$\begin{aligned}
|(101)| &\le C\Lambda^2\|\nabla\varphi_t\|^2\left(\beta^b(t) + N^{-1}\right) + \|\nabla_1 q_1\Psi_N\|^2 \\
&\le \Lambda^2 C(\|\varphi_t\|_{H^1})\left(\beta_2(t) + N^{-1}\right). \qquad (113)
\end{aligned}$$

Likewise, we estimate

$$\begin{aligned}
|(102)| &\le 2|\langle\left(N^{-1/2}\widehat{\Phi}_\kappa(x_1) - \Phi_\kappa(x_1,t)\right)p_1^{\varphi_t}\Psi_{N,t}, (-\Delta_1)q_1^{\varphi_t}\Psi_{N,t}\rangle| \\
&= 2|\langle\nabla_1\left(N^{-1/2}\widehat{\Phi}_\kappa(x_1) - \Phi_\kappa(x_1,t)\right)p_1\Psi_N, \nabla_1 q_1\Psi_N\rangle| \\
&\le \left\|\nabla_1\left(N^{-1/2}\widehat{\Phi}_\kappa(x_1) - \Phi_\kappa(x_1,t)\right)p_1\Psi_N\right\|^2 + \|\nabla_1 q_1\Psi_N\|^2. \quad (114)
\end{aligned}$$

Due to triangular inequality, $(a+b)^2 \le 2\left(a^2+b^2\right)$ and (111), this becomes

$$\begin{aligned}
|(102)| \le\ &2\left\|\left(N^{-1/2}\widehat{\Phi}_\kappa(x_1) - \Phi_\kappa(x_1,t)\right)\nabla_1 p_1\Psi_N\right\|^2 \\
&+ 2\left\|\left(N^{-1/2}(\nabla\widehat{\Phi}_\kappa)(x_1) - (\nabla\Phi_\kappa)(x_1)p_1\Psi_N\right)\right\|^2 + \|\nabla_1 q_1\Psi_N\|^2 \\
\le\ &\Lambda^4 C(\|\varphi_t\|_{H^1})\left(\beta_2(t) + N^{-1}\right). \qquad (115)
\end{aligned}$$

Next, we consider line

$$
\begin{aligned}
|(103)| &= -2\mathrm{Im}\langle \nabla_1 N^{-1/2}\widehat{\Phi}_\kappa(x_1)q_1^{\varphi_t}\Psi_{N,t}, \nabla_1 q_1^{\varphi_t}\Psi_{N,t}\rangle \\
&= -2\mathrm{Im}\langle N^{-1/2}(\nabla\widehat{\Phi}_\kappa)(x_1)q_1^{\varphi_t}\Psi_{N,t}, \nabla_1 q_1^{\varphi_t}\Psi_{N,t}\rangle \\
&\quad -2\mathrm{Im}\langle N^{-1/2}\widehat{\Phi}_\kappa(x_1)\nabla_1 q_1^{\varphi_t}\Psi_{N,t}, \nabla_1 q_1^{\varphi_t}\Psi_{N,t}\rangle.
\end{aligned} \tag{116}
$$

The scalar product in the last line is easily shown to be real. This yields

$$
\begin{aligned}
|(103)| &= -2\mathrm{Im}\langle N^{-1/2}(\nabla\widehat{\Phi}_\kappa)(x_1)q_1^{\varphi_t}\Psi_{N,t}, \nabla_1 q_1^{\varphi_t}\Psi_{N,t}\rangle \\
&= -2\mathrm{Im}\langle \left(N^{-1/2}(\nabla\widehat{\Phi}_\kappa)(x_1) - (\nabla\Phi_\kappa)(x_1,t)\right)q_1^{\varphi_t}\Psi_{N,t}, \nabla_1 q_1^{\varphi_t}\Psi_{N,t}\rangle \\
&\quad -2\mathrm{Im}\langle (\nabla\Phi_\kappa)(x_1,t)q_1^{\varphi_t}\Psi_{N,t}, \nabla_1 q_1^{\varphi_t}\Psi_{N,t}\rangle.
\end{aligned} \tag{117}
$$

and allows us to estimate

$$
\begin{aligned}
|(103)| &\leq 2|\langle \left(N^{-1/2}(\nabla\widehat{\Phi}_\kappa)(x_1) - (\nabla\Phi_\kappa)(x_1,t)\right)q_1^{\varphi_t}\Psi_{N,t}, \nabla_1 q_1^{\varphi_t}\Psi_{N,t}\rangle| \\
&\quad + 2|\langle (\nabla\Phi_\kappa)(x_1,t)q_1^{\varphi_t}\Psi_{N,t}, \nabla_1 q_1^{\varphi_t}\Psi_{N,t}\rangle| \\
&\leq \left\|\left(N^{-1/2}(\nabla\widehat{\Phi}_\kappa)(x_1) - (\nabla\Phi_\kappa)(x_1,t)\right)q_1^{\varphi_t}\Psi_{N,t}\right\|^2 + \left\|(\nabla\Phi_\kappa)(x_1,t)q_1^{\varphi_t}\Psi_{N,t}\right\|^2 \\
&\quad + 2\left\|\nabla_1 q_1^{\varphi_t}\Psi_{N,t}\right\|^2 \leq C\Lambda^4\left(\beta^b(t) + N^{-1}\right) + C\Lambda^4\|\alpha_t\|_2^2\,\beta^a(t) + 2\beta^c(t) \\
&\leq \Lambda^4 C(\|\alpha_t\|_2)\left(\beta_2(t) + N^{-1}\right).
\end{aligned} \tag{118}
$$

Here, we used (111) and the fact that

$$
\|(\nabla\Phi_\kappa)(\cdot,t)\|_\infty \leq C\Lambda^2\|\alpha_t\|_2 \tag{119}
$$

holds because of Schwarz's inequality. In total, we have

$$
|\frac{d}{dt}\beta^c(t)| \leq \Lambda^4 C(\|\varphi_t\|_{H^1}, \|\alpha_t\|)\left(\beta_2 + N^{-1}\right). \tag{120}
$$

With Lemma 8.4, this implies

$$
|\frac{d}{dt}\beta_2\left[\Psi_{N,t}, \varphi_t, \alpha_t\right]| \leq \Lambda^4 C(\|\varphi_t\|_{H^1}, \|\alpha_t\|)\left(\beta_2\left[\Psi_{N,t}, \varphi_t, \alpha_t\right] + N^{-1}\right) \tag{121}
$$

Using the shorthand notation $C(t) := C(\|\varphi_t\|_{H^1}, \|\alpha_t\|)$, we obtain

$$
\beta_2\left[\Psi_{N,t}, \varphi_t, \alpha_t\right] \leq e^{\Lambda^4\int_0^t C(s)ds}\left(\beta_2\left[\Psi_{N,0}, \varphi_0, \alpha_0\right] + N^{-1}\right) \tag{122}
$$

by means of Gronwall's lemma. □

9 Initial States

Subsequently, we are concerned with the initial states of Theorem 3.3.

Lemma 9.1. *Let* $\Psi_{N,0} \in \left(L^2_s(\mathbb{R}^{3N}) \otimes \mathcal{F}\right) \cap \mathcal{D}(\mathcal{N})$ *with* $\left\|\Psi_{N,0}\right\| = 1$ *and* $(\varphi_0, \alpha_0) \in L^2(\mathbb{R}^3) \oplus L^2(\mathbb{R}^3)$ *with* $\|\varphi_0\| = 1$. *Then*

$$\beta^a(\Psi_{N,0}, \varphi_0) \le Tr_{L^2(\mathbb{R}^3)}|\gamma_{N,0}^{(1,0)} - |\varphi_0\rangle\langle\varphi_0|| = a_N,$$

$$\beta^b(\Psi_{N,0}, \alpha_0) = N^{-1}\langle W^{-1}(\sqrt{N}\alpha_0)\Psi_{N,0}, \mathcal{N}W^{-1}(\sqrt{N}\alpha_0)\Psi_{N,0}\rangle = b_N. \quad (123)$$

Proof. The first inequality is a consequence of Lemma 7.1. Before we prove the second relation, we justify (39). Therefore, it is useful to note that the Weyl operator $(f \in L^2(\mathbb{R}^3))$

$$W(f) = \exp\left(\int d^3k\, f(k)a^*(k) - \overline{f(k)}a(k)\right) \quad (124)$$

is unitary

$$W^{-1}(f) = W^*(f) = W(-f) \quad (125)$$

and satisfies[13]

$$W^*(f)a(k)W(f) = a(k) + f(k), \quad W^*(f)a^*(k)W(f) = a^*(k) + \overline{f(k)}. \quad (126)$$

This leads to

$$\begin{aligned}
\beta^b(\Psi_{N,t}, \alpha_t) &= \int d^3k\, \left\|\left(N^{-1/2}a(k) - \alpha_t(k)\right)\Psi_{N,t}\right\|^2 \\
&= \int d^3k\, \left\|W^*(\sqrt{N}\alpha_t)\left(N^{-1/2}a(k) - \alpha_t(k)\right)W(\sqrt{N}\alpha_t)W^*(\sqrt{N}\alpha_t)\Psi_{N,t}\right\|^2 \\
&= \int d^3k\, \left\|N^{-1/2}a(k)W^*(\sqrt{N}\alpha_t)\Psi_{N,t}\right\|^2 \\
&= N^{-1}\langle W^*(\sqrt{N}\alpha_t)e^{-iH_N t}\Psi_{N,0}, \mathcal{N}W^*(\sqrt{N}\alpha_t)e^{-iH_N t}\Psi_{N,0}\rangle. \quad (127)
\end{aligned}$$

Let

$$\mathcal{U}_N(t; 0) := W^*(\sqrt{N}\alpha_t)e^{-iH_N t}W(\sqrt{N}\alpha_0) \quad (128)$$

denote the fluctuation dynamics then

$$\beta^b(\Psi_{N,t}, \alpha_t) = N^{-1}\langle \mathcal{U}_N(t; 0)W^{-1}(\sqrt{N}\alpha_0)\Psi_{N,0}, \mathcal{N}\mathcal{U}_N(t; 0)W^{-1}(\sqrt{N}\alpha_0)\Psi_{N,0}\rangle \quad (129)$$

---

[13]More information is given for instance in [12, p. 9].

follows from the unitarity of the Weyl operator. In particular, we have

$$\beta^b(\Psi_{N,0}, \alpha_0) = N^{-1}\langle W^{-1}(\sqrt{N}\alpha_0)\Psi_{N,0}, \mathcal{N}W^{-1}(\sqrt{N}\alpha_0)\Psi_{N,0}\rangle = b_N. \tag{130}$$

$\square$

In the following, we consider initial states of product form (15).

Lemma 9.2. *Let* $(\varphi_0, \alpha_0) \in H^2(\mathbb{R}^3) \oplus L_1^2(\mathbb{R}^3)$ *with* $||\varphi_0|| = 1$ *and* $\Psi_{N,0} = \varphi_0^{\otimes N} \otimes W(\sqrt{N}\alpha_0)\Omega$. *Then*

$$a_N = Tr_{L^2(\mathbb{R}^3)}|\gamma_{N,0}^{(1,0)} - |\varphi_0\rangle\langle\varphi_0|| = 0, \tag{131}$$

$$b_N = N^{-1}\langle W^{-1}(\sqrt{N}\alpha_0)\Psi_{N,0}, \mathcal{N}W^{-1}(\sqrt{N}\alpha_0)\Psi_{N,0}\rangle = 0 \ and \tag{132}$$

$$\Psi_{N,0} \in \left(L_s^2(\mathbb{R}^{3N}) \otimes \mathcal{F}\right) \cap \mathcal{D}(\mathcal{N}) \cap \mathcal{D}(\mathcal{N}H_N). \tag{133}$$

Let $(\varphi_0, \alpha_0) \in H^4(\mathbb{R}^3) \oplus L_2^2(\mathbb{R}^3)$ *with* $||\varphi_0|| = 1$ *then*

$$c_N = ||\nabla_1 q_1^{\varphi_0}\Psi_{N,0}||^2 = 0 \tag{134}$$

$$\Psi_{N,0} \in \left(L_s^2(\mathbb{R}^{3N}) \otimes \mathcal{F}\right) \cap \mathcal{D}(\mathcal{N}) \cap \mathcal{D}(\mathcal{N}H_N) \cap \mathcal{D}\left(H_N^2\right). \tag{135}$$

Proof. From the definition of the one-particle reduced density matrix and (130), we directly obtain the relations (131) and (132).
In order to show (133), we point out that

$$\Psi_{N,0}^{(n)}(X_N, K_n) = \prod_{i=1}^{N}\varphi_0(x_i)e^{-N||\alpha_0||^2/2}(n!)^{-1/2}\prod_{j=1}^{n}N^{\sqrt{1/2}}\alpha_0(k_j) \tag{136}$$

follows from the definition of the the Weyl operators [12, p. 8]. A direct calculation gives

$$\sum_{n=1}^{\infty}n^2\left|\left|\Psi_{N,0}^{(n)}\right|\right|^2 = N||\alpha_0||^2 + N^2||\alpha_0||^4. \tag{137}$$

Hence, $\Psi_{N,0}^{(n)} \in \mathcal{D}(\mathcal{N})$ (see (20)). Moreover, we have $\Psi_{N,0} \in \mathcal{D}(\sum_{i=1}^{N} -\Delta_i)$ because $\varphi_0 \in H^2(\mathbb{R}^3)$. A straightforward estimate leads to

$$\sum_{n=1}^{\infty}\int d^{3N}x\, d^{3n}k\,|\sum_{j=1}^{n}w(k_j)\Psi_{N,0}^{(n)}(X_N, K_n)|^2 \le C(N, ||\alpha_0||_{L_1^2(\mathbb{R}^3)}). \tag{138}$$

From (12), we then conclude $\Psi_{N,0}^{(n)} \in \mathcal{D}(H_f)$ and $\Psi_{N,0}^{(n)} \in \mathcal{D}(H_N) = \mathcal{D}(\sum_{i=1}^{N} -\Delta_i) \cap \mathcal{D}(H_f)$. Similarly, one derives

$$\sum_{n=1}^{N} n^2 \left\| (H_N \Psi_{N,0})^{(n)} \right\|^2 \leq C \sum_{n=1}^{\infty} n^2 \left(\left\| \sum_{j=1}^{N} \Delta_j \Psi_{N,0}^{(n)} \right\|^2 + \left\| \sum_{j=1}^{N} N^{-1/2} (\widehat{\Phi}_\kappa(x_j) \Psi_{N,0})^{(n)} \right\|^2 \right)$$

$$+ C \sum_{n=1}^{\infty} n^2 \left\| (H_f \Psi_{N,0})^{(n)} \right\|^2 \leq C(N, \Lambda, \|\varphi_0\|_{H^2(\mathbb{R}^3)}, \|\alpha_0\|_{L^2_1(\mathbb{R}^3)}).$$

$$(139)$$

and concludes $\Psi_{N,0} \in \mathcal{D}(\mathcal{N} H_N) = \{ \Psi_N \in \mathcal{D}(H_N) : H_N \Psi_N \in \mathcal{D}(\mathcal{N}) \}$.
Equation (134) holds because $\Psi_{N,0}$ is in the kernel of the projector $q_1^{\varphi_0}$.
In order to show (135), we would like to note that $(\varphi_0, \alpha_0) \in (H^4(\mathbb{R}^3), L^2_2(\mathbb{R}^3))$, $| \cdot |^2 \tilde{\eta} \in L^2(\mathbb{R}^3)$ and $\tilde{\eta} \in L^2(\mathbb{R}^3)$ imply $H_N \Psi_{N,0} \in \mathcal{D}(\sum_{i=1}^{N} -\Delta_i)$. By means of the estimate

$$\sum_{n=1}^{\infty} d^{3N} x \, d^{3n} k \, | \sum_{j=1}^{n} w(k_j) (H_N \Psi_{N,0})^{(n)} (X_N, K_n)|^2 \leq C(N, \Lambda, \|\varphi_0\|_{H^2(\mathbb{R}^3)}, \|\alpha_0\|_{L^2_2(\mathbb{R}^3)})$$

$$(140)$$

one obtains $H_N \Psi_{N,0} \in \mathcal{D}(H_f)$. In total, we have $H_N \Psi_{N,0} \in \mathcal{D}(H_N)$ and $\Psi_{N,0} \in \mathcal{D}(H_N^2)$. $\qquad \square$

10 Proof of Theorem 3.3

In order to finish the proof of Theorem 3.3, we remark that Lemma 9.1 leads to

$$\beta(\Psi_{N,0}, \varphi_0, \alpha_0) \leq a_N + b_N,$$
$$\beta_2(\Psi_{N,0}, \varphi_0, \alpha_0) \leq a_N + b_N + c_N.$$
$$(141)$$

We then choose for a given time $t \in [0, \infty)$ the number N of charged particles large enough such that the values of $\beta(\Psi_{N,t}, \varphi_t, \alpha_t)$ in (84) and $\beta_2(\Psi_{N,t}, \varphi_t, \alpha_t)$ in (108) are smaller than one and derive Theorem 3.3 by means of Lemma 7.1.

Acknowledgements We would like to thank Dirk André Deckert, Marco Falconi and David Mitrouskas for helpful discussions. N.L. gratefully acknowledges financial support by the Cusanuswerk and the European Research Council (ERC) under the European Union's Horizon 2020 research and innovation programme (grant agreement No 694227). The article appeared in slightly different form in one of the author's (N.L.) Ph.D. thesis [48].

References

1. Ammari, Z., Falconi, M.: Bohr's correspondence principle for the renormalized nelson model. SIAM J. Math. Anal. **49**(6), 5031–5095 (2017). arXiv:1602.03212 (2016)

2. Falconi, M.: Classical limit of the Nelson model with cutoff. J. Math. Phys. **54**(1), 012303 (2013). arXiv:1205.4367 (2012)
3. Frank, R.L., Schlein, B.: Dynamics of a strongly coupled polaron. Lett. Math. Phys. **104**, 911–929 (2014). arXiv:1311.5814 (2013)
4. Frank, R.L., Gang, Z.: Derivation of an effective evolution equation for a strongly coupled polaron. Anal. PDE **10**(2), 379–422 (2017). arXiv:1505.03059 (2015)
5. Ginibre, J., Nironi, F., Velo, G.: Partially classical limit of the Nelson model. Ann. H. Poincaré **7**, 21–43 (2006). arXiv:math-ph/0411046 (2004)
6. Griesemer, M.: On the dynamics of polarons in the strong-coupling limit. Rev. Math. Phys. **29**(10), 1750030 (2017). arXiv:1612.00395 (2016)
7. Leopold, N., Pickl, P.: Derivation of the Maxwell-Schrödinger Equations from the Pauli-Fierz Hamiltonian (2016). arXiv:1609.01545
8. Teufel, S.: Effective N-body dynamics for the massless Nelson Model and adiabatic decoupling without spectral gap. Ann. H. Poincaré **3**, 939–965 (2002). arXiv:math-ph/0203046 (2002)
9. Cohen-Tannoudji, C., Dupont-Roc, J., Grynberg, G.: Photons and Atoms: Introduction to Quantum Electrodynamics. John Wiley and Sons, Inc. (1997). ISBN 978-0-471-18433-1
10. Falconi, M.: Classical limit of the Nelson model, Ph.D. thesis (2012). http://user.math.uzh.ch/falconi/other/phd_thesis.pdf
11. Pickl, P.: A simple derivation of mean field limits for quantum systems. Lett. Math. Phys. **97**, 151–164 (2011). arXiv:0907.4464 (2009)
12. Rodnianski, I., Schlein, B.: Quantum fluctuations and rate of convergence towards mean field dynamics. Commun. Math. Phys. **291**(1), 31–61 (2009). arXiv:0711.3087 (2007)
13. Matulevičius, V.: Maxwell's Equations as Mean Field Equations, Master thesis (2011). http://www.mathematik.uni-muenchen.de/%7Ebohmmech/theses/Matulevicius_Vytautas_MA.pdf
14. Nelson, E.: Interaction of nonrelativistic particles with a quantized scalar field. J. Math. Phys. **5**(9), 1190–1197 (1964)
15. Spohn, H.: Dynamics of Charged Particles and their Radiation Field. Cambridge University Press, Cambridge (2004). ISBN 0-521-83697-2
16. Michelangeli, A.: Equivalent definitions of asymptotic 100% BEC. Nuovo Cimento Sec. B. **123**, 181–192 (2008)
17. Pecher, H.: Some new well-posedness results for the Klein-Gordon-Schrödinger system. Differ. Integral Equ. **25**(1–2), 117–142 (2012). arXiv:1106.2116 (2011)
18. Reed, M., Simon, B.: Methods of Modern Mathematical Physics II: Fourier Analysis, Self-Adjointness. Academic Press, Inc., New York (1975). ISBN 978-0-12-585002-5
19. Mitrouskas, D., Petrat, S., Pickl, P.: Bogoliubov corrections and trace norm convergence for the Hartree dynamics (2016). arXiv:1609.06264
20. Anapolitanos, I., Hott, M.: A simple proof of convergence to the Hartree dynamics in Sobolev trace norms. J. Math. Phys. **57**, 122108 (2016). arXiv:1608.01192 (2016)
21. Anapolitanos, I., Hott, M., Hundertmark, D.: Derivation of the Hartree equation for compound Bose gases in the mean field limit. Rev. Math. Phys. **29**, 1750022 (2017). arXiv:1702.00827 (2017)
22. Boßmann, L.: Derivation of the 1d NLS equation from the 3d quantum many-body dynamics of strongly confined bosons (2018). arXiv:1803.11011
23. Boßmann, L., Teufel, S.: Derivation of the 1d Gross-Pitaevskii equation from the 3d quantum many-body dynamics of strongly confined bosons (2018). arXiv:1803.11026
24. Jeblick, M., Pickl, P.: Derivation of the time dependent two dimensional focusing NLS equation (2017). arXiv:1707.06523
25. Jeblick, M., Pickl, P.: Derivation of the time dependent Gross-Pitaesvkii equation for a class of non purely positive potentials (2018). arXiv:1801.04799
26. Jeblick, M., Leopold, N., Pickl, P.: Derivation of the Time Dependent Gross-Pitaevskii Equation in Two Dimensions (2016). arXiv:1608.05326
27. von Keler, J., Teufel, S.: The NLS Limit for Bosons in a Quantum Waveguide. Ann. H. Poincaré **17**, 3321–3360 (2016). arXiv:1510.03243 (2015)

28. Knowles, A., Pickl, P.: Mean-field dynamics: singular potentials and rate of convergence. Commun. Math. Phys. **298**, 101–139 (2010). arXiv:0907.4313 (2009)
29. Lührmann, J.: Mean-field quantum dynamics with magnetic fields. J. Math. Phys. **53**, 022105 (2012). arXiv:1202.1065
30. Michelangeli, A., Olgiati, A.: Mean-field quantum dynamics for a mixture of Bose-Einstein condensates. Anal. Math. Phys. **7**(4), 377–416 (2017). arXiv:1603.02435 (2016)
31. Michelangeli, A., Olgiati, A.: Gross-Pitaevskii non-linear dynamics for pseudo-spinor condensates. J. Nonlinear Math. Phys. **24**, 426–464 (2017). arXiv:1704.00150
32. Olgiati, A.: Remarks on the derivation of Gross-Pitaevskii equation with magnetic Laplacian. In: Advances in Quantum Mechanics. Springer INdAM Ser. 18, pp. 257–266. Springer, Cham (2017). arXiv:1709.04841
33. Pickl, P.: Derivation of the time dependent Gross-Pitaevskii equation without positivity condition on the interaction. J. Stat. Phys. **140**, 76–89 (2010). arXiv:0907.4466 (2009)
34. Pickl, P.: Derivation of the time dependent Gross-Pitaevskii equation with external fields. Rev. Math. Phys. **27**, 1550003 (2015). arXiv:1001.4894 (2010)
35. Benedikter, N., Porta, M., Schlein, B.: Effective Evolution Equations from Quantum Dynamics. SpringerBriefs in Mathematical Physics Cambridge, Cambridge (2016). arXiv:1502.02498 (2015). ISBN 978-3-319-24898-1
36. Ginibre, J., Velo, G.: The classical field limit of scattering theory for nonrelativistic many-boson systems I and II. Commun. Math. Phys. **66**(1), 37–76 (1979); **68**(1), 45–68 (1979)
37. Hepp, K.: The classical limit for quantum mechanical correlation functions. Commun. Math. Phys. **35**, 265–277 (1974)
38. Ben Arous, G., Kirkpatrick, K., Schlein, B.: A central limit theorem in many-body quantum dynamics. Commun. Math. Phys. **321**(2), 371–417 (2013). arXiv:1111.6999 (2011)
39. Benedikter, N., de Oliveira, G., Schlein, B.: Quantitative derivation of the gross-pitaevskii equation. Commun. Pure Appl. Math. **68**(8), 1399–1482 (2015). arXiv:1208.0373 (2012)
40. Boccato, C., Cenatiempo, S., Schlein, B.: Quantum many-body fluctuations around nonlinear Schrödinger dynamics. Ann. H. Poincaré **18**(1), 113–191 (2017). arXiv:1509.03837 (2015)
41. Brennecke, C., Schlein, B.: Gross-Pitaevskii dynamics for Bose-Einstein condensates (2017). arXiv:1702.05625
42. Brennecke, C., Nam, P.T., Napiórkowski, M., Schlein, B.: Fluctuations of N-particle quantum dynamics around the nonlinear Schrödinger equation (2017). arXiv:1710.09743
43. Buchholz, S., Saffirio, C., Schlein, B.: Multivariate central limit theorem in quantum dynamics. J. Stat. Phys. **154**(1–2), 113–152 (2014). arXiv:1309.1702 (2013)
44. Chen, L., Lee, J. O., Schlein, B.: Rate of convergence towards hartree dynamics. J. Stat. Phys. **144**(4), 872–903 (2011). arXiv:1103.0948
45. Michelangeli, A., Schlein, B.: Dynamical collapse of boson stars. Commun. Math. Phys. **311**(3), 645–687 (2012). arXiv:1005.3135
46. Lieb, E.H., Thomas, L.E.: Exact ground state energy of the strong-coupling polaron. Commun. Math. Phys. **183**(3), 511–519 (1997); Erratum: ibid. **188**(2), 499–500 (1997). arXiv:cond-mat/9512112 (1995)
47. Glauber, R.J.: Coherent and incoherent states of the radiation field. Phys. Rev. **131**(6), 2766–2788 (1963)
48. Leopold, N.: Effective Evolution Equations from Quantum Mechanics, Ph.D. thesis (2018). https://edoc.ub.uni-muenchen.de/21926/

Taking Inspiration from Quantum-Wave Analogies—Recent Results for Photonic Crystals

Max Lein

Abstract Similarities between quantum systems and analogous systems for classical waves have been used to great effect in the physics community, be it to gain an intuition for quantum systems or to anticipate novel phenomena in classical waves. This proceeding reviews recent advances in putting these quantum-wave analogies on a mathematically rigorous foundation for classical electromagnetism. Not only has this *Schrödinger formalism of electromagnetism* led to new, interesting mathematical problems for the so-called Maxwell-type operators, but has also improved the understanding of the physics of topological phenomena in electromagnetic media. For example, it enabled us to classify electromagnetic media by their material symmetries, and explained why "fermionic time-reversal symmetries" that were conjectured to exist in the physics literature are, in fact, forbidden.

The idea behind quantum-wave analogies is that two effects, one in a quantum system and another in a classical wave, are, in fact, *different manifestations of the same underlying physical principles*. They can be read both ways: The founding fathers of modern quantum mechanics relied on them to transfer some of their understanding of electromagnetism to their fledgling theory. Nowadays, it is usually the other way around, quantum mechanics is an established theory in its own right, and quantum-wave analogies can help guide one's intuition in the search for new phenomena in *classical* waves. To name but one example, on the basis of the analogy to crystalline solids, Yablonovitch [84] and John [33] proposed the concept of *photonic crystals*, man-made electromagnetic media whose periodic structure endows them with peculiar light conduction properties. This deceptively simple idea gave birth to several vibrant subcommunities in physics. Despite very exciting experiments such

M. Lein (✉)
Advanced Institute of Materials Research, Tohoku University, 2-1-1 Katahira,
Aoba-ku, Sendai 980-8577, Japan
e-mail: max.lein@tohoku.ac.jp

© Springer Nature Switzerland AG 2018
D. Cadamuro et al. (eds.), *Macroscopic Limits of Quantum Systems*,
Springer Proceedings in Mathematics & Statistics 270,
https://doi.org/10.1007/978-3-030-01602-9_10

as the realization of topological effects on a truly macroscopic scale [72, 77] or the experimental observation of Weyl points [44], only very few mathematicians are systematically working on classical wave equations with the help of quantum-wave analogies. That is a bit of a pity as new insights into the physics of classical waves open up very challenging mathematical questions, which in some cases have a direct impact on the physics community.

Least of which, not all of the quantum-wave analogies physicists have identified stand on a solid mathematical foundation, though, and it is not always clear whether and to what extent they can be made rigorous. Similarly, *wave*-wave analogies are often not derived from first principles, because these different equations are usually not considered in a *unified mathematical framework*. This proceeding will detail a mathematical framework that applies to a host of classical wave equations and discuss specific instances of quantum-wave analogies that were the subject of recent publications [12–15] and preprints [17, 18]. Among other wave equations, the formalism applies to *Maxwell's equations for linear dielectrics*,

$$
\begin{pmatrix} \varepsilon & 0 \\ 0 & \mu \end{pmatrix} \frac{\partial}{\partial t} \begin{pmatrix} \mathbf{E}(t) \\ \mathbf{H}(t) \end{pmatrix} = \begin{pmatrix} +\nabla \times \mathbf{H}(t) \\ -\nabla \times \mathbf{E}(t) \end{pmatrix} - \begin{pmatrix} \mathbf{J}(t) \\ 0 \end{pmatrix}, \tag{1a}
$$

$$
\begin{pmatrix} \nabla \cdot \varepsilon \mathbf{E}(t) \\ \nabla \cdot \mu \mathbf{H}(t) \end{pmatrix} = \begin{pmatrix} \rho(t) \\ 0 \end{pmatrix}, \tag{1b}
$$

$$
\partial_t \rho(t) + \nabla \cdot \mathbf{J}(t) = 0, \tag{1c}
$$

certain *linearized magneto-hydrodynamic equations* [18, Section 6.2.1],

$$
\frac{d\rho_1}{dt} = -\nabla \cdot (\rho_0 \mathbf{v}), \tag{2a}
$$

$$
\rho_0 \frac{d\mathbf{v}}{dt} = -\nabla \wp_1 - \mathbf{g} \rho_1 + (\mathbf{J}_0 \times \mathbf{B}_1 + \mathbf{J}_1 \times \mathbf{B}_0), \tag{2b}
$$

$$
\frac{d\mathbf{B}_1}{dt} = \nabla \times (\mathbf{v} \times \mathbf{B}_0), \tag{2c}
$$

$$
\wp_1 = \Gamma(\rho_1, \mathbf{v}), \tag{2d}
$$

and linearized *equations for spin propagating in a periodic medium* [66, equation (21)],

$$
\mathrm{i} \frac{\partial}{\partial t} \begin{pmatrix} \beta(t, k) \\ \beta(t, -k) \end{pmatrix} = \sigma_3 H(k) \begin{pmatrix} \beta(t, k) \\ \beta(t, -k) \end{pmatrix}. \tag{3}
$$

These equations share certain basic features: (1) They are first-order in time, (2) they have a product structure, and (3) the waves are *real*- as opposed to complex-valued. It turns out that all of these wave equations can be recast as a Schrödinger equation.

1 The Physics of Photonic Crystals

The big appeal of photonic crystals and periodic media for other classical waves is that they can be *engineered* to have certain properties such as (photonic) band gaps and crystallographic symmetries. Here, physicists have much wider latitude than with condensed matter, not least because they are able to choose different spatial scales by varying the wavelength (e. g., infrared or microwaves).

The birth paper of the field of photonic crystals is the work by Yablonovitch [84], who proposed a dielectric with three-dimensional periodicity lattice and a *photonic band gap* as an "omnidirectional quarter wave plate". This inspired John [33] to suggest that also light can be Anderson localized in a photonic crystal with band gap in the presence of moderate disorder. These ideas were quickly generalized to other classical waves, and nowadays there are phononic crystals [8, 22, 53, 63], magnonic crystals [66], and periodic arrays of coupled oscillators [72, 73], whose physics is in many respects similar to that of crystalline solids.

1.1 Photonic Crystals with Band Gaps

Initially, the focus of theoretical research centered around proposing dielectric structures which (1) exhibited a photonic band gap and (2) were thought to be realizable in experiment. Also, robust numerical tools for band structure calculations have been developed [34].

Within the next decade or so, experimentalists pushed along two directions, namely improving the fidelity (absence of defects) and size (measured in numbers of unit cells, for instance) on the one hand, and the invention and improvement of manufacturing techniques for structures at smaller scales on the other (cf. [41, 69] and [57, Chapter 6]). This means photonic crystals can now be designed and manufactured to have band gaps in specific, predetermined frequency ranges [32].

1.2 Topological Photonic Crystals

Late in 2005, Raghu and Haldane proposed in two seminal works [29, 59] to consider topological effects analogous to the Integer Quantum Hall Effect in adiabatically perturbed photonic crystals. Such slow modulations can be the result of strain [80] or a thermal gradient [19, 20]. Of particular interest are photonic crystals where the material weights $W = \left(\begin{smallmatrix} \varepsilon & 0 \\ 0 & \mu \end{smallmatrix} \right) \neq \overline{W}$ are complex; this is analogous to magnetic quantum systems with broken time-reversal symmetry. Raghu and Haldane realized that the geometric (Berry) connection can be defined just like in crystalline solids and proposed a photonic analog of the *Peierls substitution*: for states associated to an isolated band ω_n "semiclassical" ray optics equations of motion for r and k

approximate the full light dynamics. Topological effects enter through the Berry curvature $\Omega_n := \nabla_k \times i\langle \varphi_n, \nabla_k \varphi_n \rangle_W$. Based on this analogy, Haldane proposed the following *photonic bulk-edge correspondence* [59, p. 7]:

> [...] if the Chern number of a band changes at an interface, the net number of unidirectionally moving modes localized at the interface is given by the difference between the Chern numbers of the band at the interface (Fig. 2).

Initially, parts of the physics community were very skeptical that topological phenomena may present themselves in classical waves. Hence, it was only after the groups of Joannopoulos and Soljačić at MIT provided conclusive experimental evidence [76, 77] for the existence of unidirectional, backscattering-free edge modes, that Haldane was able to publish his two works.

Haldane's insight has had a tremendous impact, it kickstarted the search for novel topological phenomena in photonic crystals [37] and other classical waves [8, 22, 53, 63, 72, 73]. To give but one example, Lu et al. carefully designed [43] and realized [44] a photonic crystal with gyroid structures that exhibited Weyl points. Indeed, theirs was the first experimental observation of Weyl points (concurrently with a group at Princeton [83] in a condensed matter system). For a comprehensive review of the physics, we point the interested reader to [48] or [36].

All of these tabletop experiments used microwaves. These are very convenient to work with for proof-of-principle experiments as their wavelength is in the millimeter to centimeter range: with yttrium iron garnet (YIG) there is a medium that is an *insulator* and where the electric permittivity is *complex* [56]. Hence, we may regard the medium as being lossless for microwaves and $\varepsilon \neq \bar{\varepsilon}$ implies time-reversal symmetry is broken. Moreover, the degree to which it is broken can be tuned by applying a constant magnetic field. A second advantage is that these media can literally be machined [44] or, potentially, 3d printed [35].

1.3 Topological Phenomena in Media with Time-Reversal Symmetry

However, for many applications is it desirable to use light at infrared or optical frequencies (e. g., to realize "one-way wave guides" for photonic CMOS chips [26]). Fortunately, Maxwell's equations are scale-independent, and, therefore, the existence of topological phenomena for microwaves implies their existence for *all other* frequencies—provided we are able to find media with the same material weights at those frequencies.

This is where mathematics and physics part ways: currently, there are no known examples of *lossless* media with complex material weights at optical or infrared frequencies. Consequently, it is at present not feasible to miniaturize the experiments by Wang et al. [76, 77] or Lu et al. [44]. Moreover, if one wants to deploy existing mass manufacturing techniques for chips, we would have to deal with time-reversal symmetric media like silicon or gallium arsenide.

Wu and Hu [81] proposed a way to realize topological effects in ordinary, time-reversal symmetric media with the help of lattice symmetries. The idea is that the presence of the extra lattice symmetry gives rise to an "isospin" degree of freedom, and we may consider isospin-↑ and isospin-↓ systems separately. And as time-reversal symmetry flips the isospin, time-reversal symmetry is broken *on each sub-system*. Thus, if edge modes exist, they come in counter-propagating pairs of opposite isospin. When states of a given isospin are selectively excited, then they feature the same hallmarks of topological protection as those observed in [77]. However, these unidirectional edge modes are only robust against perturbations, which *preserve the relevant crystallographic symmetries*. Their idea has since been realized in a variety of experiments [82, 86]. The significance here is that time-reversal symmetric media are plentiful at many wavelengths, and exploiting crystallographic symmetries provide a viable avenue to engineer media with topological phenomena based on readily available materials.

1.4 Periodic Waveguide Arrays

Another class of experiments with optical and infrared light is performed with periodic waveguide arrays, where topologically protected edge states have also been observed [25, 28, 54, 55]. Here, the waveguides are formed by tubular regions where the refractive index n of the substrate material (usually silica where $n \approx 1.45$) is increased by $\Delta n \approx 10^{-3} \sim 10^{-4}$. For light of specific wavelengths, these regions of increased refractive index act as *evanescently coupled waveguides*. By carefully designing these waveguide arrays, experimentalists can, to good approximation, realize a variety of tight-binding Hamiltonians with nearest and next-nearest neighbor hopping, e. g., "photonic graphene" [55] or a Lieb lattice [28]. The wave guide array geometry can be modified in other ways, say with helically wound waveguides and "squeezed" configurations [60], and its effect on light conduction properties have been investigated. Physicists propose the paraxial wave equation

$$\mathrm{i}\partial_z \psi(x, z) = -\frac{1}{2k_0} \Delta_x \psi(x, z) - \frac{k_0\, \Delta n(x)}{n_0} \psi(x, z) \qquad (4)$$

as the equation which governs the "dynamics". Here, z is the coordinate along the fiber direction which takes the role of time, $x = (x_1, x_2)$ are the coordinates orthogonal to the z-axis, and ψ is the envelope of the electromagnetic field, i. e., $\mathbf{E}(t, x, z) = \mathrm{Re}\, \psi(x, z)\, \mathrm{e}^{+\mathrm{i}(k_0 z - \omega t)}$. The operator which appears on the right-hand side of (4) has the same form as a regular Schrödinger operator; sometimes a pseudomagnetic field and other potential terms are added to phenomenologically account for geometric effects [42, 61]. These waveguide arrays can be endowed with "artificial dimensions" by making them parameter-dependent; in recent experiments [49, 87] topological phenomena have been found in two-dimensional boundaries to effectively four-dimensional waveguide arrays.

Seeing as the effective equation (4) describing waveguide arrays is different from Maxwell's equations, we will not cover them any further in these proceedings. Instead, we refer to [48, Section III.A.2] for details.

2 The Schrödinger Formalism for Classical Waves

When one wants to make a specific quantum-wave analogy rigorous, rewriting wave equations such as (1)–(3) in the form

$$i\partial_t \Psi(t) = M\Psi(t) - iJ(t), \qquad \Psi(t_0) = \Phi \in \mathcal{H}_W, \qquad (5)$$

is a natural preliminary step that necessarily precedes anything else. Here, $\Psi(t)$ is a *complex* wave that *represents* the *real*-valued physical wave $u(t) = 2\operatorname{Re}\Psi(t)$ and belongs to a *complex* Hilbert space $\mathcal{H}_W$; the self-adjoint operator M takes the place of the quantum Hamiltonian; and $J(t)$ is a current. The derivation of all the pieces in Eq. (5) consists of two distinct steps:

(1) We complexify the wave equation and obtain an equation that looks like (5); this gives us access to many of the powerful tools developed for self-adjoint operators.
(2) We reduce the complexified equations to waves with frequencies $\omega \geq 0$. Then, real-valued physical waves $u = 2\operatorname{Re}\Psi$ are *uniquely represented* by a complex wave Ψ composed solely of non-negative frequencies.

2.1 Maxwell-Type Operators for the Complexified Equations

What makes these *Maxwell-type operators* $M^{\mathbb{C}} = W_L D W_R$, which arise in the context of wave equations, mathematically interesting is their *product structure*.

Definition 1 (Complexified Maxwell-type operator). A linear operator

$$M^{\mathbb{C}} := W_L D W_R \qquad (6)$$

with product structure on a complex Hilbert space $\mathcal{H}^{\mathbb{C}}$ is a complexified Maxwell-type operator if it has the following properties:

(a) D is a (possibly unbounded) self-adjoint operator on $\mathcal{H}^{\mathbb{C}}$ with domain $\mathcal{D}_0$.
(b) $W_L, W_R \in \mathcal{B}(\mathcal{H}^{\mathbb{C}})$ are bounded, self-adjoint, commuting operators with bounded inverses, i. e., $[W_L, W_R] = 0$ and $W_L^{-1}, W_R^{-1} \in \mathcal{B}(\mathcal{H}^{\mathbb{C}})$.
(c) The product $W_R W_L^{-1}$ is self-adjoint and bounded away from 0, i. e., there exists a positive constant $c > 0$ so that $W_R W_L^{-1} \geq c \, \mathbb{1}_{\mathcal{H}^{\mathbb{C}}}$.
(d) $M^{\mathbb{C}}$ is endowed with the domain $\mathcal{D}_R := W_R^{-1} \mathcal{D}_0$.

(e) $M^{\mathbb{C}}$ anticommutes with an even antiunitary conjugation C (which we will refer to as *complex conjugation*), i. e., $C \, \mathscr{D}_R = \mathscr{D}_R$ is left invariant and $M^{\mathbb{C}}$ satisfies

$$C \, M^{\mathbb{C}} \, C = -M^{\mathbb{C}}. \tag{7}$$

Straightforward arguments show that Maxwell-type operators are necessarily closed [18, Lemma B.1]; however, they generally fail to be symmetric on $\mathscr{H}^{\mathbb{C}}$ and, therefore, cannot be self-adjoint. This apparent lack of self-adjointness can be cured by endowing the *vector* space $\mathscr{H}^{\mathbb{C}}$ with the *weighted* scalar product

$$\langle \Phi, \Psi \rangle_W := \langle \Phi, W_L^{-1} \, W_R \, \Psi \rangle \tag{8}$$

defined in terms of $\mathscr{H}^{\mathbb{C}}$'s scalar product $\langle \, \cdot \, , \, \cdot \, \rangle$. We will denote the resulting Hilbert space with $\mathscr{H}_W^{\mathbb{C}}$; due to our assumptions on W_L and W_R, the vector spaces $\mathscr{H}_W^{\mathbb{C}}$ and $\mathscr{H}^{\mathbb{C}}$ agree as *Banach* spaces [18, Section 6.1.1]. A quick computation shows that Maxwell-type operators are indeed symmetric on $\mathscr{H}_W$,

$$\begin{aligned} \langle \psi, M\varphi \rangle_W &= \langle \psi, W_R \, D \, W_R \varphi \rangle = \langle W_R \, D \, W_R \psi, \varphi \rangle \\ &= \langle W_L^{-1} \, W_L \, D \, W_R \psi, W_R \varphi \rangle = \langle M\psi, \varphi \rangle_W, \end{aligned}$$

and elementary arguments, in fact, lead us to conclude that M coincides with its $\langle \, \cdot \, , \, \cdot \, \rangle_W$-adjoint M^{*w} [18, Proposition 6.2]. Thanks to that, we can employ functional calculus, e. g., to define the *unitary* evolution group $\mathrm{e}^{-\mathrm{i}t M^{\mathbb{C}}}$. Note that the unitarity leads to the existence of at least one conserved quantity, $\mathscr{E}(\psi) = \frac{1}{2} \langle \psi, \psi \rangle_W$.

The role of the symmetry relation (7) is to ensure the existence of real solutions: $C \, M^{\mathbb{C}} \, C = -M^{\mathbb{C}}$ implies that complex conjugation C and therefore, the associated real part operator $\mathrm{Re} = \frac{1}{2}(\mathbb{1} + C)$, commute with the unitary evolution group,

$$\mathrm{Re} \; \mathrm{e}^{-\mathrm{i}t M^{\mathbb{C}}} = \mathrm{e}^{-\mathrm{i}t M^{\mathbb{C}}} \, \mathrm{Re} \, .$$

Put succinctly, the fact that the original equations describe real waves translates to the presence of the "symmetry" C. Moreover, it gives us a way to represent real waves as the real part of complex waves: if $\Psi \in \mathscr{H}_W^{\mathbb{C}}$ is a complex wave with $u = 2\mathrm{Re} \, \Psi$, then we may either evolve u or evolve the complex wave Ψ and then take $2\mathrm{Re}$ afterwards.

2.2 Reduction to Complex Waves of Non-negative Frequencies

This insight is crucial for the second step where we establish a one-to-one correspondence between real, physical waves, and complex waves composed only of frequencies with $\omega \geq 0$. In principle, there are many different one-to-one mappings $u \mapsto \Psi$ with $u = 2\mathrm{Re} \, \Psi$, but spectral conditions are particularly easy to impose and so an especially convenient choice. For the *in vacuo* Maxwell equations, this strategy is a part and parcel of every course on electromagnetism, but it is worth generalizing them to cases where solutions to Maxwell's equations can no longer be explicitly expanded in terms of pseudo-eigenfunctions.

Any real field $u = \mathrm{Re}\, u \in \mathscr{H}_W^{\mathbb{C}}$ can be decomposed via the maps

$$Q_\pm = 1_{(0,\infty)}(\pm M^{\mathbb{C}}) + \tfrac{1}{2} 1_{\{0\}}(M^{\mathbb{C}}) = 1 - Q_\mp = C\, Q_\mp\, C \tag{9}$$

into a non-negative and a non-positive frequency component,

$$u = \Psi_+ + \Psi_- := Q_+ u + Q_- u.$$

Thus, for real waves u the non-positive frequency part $\Psi_- = Q_- u$ can be reconstructed from the non-negative frequency contribution

$$\Psi_- = Q_- u = \overline{\Psi_+} = C\, Q_+ u \tag{10}$$

by taking the complex conjugate. This *phase locking condition* tells us that Ψ_+ and Ψ_- are *not independent degrees of freedom*, and that the real-valued physical fields

$$u = 2\mathrm{Re}\,\Psi_\pm$$

can be recovered from just $\Psi_\pm \in \mathrm{ran}\, Q_\pm$ alone. Conversely, u uniquely determines $\Psi_\pm = Q_\pm u$, and a straightforward adaption of the arguments that led to [18, Proposition 3.3] yields

Proposition 1 (1-to-1 correspondence between real and complex $\omega \geq 0$ waves). *Given a complexified Maxwell operator $M^{\mathbb{C}} = W_L\, D\, W_R$, let $Q_\pm$ denote the maps from Eq. (9). Then for real states $2\mathrm{Re}$ is an inverse of $Q_\pm$ in the sense that*

$$2\mathrm{Re}\, Q_\pm \mathrm{Re} = \mathrm{Re} : \mathscr{H}_W^{\mathbb{C}} \longrightarrow \mathscr{H}_W^{\mathbb{C}}$$

holds. Therefore, there is a one-to-one correspondence between the real wave $u = \mathrm{Re}\, u = 2\mathrm{Re}\,\Psi_\pm \in \mathscr{H}_W^{\mathbb{C}}$ and the complex wave $\Psi_\pm = Q_\pm u \in \mathscr{H}_W^{\mathbb{C}}$.

As a matter of convention, real waves are represented as complex waves comprised of non-negative frequencies. Due to this correspondence, we can describe the real wave equation for $u = 2\mathrm{Re}\,\Psi_+$ as a Schrödinger-type equation for the complex wave Ψ_+,

$$\mathrm{i}\partial_t \Psi_+(t) = M_+ \Psi_+(t) - \mathrm{i}J_+(t), \qquad \Psi_+(t_0) = Q_+ u(t_0) \in \mathscr{H}_W, \tag{11}$$

where the Hilbert space $\mathscr{H}_W = Q_+\big[\mathscr{H}_W^{\mathbb{C}}\big]$ is the *non-negative frequency subspace of the Hilbert space $\mathscr{H}_W^{\mathbb{C}}$ that inherits the weighted scalar product* (8), and the Maxwell-type operator

$$M_+ = W_L D W_R \big|_{\omega \geq 0} := M^{\mathbb{C}} \big|_{\mathscr{H}_W} \tag{12}$$

is the restriction of the complexified Maxwell operator to the spectral subspace $\omega \geq 0$. Note that all essential properties transfer from $M^{\mathbb{C}}$, e. g., as a restriction of a self-adjoint operator to a spectral subspace, it is automatically self-adjoint.

The physical, real-valued wave

$$u(t) = 2\mathrm{Re}\,\Psi_+(t)$$

is recovered by taking twice the real part.

Note that the self-adjointess of M_+ leads to the existence of at least one conserved quantity in the absence of sources, the square of the weighted norm,

$$\mathscr{E}\big(u(t)\big) := \big\langle \mathrm{e}^{-\mathrm{i}t M_+} Q_+ u,\ \mathrm{e}^{-\mathrm{i}t M_+} Q_+ u \big\rangle_W = \big\langle Q_+ u,\ Q_+ u \big\rangle_W = \mathscr{E}\big(u(0)\big).$$

In case of Maxwell's equations, this conserved quantity is nothing but the total field energy.

2.3 Changes of Representation

One of the concepts that directly transfer from quantum mechanics is that of *representations*; this can be useful to exploit extra structures, symmetries or other relations. Unitary operators between $\mathscr{H}_W$ and other Hilbert spaces facilitate changes of representation. The two most common examples are the continuous Fourier transform to analyze homogeneous media or a version of the discrete Fourier transform for periodic media.

Another relevant example is $W_R : \mathscr{H}_W \longrightarrow \mathscr{H}_{W'}$, viewed as a *unitary* operator between $\mathscr{H}_W$ and the Hilbert space $\mathscr{H}_{W'} = W_R[\mathscr{H}_W]$, where the latter is endowed with the scalar product weighted by the operator $W' := W_L^{-1} W_R^{-1}$,

$$\big\langle \Phi, \Psi \big\rangle_{W'} = \big\langle \Phi,\ W_L^{-1} W_R^{-1} \Psi \big\rangle.$$

The advantage here is that W_R transforms a Maxwell-type operator of the form (6) to one where the weights are all on the left,

$$M = W'^{-1} D.$$

Consequently, we may take this to be the canonical form of Maxwell-type operators.

2.4 The Schrödinger Formalism for Maxwell's Equations for Linear, Dispersionless Media

Let us now make the construction explicit for the case of Maxwell's equations that govern the propagation of electromagnetic fields $\big(\mathbf{E}(t), \mathbf{H}(t)\big) \in L^2(\mathbb{R}^3, \mathbb{R}^6)$ in a non-gyrotropic dielectric. Here, the material weights are constructed from the electric permittivity ε and the magnetic permeability μ,

$$W := \begin{pmatrix} \varepsilon & 0 \\ 0 & \mu \end{pmatrix}.$$

Non-gyrotropic dielectrics are those where the weights $W = \overline{W}$ are real and satisfy

Assumption 1 (Material weights) *The medium described by the material weights* $W \in L^\infty\big(\mathbb{R}^3, \mathrm{Mat}_{\mathbb{C}}(6)\big)$ *has the following properties:*

(a) *The medium is* lossless, *i. e.,* $W(x) = W(x)^*$ *takes values in the hermitian matrices.*

(b) *The medium is* not a negative index material, *i. e., for some* $0 < c < C < 0$ *the weights satisfy*

$$0 < c\,\mathbb{1} \leq W \leq C\,\mathbb{1} < \infty.$$

2.4.1 Complexified Equations

Under these conditions, we can multiply both sides of the dynamical Maxwell equation (1a) with $i\,W^{-1}$ and obtain the Schrödinger-type equation

$$i\,\partial_t \Psi(t) = M^{\mathbb{C}}\,\Psi(t) - i\,W^{-1}\big(\mathbf{J}(t), 0\big) \tag{13}$$

with the initial condition

$$\Psi(t_0) = \big(\mathbf{E}(t_0), \mathbf{H}(t_0)\big) \in L^2(\mathbb{R}^3, \mathbb{R}^6) \subset L^2(\mathbb{R}^3, \mathbb{C}^6).$$

Here, $\Psi(t) = \big(\mathbf{E}(t), \mathbf{H}(t)\big)$ is just a new label for the electromagnetic field and the complexified Maxwell operator

$$M^{\mathbb{C}} = W^{-1}\,\mathrm{Rot} := \begin{pmatrix} \varepsilon & 0 \\ 0 & \mu \end{pmatrix}^{-1} \begin{pmatrix} 0 & +i\nabla^\times \\ -i\nabla^\times & 0 \end{pmatrix}$$

is a succinct way to write the right-hand side of (13). The *free Maxwell operator* Rot is defined in terms of the curl $\nabla^\times \mathbf{E} := \nabla \times \mathbf{E}$. On the usual $L^2(\mathbb{R}^3, \mathbb{C}^6)$ the complexified Maxwell operator is closed, but not self-adjoint; however, if we endow this Banach space with the (weighted) *energy scalar product*

$$\langle \Phi, \Psi \rangle_W := \int_{\mathbb{R}^3} dx \, \Phi(x) \cdot W(x) \Psi(x),$$

and denote the resulting Hilbert space with $\mathscr{H}_W^{\mathbb{C}} := L_W^2(\mathbb{R}^3, \mathbb{C}^6)$, then $M^{\mathbb{C}}$ endowed with the domain of Rot indeed defines a self-adjoint operator on $\mathscr{H}_W^{\mathbb{C}}$ [18, Proposition 6.2]; note that complex conjugation in the above expression is contained in the Euclidean scalar product $a \cdot b := \sum_{j=1}^6 \overline{a_j} b_j$ of $\mathbb{C}^6$.

This takes care of the dynamical equation, but we must not forget about the *constraint equation* (1b). To see that this is also satisfied, we introduce a Helmholtz splitting

$$\mathscr{H}^{\mathbb{C}} = \mathscr{J} \oplus \mathscr{G}$$

into transversal and longitudinal waves that is adapted to the medium. Here, longitudinal waves are those which are static, i. e., gradient fields

$$\mathscr{G} := \left\{ (\nabla \varphi^E, \nabla \varphi^H) \in L^2(\mathbb{R}^3, \mathbb{C}^6) \mid \varphi^E, \varphi^H \in L_{loc}^2(\mathbb{R}^3) \right\} = \ker \text{Rot} = \ker M^{\mathbb{C}},$$

whereas transversal states are those which are $\langle \cdot, \cdot \rangle_W$-orthogonal to them,

$$\mathscr{J} := \mathscr{G}^{\perp_W} = \left\{ \Psi \in L^2(\mathbb{R}^3, \mathbb{C}^6) \mid \text{Div } W\Psi = 0 \right\} = \text{ran } M^{\mathbb{C}},$$

where $\text{Div}(\mathbf{E}, \mathbf{H}) := (\nabla \cdot \mathbf{E}, \nabla \cdot \mathbf{H})$ consists of two copies of the usual divergence operator. Transversal states $\Psi_{\perp} \in \mathscr{J}$ are exactly those that satisfy the constraint equation (1b) in the distributional sense for $\rho = 0$.

Applying this decomposition to the Schrödinger-type equation yields that the longitudinal part $\Psi_{\parallel} \in \mathscr{G}$ is fixed by the constraint equation (1b) (cf. [18, Section 3.2.1]), and only the transversal part $\Psi_{\perp} \in \mathscr{J}$ has non-trivial dynamics. Consequently, given sources $\rho(t)$ and $\mathbf{J}(t)$ which satisfy local charge conservation (1c) and an initial condition $(\mathbf{E}(t_0), \mathbf{H}(t_0))$ that fulfills the constraint equation (1b), then also the solution

$$(\mathbf{E}(t), \mathbf{H}(t)) = e^{-i(t-t_0)M^{\mathbb{C}}} (\mathbf{E}(t_0), \mathbf{H}(t_0)) - i \int_{t_0}^t ds \, e^{-i(t-s)M^{\mathbb{C}}} \, W^{-1} (\mathbf{J}(t), 0)$$

to the Schrödinger equation (13) satisfies (1b) for all times.

2.4.2 Reduction to $\omega \geq 0$ and One-to-One Correspondence

Because this operator is self-adjoint, the operator Q_+ defined through (9) makes sense and can be used to define the non-negative frequency subspace

$$\mathscr{H}_W := Q_+ [L_W^2(\mathbb{R}^3, \mathbb{C}^6)]$$

as well as the one-to-one correspondence between real fields and complex fields with $\omega \geq 0$ from Proposition 1,

$$L^2(\mathbb{R}^3, \mathbb{R}^6) \ni (\mathbf{E}, \mathbf{H}) = 2\operatorname{Re}\Psi_+ \quad \longleftrightarrow \quad \Psi_+ = Q_+(\mathbf{E}, \mathbf{H}) \in \mathscr{H}_W. \quad (14)$$

That means Maxwell's equations (1) are equivalent to the Schrödinger equation

$$\mathrm{i}\,\partial_t \Psi_+(t) = M_+\Psi_+(t) - \mathrm{i}\,J_+(t), \qquad \Psi_+(t_0) = Q_+\bigl(\mathbf{E}(t_0), \mathbf{H}(t_0)\bigr), \quad (15)$$

where $J_+(t) := Q_+\,W^{-1}\bigl(\mathbf{J}(t), 0\bigr)$ is the non-negative frequency contribution to the current density and the *Maxwell operator*

$$M_+ := M^{\mathbb{C}}\big|_{\mathscr{H}_W}$$

is the restriction of the complexified Maxwell operator to non-negative frequencies. The equivalence of the dynamical equations (15) and (1a) follows from the one-to-one correspondence (14) and the equivalence of the complexified equations. Similarly, if the initial state $\bigl(\mathbf{E}(t_0), \mathbf{H}(t_0)\bigr)$ satisfies the constraint equation and charge is conserved, then $2\operatorname{Re}\Psi_+(t)$ also satisfies the constraint equation (1b).

Lastly, the self-adjointness of M_+ translates to conservation of total field energy, which is the reason why in the context of electromagnetism we refer to the weighted scalar product (8) as *energy* scalar product.

2.4.3 Extension to Media with Complex Material Weights

For media with real weights, the (complexified) Schrödinger formalism for Maxwell's equations in media as presented above, was well known since at least the 1960s, one of the earliest references we are aware of is due to Wilcox [79] which predates [7] by two decades. However, it seems that Maxwell's equations for media with *complex* weights

$$W = \begin{pmatrix} \varepsilon & \chi \\ \chi^* & \mu \end{pmatrix}$$

that satisfy Assumption 1 have not been derived and studied prior to [18].

The main difficulty is to obtain physically meaningful Maxwell equations in the first place; these have to be derived from Maxwell's equations for a linear *dispersive* medium [18, Section 2]. However, even if it is not at all obvious, it turns out that at the end of the day the $\omega \geq 0$ Schrödinger equation (15) for media with complex material weights is exactly the same as that in the real case [18, Section 3.3].

3 Rigorous Quantum-Wave Analogies in Photonic Crystals

The Schrödinger formalism for classical waves that we have summarized in the last section becomes the starting point for the actually interesting part—investigating whether and to what extent specific quantum-wave analogies hold. Examples that have been covered by other authors include Anderson localization [21, 33], effective dynamics in weakly nonlinear photonic crystals in the form of nonlinear Schrödinger equations [4, 6], a proof of the Bethe-Sommerfeld conjecture in two dimensions [75] as well as scattering theory for asymptotically homogeneous electromagnetic media [62, 65, 78].

Our contributions to the subject [12–15, 17, 18] work toward a proof of *photonic bulk-boundary correspondences*, which link specific physical phenomena in photonic crystals to topology; the paradigmatic example is the one proposed by Haldane in [59], the seminal paper that kickstarted the search for topological effects in classical waves.

3.1 Fundamental Properties of Periodic Maxwell Operators

The first step was to better understand the mathematical properties of *photonic crystals* [14, 39], an idea Yablonovitch [84] and John [33] independently came up with. Yablonovitch envisioned that periodic patterning would allow one to create a medium that acts as an "omnidirectional quarter wave plate"; in the parlance of condensed matter physics, the medium should have a *photonic band gap*. What makes photonic crystals so interesting is that compared to crystalline solids physicists have much wider latitude when *engineering* them for a specific purpose. Indeed, a few years after the concept was proposed, Yablonovitch successfully designed and manufactured a photonic crystal with a band gap [85].

Mathematically speaking, photonic crystals are described by Maxwell operators with periodic weights.

Assumption 2 (Periodic weights) *W is periodic with respect to some lattice $\Gamma \cong \mathbb{Z}^3$.*

Many of the standard techniques for periodic operators [38, 39] apply directly, and it is not surprising that periodic Maxwell operators admit a frequency band picture [14, Theorem 1.4]: the usual Bloch-Floquet transform decomposes periodic Maxwell operators

$$M_+ \cong \int_{\mathbb{T}^*}^{\oplus} dk \, M_+(k)$$

into a family of Maxwell operators acting on the Hilbert space $\mathscr{H}_W(k) \subset L^2_W(\mathbb{T}^3, \mathbb{C}^6)$ of the unit cell, that has been endowed with a weighted scalar product akin to (8).

Except for the infinitely degenerate eigenvalue $\omega = 0$ due to longitudinal gradient fields, the spectrum $\sigma\left(M_+(k)\right) \subseteq [0, \infty)$ is purely discrete. Just like with periodic Schrödinger operators, band functions are continuous and locally analytic away from band crossings; Bloch functions are locally analytic away from band crossings.

Periodic Maxwell operators do possess characteristics that set them apart from periodic Schrödinger operators. In a departure from quantum theory, Maxwell operators necessarily feature two "ground state bands" with approximately linear dispersion around $k = 0$ and $\omega = 0$. The presence of these ground state bands can be easily inferred from heuristic arguments that can be made rigorous (cf. Lemma 3.7 and Theorem 1.4 in [14]): low-frequency waves have very long wavelengths, which eventually become longer than the lattice length. Then to leading order, these waves are subjected to the unit cell averaged weights W_{avg}, and consequently, ground state frequency band and Bloch functions near $k = 0$ are well-approximated by those of the homogeneous medium with weights W_{avg}.

3.2 Effective Tight-Binding Operators for Adiabatically Perturbed Photonic Crystals

The main purpose of the technical paper [14] was to show that the class of Maxwell operators that models slowly (i. e., *adiabatically*) modulated photonic crystals can be viewed as pseudodifferential operators, a necessary preliminary step for [12]. This is the *photonic analog* of a work by Panati, Spohn and Teufel [51] who studied a Bloch electron subjected to slowly varying, external electromagnetic fields via *space-adiabatic perturbation theory*.

3.2.1 Original Result

Unlike in quantum mechanics where the potentials due to the external fields are *added*, perturbations naturally act *multiplicatively* in photonics, so that the perturbed weights take the form

$$W_\lambda(x) = S(\lambda x)^{-2}\, W(x) = \begin{pmatrix} \tau_\varepsilon^{-2}(\lambda x) & 0 \\ 0 & \tau_\mu^{-2}(\lambda x) \end{pmatrix} \begin{pmatrix} \varepsilon(x) & 0 \\ 0 & \mu(x) \end{pmatrix},$$

where W are the periodic weights and $S(\lambda x)$ is a sufficiently regular perturbation (see, e. g., [14, Assumption 1.2] or [12, Assumption 1]). The dimensionless small parameter $\lambda \ll 1$ of this perturbation problem is the ratio of the lattice length to the slow length scale on which the modulation varies.

Then the associated Maxwell operator $M_\lambda = W_\lambda^{-1}$ Rot (without frequency restriction) acting on the λ-dependent Hilbert space $\mathscr{H}_\lambda := L^2_{W_\lambda}(\mathbb{R}^3, \mathbb{C}^6)$ is a pseudodifferential operator [14, Theorem 1.3], and therefore *space-adiabatic perturbation theory* developed by Panati, Spohn, and Teufel [51, 52] can be adapted.

This perturbation scheme allowed us to approximate the full dynamics e^{-itM_λ} by simpler, effective dynamics $e^{-itM_{\text{eff},\lambda}}$ for states from a given fixed frequency range of interest (see [12, Theorem 1] for the precise mathematical statement),

$$e^{-itM_\lambda}\, \Pi_\lambda = e^{-itM_{\text{eff},\lambda}}\, \Pi_\lambda + \mathcal{O}_{\|\cdot\|}(\lambda^\infty), \qquad (16)$$

where $\Pi_\lambda \asymp \sum_{n=0}^\infty \lambda^n\, \Pi_n$ is the projection onto the almost invariant subspace associated to this spectral region; we will make this more precise below. All of these operators admit asymptotic series in λ, where each term can in principle be computed explicitly.

Physicists commonly use the reasoning that is behind (16) to *justify effective tight-binding operators* that encapsulate the physics near interesting points in the band spectrum.

The first order of business is to pick a *finite frequency range of interest*. For the unperturbed, perfectly periodic photonic crystal, this is equivalent to selecting a finite family of frequency bands $\sigma_{\text{rel}}(k) = \bigcup_{n\in\mathscr{I}}\{\omega_n(k)\}$, $\mathscr{I} \subset \mathbb{N}$. However, for this spectral subspace to decouple, it is necessary to assume that $\sigma_{\text{rel}}(k)$ is separated from all other frequency bands by a local *spectral gap*; moreover, for physical as well as technical reasons, we need to exclude the ground state bands.

Now let us turn on the perturbation. To quantify the error when comparing dynamics via (16), we have to choose a norm. This is not as immediate as it is in quantum mechanics, since for $\lambda \neq \lambda'$ the operators M_λ and $M_{\lambda'}$ are defined on *different* Hilbert spaces and the norms of $\mathscr{H}_\lambda$ and $\mathscr{H}_{\lambda'}$ *depend on the perturbation parameter*. Instead, we represent all of the operators on a λ-*independent reference space*, namely that on which the periodic operator lives. This way the norm we actually use for the estimates is independent of the perturbation parameter, and all norm estimates carry over to other representations even if the unitaries themselves depend on λ.

In the end, we were able to mimic the construction of Panati, Spohn, and Teufel. For example, we constructed the projection $\Pi_\lambda \asymp \sum_{n=0}^\infty \lambda^n\, \Pi_n$ from Eq. (16) [12, Proposition 1]. This projection is uniquely determined up to $\mathcal{O}(\lambda^\infty)$ by three data: (1) Π_λ is an orthogonal projection; (2) Π_λ commutes with M_λ up to $\mathcal{O}(\lambda^\infty)$; and (3) the leading-order term $\Pi_0 \cong \sum_{n\in\mathscr{I}} |\varphi_n(\hat{k})\rangle\langle\varphi_n(\hat{k})|$ is unitarily equivalent to the projection onto the states associated to the relevant bands $\sigma_{\text{rel}}(k) = \bigcup_{n\in\mathscr{I}}\{\omega_n(k)\}$. Moreover, we have verified that modulo an $\mathcal{O}(\lambda^\infty)$ error, its range ran Π_λ is comprised of transversal waves [12, Proposition 7], so that states from ran Π_λ satisfy the constraint (1b) for $\rho = 0$.

3.2.2 Incorporating the Restriction to $\omega \geq 0$

Strictly speaking [12] uses unphysical equations as it predates our more recent works [15, 18] that were the first to start the analysis with the correct equations. Compared to our newer works, the Maxwell operator $M_\lambda = W_\lambda^{-1}\,\text{Rot}$ from the older paper [12] lacks the restriction to non-negative frequencies.

Nevertheless, this can be remedied with simple, straightforward arguments. First of all, instead of picking bands "symmetrically"as discussed in [12, Section 4.1.1], we only choose *positive* frequency bands in the construction of Π_λ.

To show we can replace the evolution generated by $M_{+,\lambda}$ with that of M_λ on ran Π_λ, we just combine a Duhamel argument with

Lemma 1. *In the setting of [12, Theorem 1], we have*

$$M_{+,\lambda}\,\Pi_\lambda = M_\lambda\,\Pi_\lambda + \mathcal{O}_{\|\cdot\|}(\lambda^\infty). \tag{17}$$

This lemma can be shown by modifying the proof of [12, Proposition 7].

Proof (Sketch). First, we pick a small enough $\omega_0 > 0$ and smooth function $\chi : \mathbb{R} \longrightarrow [0, 1]$ whose derivatives are compactly supported and for which

$$\chi(\omega)\,1_{[\omega_0,\infty)}(\omega) = 1_{[\omega_0,\infty)}(\omega)$$

holds. This gives rise to a "smoothened" version $\chi(M_\lambda)$ of the spectral projection $1_{[\omega_0,\infty)}(M_\lambda)$ onto frequencies $\omega \geq \omega_0 > 0$. Since χ is regular enough, $\chi(M_\lambda)$ is, in fact, a *pseudodifferential* operator and we obtain for $\omega_0 > 0$ small enough

$$\Pi_\lambda\,\chi(M_\lambda) = \Pi_\lambda + \mathcal{O}_{\|\cdot\|}(\lambda^\infty) = \chi(M_\lambda)\,\Pi_\lambda + \mathcal{O}_{\|\cdot\|}(\lambda^\infty)$$

via Weyl calculus and the Helffer-Sjöstrand formula for operator-valued symbols.

The Helffer-Sjöstrand formula on the level of operators, on the other hand, yields

$$\chi(M_\lambda) = \chi(M_\lambda)\,1_{[0,\infty)}(M_\lambda) = 1_{[0,\infty)}(M_\lambda)\,\chi(M_\lambda).$$

So starting from the right-hand side of (17), we can insert $1_{[0,\infty)}(M_\lambda)\,\chi(M_\lambda)$ at the expense of an $\mathcal{O}_{\|\cdot\|}(\lambda^\infty)$ error and obtain the claim,

$$\begin{aligned}
M_\lambda\,\Pi_\lambda &= M_\lambda\,1_{[0,\infty)}(M_\lambda)\,\chi(M_\lambda)\,\Pi_\lambda + \mathcal{O}_{\|\cdot\|}(\lambda^\infty)\\
&= M_{+,\lambda}\,1_{[0,\infty)}(M_\lambda)\,\chi(M_\lambda)\,\Pi_\lambda + \mathcal{O}_{\|\cdot\|}(\lambda^\infty)\\
&= M_{+,\lambda}\,\Pi_\lambda + \mathcal{O}_{\|\cdot\|}(\lambda^\infty).
\end{aligned}$$

Combining this Lemma with [12, Theorem 1] gives us an analog of [12, Theorem 1] for the physically meaningful equations.

Theorem 3 (Effective tight-binding dynamics). *Suppose the adiabatically perturbed weights are of the form in [12, Theorem 1] and $\sigma_{\mathrm{rel}}(k) = \bigcup_{n\in\mathscr{I}}\{\omega_n(k)\}$ are a finite family of bands of the unperturbed operator $M_{+,0}$ that are separated from the others by a local spectral gap (cf. [12, Assumption 3]) and $0 \notin \sigma_{\mathrm{rel}}(0)$. Moreover, we assume that the Bloch bundle (cf. [17, Section 4.2]) is trivial.*

Then there exists an orthogonal projection $\Pi_\lambda \asymp \sum_{n=0}^\infty \lambda^n \Pi_n$ onto an almost invariant subspace associated to σ_{rel} and an effective Maxwell operator $M_{\mathrm{eff},\lambda} \asymp \sum_{n=0}^\infty \lambda^n M_{\mathrm{eff},n}$ so that

$$\mathrm{e}^{-\mathrm{i} t M_{+,\lambda}} \, \Pi_\lambda = \mathrm{e}^{-\mathrm{i} t M_{\mathrm{eff},\lambda}} \, \Pi_\lambda + \mathscr{O}_{\|\cdot\|}(\lambda^\infty)$$

holds. These operators are in fact pseudodifferential operators that can be computed to any order in λ (see [12, equations (8)-(10)]).

Unfortunately, the price we have to pay is that our work inherits the same restriction as [10, 51]: the Bloch bundle associated to $\sigma_{\mathrm{rel}}(k)$ needs to be trivial ([51, Assumption A_2] or [10, Assumption 3.2]). However, a more recent work of Freund and Teufel [23] shows how we might go about extending Theorem 3 to the topologically nontrivial case in the future. Their elegant approach develops a pseudodifferential calculus on vector bundles that crucially also works when the Bloch vector bundle is non-trivial.

3.3 Topological Classification of Electromagnetic Media

While our previous works were mostly focussed on Haldane's *photonic bulk-edge conjecture* [59], a natural and for physicists perhaps more interesting question is whether there exist electromagnetic media with as-of-yet undiscovered topological phenomena. Simply put, we can say that a phenomenon is of topological origin if there is a physical observable

$$O(t) \approx T$$

that is approximately given by a topological invariant T; for the Quantum Hall Effect that observable is the transverse conductivity and T is the Chern number.

Since the *types* of topological invariants supported by a physical system depend on its dimensionality as well as its topological class, and the topological class is determined by the *number and nature of its discrete symmetries*, this question can be answered by applying the standard classification tool for topological insulators, the Cartan–Altland–Zirnbauer scheme [1, 9], to electromagnetic media.

3.3.1 The Cartan–Altland–Zirnbauer Classification of Topological Insulators

The relevant symmetries for the topological classification are unitaries or antiunitaries V that *square to* $\pm \mathbb{1}$, and *either commute or anticommute* with the quantum Hamilton operator,

$$V H V^{-1} = \pm H.$$

In the parlance of topological insulators, unitary, commuting symmetries are referred to as ordinary symmetries; unitary, *anti*commuting symmetries are chiral symmetries; *anti*unitary, commuting symmetries are time-reversal symmetries; and *anti*unitary, *anti*commuting symmetries are particle-hole symmetries. The two antiunitary symmetries come in an even and an odd variety depending on whether $V^2 = \pm 1$.

Before proceeding with the classification, we need to block-diagonalize $H = \begin{pmatrix} H_+ & 0 \\ 0 & H_- \end{pmatrix}$ with respect to unitary, commuting symmetries—if it has any—until none are left. Note that the resulting block operators $H_\pm$ may lose other symmetries along the way, as the other symmetries $V \neq \begin{pmatrix} V_+ & 0 \\ 0 & V_- \end{pmatrix}$ need not be block-diagonal. For example, this is why some time-reversal symmetric electromagnetic media with a certain crystallographic symmetry have topologically protected edge modes with fixed isospin (e. g., [81] and the discussion in [17, Section 5.2.2]).

3.3.2 Material Symmetries of Electromagnetic Media

When experimentalists design a topological photonic crystal, they have two axes to explore: (1) they can select different materials from which to build the photonic crystal (with *material symmetries*) and (2) then decide how to periodically arrange them (*crystallographic symmetries*). The primary focus of [13, 17] was to obtain a classification in terms of *material symmetries*. Those relate **E** and **H**, and are of the following form:

$$U_n := \sigma_n \otimes 1, \qquad\qquad n = 1, 2, 3, \qquad\qquad (18a)$$
$$T_n := (\sigma_n \otimes 1)\, C, \qquad\qquad n = 0, 1, 2, 3, \qquad\qquad (18b)$$

where the tensor product refers to the $(\mathbf{E}, \mathbf{H})$ splitting, $\sigma_0 = 1$ is the identity and σ_1, σ_2 and σ_3 are the three Pauli matrices. The usual time-reversal symmetry operator $T_3(\psi^E, \psi^H) = (\overline{\psi^E}, -\overline{\psi^H})$, for example, complex conjugates the fields and flips the sign of the magnetic component. The form of the material symmetries is suggested by the free Maxwell operator

$$\text{Rot} = \begin{pmatrix} 0 & +i\nabla^\times \\ -i\nabla^\times & 0 \end{pmatrix} = -\sigma_2 \otimes \nabla^\times$$

that can be written in terms of the Pauli matrix σ_2. Thus, symmetries of the form (18) either commute or anticommute with Rot.

Initially, we performed a symmetry analysis on the operator $M = W^{-1} \text{Rot}$ that lacked the restriction to non-negative frequencies [13]. While all of the arguments in that paper are mathematically correct, the equations we studied there are unphysical.

One of the mistakes we have made is that we have not taken into account that positive and negative frequency states are not independent degrees of freedom. Indeed, some of the symmetries we have considered in our earlier work mix the two—which is inadmissible.

Put another way, symmetries $V = U_n, T_n$ of the form (18) need to be compatible with the restriction to $\omega \geq 0$. This admissibility condition requires that V has to *commute* with the Maxwell operator $M = W^{-1}$ Rot that lacks the frequency restriction to $\omega \geq 0$,

$$V M V^{-1} = +M,$$

for otherwise V maps $\omega \geq 0$ states onto $\omega \leq 0$ states. Only then does V restrict to a symmetry of $M_+ = M\,|_{\omega \geq 0}$.

The second requirement is that $V = U_n, T_n$ needs to be also an (anti)unitary with respect to the *weighted* scalar product (8), which translates to

$$V W V^{-1} = +W.$$

Combining these two commutativity conditions with the product structure of M_+, we see that the only symmetries that *may* play a role are those three which commute with the free Maxwell operators, namely T_1, U_2 and T_3 [17, Lemma 3.1].

3.3.3 Classification of Electromagnetic Media

Of those three symmetries, U_2 is a unitary, commuting symmetry and, therefore, plays no role in the classification scheme. That gives us *four topologically distinct* electromagnetic media, each is characterized by which of the even time-reversal symmetries, T_1 or T_3, is present or broken.

Theorem 4 (Symmetry classification of media [17, Theorem 1.4]**).** *Suppose the material weights*

$$W(x) = \begin{pmatrix} \varepsilon(x) & \chi(x) \\ \chi(x)^* & \mu(x) \end{pmatrix} = \begin{pmatrix} w_0(x) + w_3(x) & w_1(x) - iw_2(x) \\ w_1(x) + iw_2(x) & w_0(x) - w_3(x) \end{pmatrix},$$

expressed in terms of four Hermitian 3×3 matrices $w_j(x) = w_j(x)^*$, $j = 0, 1, 2, 3$, are lossless and have strictly positive eigenvalues that are bounded away from 0 and ∞, i. e., they satisfy Assumption 1. Moreover, we assume M_+ has no additional unitary commuting symmetries. Then there are four topologically distinct types of electromagnetic media:

Material	Conditions on W	Symmetries	CAZ class
Dual-symmetric materials & vacuum	$w_0 = \operatorname{Re} w_0,$ $w_3 = 0,$ $w_1 = 0,$ $w_2 = \operatorname{Re} w_2$	T_1, T_3, U_2	$2 \times$ AI
Non-dual symmetric & non-gyrotropic	$w_0 = \operatorname{Re} w_0,$ $w_3 = \operatorname{Re} w_3,$ $w_1 = \operatorname{i} \operatorname{Im} w_1,$ $w_2 = \operatorname{Re} w_2$	T_3	AI
Magnetoelectric	$w_0 = \operatorname{Re} w_0,$ $w_3 = \operatorname{i} \operatorname{Im} w_3,$ $w_1 = \operatorname{Re} w_1,$ $w_2 = \operatorname{Re} w_2$	T_1	AI
Gyrotropic	None	None	A

The conditions on W in each row are exclusive, meaning that, e. g., non-gyrotropic materials must violate at least one of the conditions that single out magnetoelectric materials.

For three of these types of media, the topological classification is well known in the literature [11, 45, 50].

The case of dual-symmetric, non-gyrotropic media falls outside of standard theory and we had to perform that classification ourselves [17, Section 4.2.4]. It turns out that after reducing out the unitary symmetry U_2, i. e., considering left- and right-handed circularly polarized waves separately, only *one* of the two time-reversal symmetries survives; each helicity component separately is then of class AI (cf. [17, Section 4.2.4]).

Theorem 5 (Topological bulk classification [17, Theorem 1.5]). *Suppose the material weights are periodic and satisfy Assumption 1.*

*(1) **Class A: Gyrotropic media***
  Phases are labeled by $\mathbb{Z}$-valued Chern numbers, in
  $d = 1$ by none (topologically trivial),
  $d = 2$ by a single first Chern number ($\mathbb{Z}$),
  $d = 3$ by three first Chern numbers ($\mathbb{Z}^3$),
  $d = 4$ by six first and one second Chern number ($\mathbb{Z}^6 \oplus \mathbb{Z}$).
*(2) **Class AI: Non-dual symmetric, non-gyrotropic and magnetoelectric media***
  In $d = 1, 2, 3$ these media are topologically trivial, i. e., there is a single phase.
  In $d = 4$, phases are labeled by a single second Chern number ($\mathbb{Z}$).
*(3) **Dual-symmetric, non-gyrotropic media***
  In $d = 1, 2, 3$ these media are topologically trivial, i. e., there is a single phase.
  In $d = 4$, phases are labeled by two second Chern numbers ($\mathbb{Z}^2$).

Our classification result has two important consequences for physics: first of all, *gyrotropic photonic crystals are indeed in the same topological class*, class A, *as quantum systems exhibiting the Quantum Hall Effect.* This is consistent with Haldane's conjecture that the edge modes predicted in [59] and later observed by experimentalists [76] *are indeed a photonic analog of the Quantum Hall Effect.* Given the wealth of experimental and theoretical evidence, though, this is not a surprising finding for physicists.

What *is* new is the insight that in $d \leq 3$ *only gyrotropic materials are topologically non-trivial,* provided there are no other unitary, commuting symmetries. And *the only* topological invariants supported in that situation are Chern numbers. In particular, because all potential time-reversal symmetries are *even,* there are no electromagnetic media of class AII and consequently, there exists *no photonic analog of the Quantum Spin Hall Effect* (cf. we refer to [17, Section 5.2.2] for a detailed discussion of the literature, including works on spin-momentum locking).

4 Conclusion and Outlook

In summary, the Schrödinger formalism of electromagnetism and other classical waves opens the door to systematically adapting techniques from quantum mechanics to classical waves. Two specific cases were covered here, effective dynamics in adiabatically perturbed photonic crystals (Theorem 3) and the topological classification of electromagnetic media (Theorem 4). This is not just relevant to mathematical physicists, but at least the latter result is new for and of immediate interest to physicists.

Going forward, quantum-wave analogies will continue to serve as inspiration for mathematical, theoretical and experimental physicists. Experimentalists enjoy the much wider latitude with which media for classical waves can be engineered. Depending on the circumstances, they may choose the most suitable wave (acoustic, electromagnetic, etc.) and wavelength regime. For instance, the dynamics of spin wave packets can be "filmed" because their propagation speed is much lower than that of light [64]—something that would be very hard to impossible to realize with a quantum system. Theoreticians can rely on quantum-wave or *wave*-wave analogies to transfer insights from one physical system to another and to propose novel experiments. And mathematicians can find a whole host of interesting and non-trivial problems that are of immediate relevance to physics; I close this review by discussing a select few.

4.1 Systematically Developing Mathematical Techniques for Maxwell-Type Operators

One asset mathematicians bring to the table is the tendency to identify and exploit systematic commonalities: a whole host of wave equations (see, e. g., [18, Section 6],

[65, 79] or [47]) can be phrased in the form of a Schrödinger-type equation, where a Maxwell-type operator (cf. Definition 1) takes the place of a quantum Hamiltonian. Thus, instead of treating Maxwell's equations, acoustic equations, and linearized MHD equations separately, we may study all of them simultaneously by analyzing Maxwell-type operators instead. Simply put, we aim to connect properties of the weights to properties of Maxwell-type operators. When adapting methods and arguments from quantum mechanics, a few differences must be considered.

4.1.1 Native Techniques for Operators with Product Rather than Sum Structure

The strategy of many earlier efforts to rigorously analyze Maxwell-type operators (e. g., [40, Theorem 20]) was to transform Maxwell-type operators $M = W^{-1} D \mid_{\omega \geq 0}$ (with product structure) to operators of the form $\tilde{M} + V$, where the "potential" V is connected to the commutator of the differential operator D and a multiplication operator connected to the weights. To ensure that V is "well-behaved", additional regularity conditions on W need to be imposed (typically $\mathscr{C}^k$ with bounded derivatives up to kth order). However, commonly the material weights are piecewise constant functions, i. e., they are only L^∞ and not even continuous, as many media are fabricated by alternating two or more materials. Therefore, it will be necessary to develop more techniques, which are "native" to operators with product structure.

4.1.2 The Hilbert Space Depends on the Weights

Maxwell-type operators are naturally seen as self-adjoint operators on a Hilbert space that explicitly depends on the weights in two ways: first of all, the *scalar product* and, therefore, the norm depends on the weights. This is important, for e. g., perturbation techniques, where we would like to compare the evolution groups of the perturbed $M_\lambda = W_\lambda^{-1} D \mid_{\omega \geq 0}$ and unperturbed operators $M_0 = W_0^{-1} D \mid_{\omega \geq 0}$ with one another. But because the norm on $\mathscr{H}_\lambda$ explicitly depends on the perturbation parameter λ, it is not immediately clear how to quantify the error in "power series expansions" in λ. Depending on the situation, conceptually different choices are sensible (compare e. g., the approaches in [14, Section 2.2] and [65, Section 4]). Second, it is not at all clear whether e. g., $e^{-it M_0}$ is even well-defined as an operator on $\mathscr{H}_\lambda$ in the first place: a priori we do not know *whether it maps non-negative frequency states of M_λ onto non-negative frequency states*. This consideration was crucial in identifying how to represent material symmetries on the complex Hilbert space in Sect. 3.3.2.

4.1.3 Finding Physically Meaningful Mathematical Problems and Interpreting Mathematical Results Physically

A third consideration concerns conceptual differences in the physical interpretations. Mathematical physicists working on quantum systems have internalized a lot of concepts that do not always transfer as-is to classical waves. This concerns both directions, namely when giving physical meaning to mathematical statements and when translating physics into concrete mathematical problems.

Mathematically, it is completely legitimate to study the operator $M_\Phi = W^{-1} D + \sum_{j=1}^{3} \Phi_j x_j$ in analogy to the Stark operator [46], and apply techniques from linear response theory [16]. However, adding a "potential" makes no physical sense in the context of Maxwell's equations. Instead, in experiment there are two ways to "drive the transmission" of electromagnetic fields: we can either insert sources such as an antenna [6]; or we can modulate the weights (as was the case in [12, 59]). Depending on the precise physical setting, either approach may make sense.

Also in the context of topological "insulators" a tailor-made interpretation of the results is important. For quantum mechanical models for systems that exhibit the Integer Quantum Hall Effect, the Fermi projection $P_F = 1_{(-\infty, E_F]}(H)$ is to be interpreted as the state of the system at zero temperature. Therefore, the bulk-boundary correspondence that furnishes an explanation for the Quantum Hall Effect [30, 31, 58, 74] is a statement about how the states at the boundary relate to states in the bulk. However, in electromagnetism the "Fermi projection" $P_F = 1_{(-\infty, \omega_F]}(M)$, a perfectly well-defined operator, is not linked to a state and enters Haldane's photonic bulk-boundary correspondence merely as an *auxiliary quantity*.

4.2 Dispersive Media

Another important feature that distinguishes classical waves from quantum systems is dispersion. To be precise here, we are not referring to the characteristic broadening of wave packets under the time evolution that also occurs in quantum systems. Instead, the constitutive relations that express the auxiliary fields $(\mathbf{D}, \mathbf{B})$ at time t in terms of the physical fields $(\mathbf{E}, \mathbf{H})$ not only depend on the instantaneous field configuration at time t, but also on the field configuration in the past—the medium has a "memory"(see [18, equation (2.3)] for an explicit formula). This type of dispersion implies that initial values to Maxwell's equations for dispersive media are then *past trajectories* in contrast to the setting covered here where only the instantaneous field configuration at $t = t_0$ matters.

Within the context of topological insulators, Silveirinha has proposed to use dispersion in homogeneous instead of periodic patterning to *open spectral gaps*, and explores in a systematic fashion [24, 67, 68] whether topologically non-trivial bulk band gaps exist. While there is no experimental evidence at this point and the mathematics is not well-understood, this is certainly an intriguing idea *with no quantum analog* that merits further study.

4.3 Non-linear Media

The fundamental equation of quantum mechanics, the Schrödinger equation, is manifestly linear; non-linear equations like Gross–Pitaevskii [70, 71] or Hartree–Fock equations [27, Chapter IV.5] often emerge as *effective* single particle equations from linear many-body quantum mechanics. On the other hand, many media of classical waves are *manifestly non-linear* and it is the *linear* equations that are approximations.

That raises the question whether certain non-linear media allow us to gain insights into many-body quantum systems. For example, Babin and Figotin have shown that in a particular scaling the non-linear Maxwell equations can be approximated by a *non-linear Schrödinger equation* [6], in essence allowing physicists to engineer specific non-linearities.

Babin and Figotin's works [2–6] precede that of Raghu and Haldane [59], so topological phenomena were not on their radar at the time. Hence, the question whether novel *non-linear* topological phenomena exist is also still open.

Acknowledgements The author would like to thank his long-term collaborator Giuseppe De Nittis for tackling this field together. Moreover, the author's work has been supported by JSPS through a WAKATE B grant (grant number 16K17761).

References

1. Altland, A., Zirnbauer, M.R.: Non-standard symmetry classes in mesoscopic normal-superconducting hybrid structures. Phys. Rev. B **55**, 1142–1161 (1997). https://doi.org/10.1103/PhysRevB.55.1142
2. Babin, A., Figotin, A.: Nonlinear photonic crystals I. Quadratic nonlinearity. Waves Random Media **11**, R31–R102 (2001). https://doi.org/10.1088/0959-7174/11/2/201
3. Babin, A., Figotin, A.: Nonlinear photonic crystals: II. Interaction classification for quadratic nonlinearities. Waves in Random Media **12**, R25–R52 (2002). https://doi.org/10.1088/0959-7174/12/4/202
4. Babin, A., Figotin, A.: Nonlinear Maxwell equations in inhomogeneous media. Commun. Math. Phys. **241**(2–3), 519–581 (2003). https://doi.org/10.1007/s00220-003-0939-9
5. Babin, A., Figotin, A.: Nonlinear photonic crystals: III. Cubic nonlinearity. Waves Random Media **13**, R41–R69 (2003). https://doi.org/10.1088/0959-7174/13/4/201
6. Babin, A., Figotin, A.: Nonlinear photonic crystals: IV. Nonlinear Schrödinger equation regime. Waves Random Complex Media **15**(2), 145–228 (2005). https://doi.org/10.1080/17455030500196929
7. Birman, M.S., Solomyak, M.Z.: L_2-Theory of the Maxwell operator in arbitrary domains. Uspekhi Mat. Nauk **42**(6), 61–76 (1987). https://doi.org/10.1070/RM1987v042n06ABEH001505
8. Chen, Z.G., Zhao, J., Mei, J., Wu, Y.: Acoustic frequency filter based on anisotropic topological phononic crystals. 1–7 (2017). arXiv:1706.07283
9. Chiu, C.K., Teo, J.C., Schnyder, A.P., Ryu, S.: Classification of topological quantum matter with symmetries. Rev. Mod. Phys. **88**, 035,005 (2016). https://doi.org/10.1103/RevModPhys.88.035005
10. De Nittis, G., Lein, M.: Applications of magnetic Ψ DO techniques to SAPT - beyond a simple review. Rev. Math. Phys. **23**, 233–260 (2011). https://doi.org/10.1142/S0129055X11004278

11. De Nittis, G., Lein, M.: Exponentially localized Wannier functions in periodic zero flux magnetic fields. J. Math. Phys. **52**, 112,103 (2011). https://doi.org/10.1063/1.3657344
12. De Nittis, G., Lein, M.: Effective light dynamics in perturbed photonic crystals. Commun. Math. Phys. **332**, 221–260 (2014). https://doi.org/10.1007/s00220-014-2083-0
13. De Nittis, G., Lein, M.: On the role of symmetries in photonic crystals. Ann. Phys. **350**, 568–587 (2014). https://doi.org/10.1016/j.aop.2014.07.032
14. De Nittis, G., Lein, M.: The perturbed Maxwell operator as pseudodifferential operator. Doc. Math. **19**, 63–101 (2014)
15. De Nittis, G., Lein, M.: Derivation of ray optics equations in photonic crystals via a semiclassical limit. Ann. Henri Poincaré **18**, 1789–1831 (2017). https://doi.org/10.1007/s00023-017-0552-7
16. De Nittis, G., Lein, M.: Linear Response Theory. Springer Briefs in Mathematical Physics, vol. 21. Springer, Berlin (2017)
17. De Nittis, G., Lein, M.: Symmetry classification of topological photonic crystals. 1–49 (2017). arXiv:1710.08104
18. De Nittis, G., Lein, M.: The Schrödinger formalism of electromagnetism and other classical waves—how to make quantum-wave analogies rigorous. Ann. Phys. **396**:579–617 (2018). https://doi.org/10.1016/j.aop.2018.02.019
19. Dündar, M.A., Wang, B., Nötzel, R., Karouta, F., van der Heijden, R.W.: Optothermal tuning of liquid crystal infiltrated InGaAsP photonic crystal nanocavities. J. Opt. Soc. Am. B **28**(6), 1514–1517 (2011). https://doi.org/10.1364/JOSAB.28.001514
20. van Driel, H.M., Leonard, S.W., Tan, H.W., Birner, A., Schilling, J., Schweizer, S.L., Wehrspohn, R.B., Gosele, U.: Tuning 2D photonic crystals. In: Fauchet, P.M., Braun, P.V. (eds.) Tuning the Optical Response of Photonic Bandgap Structures, vol. 5511, pp. 1–9 (2004). https://doi.org/10.1117/12.559914
21. Figotin, A., Klein, A.: Localization of classical waves II: electromagnetic waves. Commun. Math. Phys. **184**, 411–441 (1997)
22. Fleury, R., Sounas, D.L., Sieck, C.F., Haberman, M.R., Alù, A.: Sound isolation and giant linear nonreciprocity in a compact acoustic circulator. Science **343**, 516–519 (2014). https://doi.org/10.1126/science.1246957
23. Freund, S., Teufel, S.: Peierls substitution for magnetic Bloch bands. Anal. PDE **9**(7):773–811 (2016). https://doi.org/10.2140/apde.2016.9.773
24. Gangaraj, S.A.H., Silveirinha, M.G., Hanson, G.W.: Berry phase, Berry connection, and Chern number for a continuum bianisotropic material from a classical electromagnetics perspective. IEEE J. Multiscale Multiphysics Comput. Tech. **2**, 3–17 (2017). https://doi.org/10.1109/JMMCT.2017.2654962
25. Garanovich, I.L., Longhi, S., Sukhorukov, A.A., Kivshar, Y.S.: Light propagation and localization in modulated photonic lattices and waveguides. Phys. Rep. **518**(1–2), 1–79 (2012). https://doi.org/10.1016/j.physrep.2012.03.005
26. Gill, D.M., Xiong, C., Proesel, J.E., Rosenberg, J.C., Orcutt, J., Khater, M., Ellis-Monaghan, J., Viens, D., Vlasov, Y., Haensch, W., Green, W.M.J.: Demonstration of error free operation up to 32 Gb/s from a CMOS integrated monolithic nano-photonic transmitter. In: 2015 Conference on Lasers and Electro-Optics (CLEO 2015), vol. 2015. Institute of Electrical and Electronics Engineers Inc (2015)
27. Grosso, G., Parravicini, G.P.: Solid State Physics. Academic Press, New York (2003)
28. Guzmán-Silva, D., Mejía-Cortés, C., Bandres, M.A., Rechtsman, M.C., Weimann, S., Nolte, S., Segev, M., Szameit, A., Vicencio, R.A.: Experimental observation of bulk and edge transport in photonic Lieb lattices. New J. Phys. **16**, 063,061 (2014). https://doi.org/10.1088/1367-2630/16/6/063061
29. Haldane, F.D.M., Raghu, S.: Possible realization of directional optical waveguides in photonic crystals with broken time-reversal symmetry. Phys. Rev. Lett. **100**, 013,904 (2008). https://doi.org/10.1103/PhysRevLett.100.013904
30. Hatsugai, Y.: Chern number and edge states in the integer quantum Hall effect. Phys. Rev. Lett. **71**, 3697–3700 (1993). https://doi.org/10.1103/PhysRevLett.71.3697

31. Hatsugai, Y.: Edge states in the integer quantum Hall effect and the Riemann surface of the Bloch function. Phys. Rev. B **48**, 11851–11862 (1993). https://doi.org/10.1103/PhysRevB.48. 11851

32. Joannopoulos, J.D., Johnson, S.G., Winn, J.N., Meade, R.D.: Photonic Crystals. Princeton University Press, Princeton (2008)

33. John, S.: Strong localization of photons in certain disordered dielectric superlattices. Phys. Rev. Lett. **58**(23), 2486–2489 (1987). https://doi.org/10.1103/PhysRevLett.58.2486

34. Johnson, S.G., Joannopoulos, J.D.: Block-iterative frequency-domain methods for Maxwell's equations in a planewave basis. Opt. Express **8**(3), 173–190 (2001). https://doi.org/10.1364/OE.8.000173

35. Kadic, M., Bückmann, T., Schittny, R., Wegener, M.: Metamaterials beyond electromagnetism. Rep. Prog. Phys. **76**, 126,501 (2013). https://doi.org/10.1088/0034-4885/76/12/126501

36. Khanikaev, A., Lein, M.: Understanding topological photonic crystals from first principles—a pedagogical review, in preparation (2018)

37. Khanikaev, A.B., Mousavi, S.H., Tse, W.K., Kargarian, M., MacDonald, A.H., Shvets, G.: Photonic topological insulators. Nat. Mater. **12**, 233–239 (2013). https://doi.org/10.1038/nmat3520

38. Kuchment, P.: Floquet Theory for Partial Differential Equations. Operator Theory, Advances and Applications. Birkhäuser, Basel (1993)

39. Kuchment, P.: The mathematics of photonic crystals. Frontiers in Applied Mathematics, vol. 22, pp. 207–272. SIAM, Philadelphia (2001)

40. Kuchment, P., Levendorskiĭ, S.: On the structure of spectra of periodic elliptic operators. Trans. Am. Math. Soc. **354**(2), 537–569 (2001)

41. Kuramochi, E.: Fabrication of 2D and 3D Photonic Crystals, pp. 479–499. World Scientific, Singapore (2011)

42. Longhi, S.: Quantum-optical analogies using photonic structures. Laser Photon. Rev. **3**, 243–261 (2009). https://doi.org/10.1002/lpor.200810055

43. Lu, L., Fu, L., Joannopoulos, J.D., Soljačić, M.: Weyl points and line nodes in gapless gyroid photonic crystals. Nat. Photon. **7**, 294–299 (2013). https://doi.org/10.1038/NPHOTON.2013. 42

44. Lu, L., Wang, Z., Ye, D., Ran, L., Fu, L., Joannopoulos, J.D., Soljačić, M.: Experimental observation of Weyl points. Science **349**(6248), 622–624 (2015). https://doi.org/10.1126/science. aaa9273

45. Nenciu, G.: Existence of the exponentially localised Wannier functions. Commun. Math. Phys. **91**, 81–85 (1983). https://doi.org/10.1007/BF01206052

46. Nenciu, G.: Dynamics of band electrons in electric and magnetic fields: rigorous justification of the effective Hamiltonians. Rev. Mod. Phys. **63**(1), 91–127 (1991). https://doi.org/10.1103/RevModPhys.63.91

47. Nenciu, G., Nenciu, I.: On essential self-adjointness for first order differential operators on domains in $\mathbb{R}^d$. 1–19 (2018). arXiv:1803.08106

48. Ozawa, T., Price, H.M., Amo, A., Hafezi, M., Lu, L., Rechtsman, M.C., Schuster, D., Simon, J., Zilberberg, O., Carusotto, I.: Topological photonics. 1–83 (2018). arXiv:1802.04173

49. Ozawa, T., Price, H.M., Goldman, N., Zilberberg, O., Carusotto, I.: Synthetic dimensions in integrated photonics: from optical isolation to four-dimensional quantum Hall physics. Phys. Rev. A **93**, 043,827 (2016). https://doi.org/10.1103/PhysRevA.93.043827

50. Panati, G.: Triviality of Bloch and Bloch-Dirac bundles. Ann. Henri Poincaré **8**, 995–1011 (2007). https://doi.org/10.1007/s00023-007-0326-8

51. Panati, G., Spohn, H., Teufel, S.: Effective dynamics for Bloch electrons: Peierls substitution and beyond. Commun. Math. Phys. **242**, 547–578 (2003). https://doi.org/10.1007/s00220-003-0950-1

52. Panati, G., Spohn, H., Teufel, S.: Space-adiabatic perturbation theory. Adv. Theor. Math. Phys. **7**(1), 145–204 (2003). https://doi.org/10.4310/ATMP.2003.v7.n1.a6

53. Peano, V., Brendel, C., Schmidt, M., Marquardt, F.: Topological phases of sound and light. Phys. Rev. X **5**, 031,011 (2015). https://doi.org/10.1103/PhysRevX.5.031011

54. Plotnik, Y., Rechtsman, M.C., Song, D., Heinrich, M., Szameit, A., Malkova, N., Chen, Z., Segev, M.: Observation of dispersion-free edge states in honeycomb photonic lattices. In: 2012 Conference on Lasers and Electro-Optics, p. QF2H.6. Optical Society of America (2012). https://doi.org/10.1364/QELS.2012.QF2H.6, http://www.opticsinfobase.org/abstract.cfm?URI=QELS-2012-QF2H.6

55. Plotnik, Y., Rechtsman, M.C., Song, D., Heinrich, M., Zeuner, J.M., Nolte, S., Lumer, Y., Malkova, N., Xu, J., Szameit, A., Chen, Z., Segev, M.: Observation of unconventional edge states in photonic graphene. Nat. Mater. **13**, 57–62 (2014). https://doi.org/10.1038/nmat3783

56. Pozar, D.M.: Microwave Engineering. Wiley, New York (2011)

57. Prather, D.W., Shi, S., Sharkawy, A., Murakowski, J., Schneider, G.J.: Photonic Crystals: Theory, Applications and Fabrication. Wiley Series in Pure and Applied Optics. Wiley, New York (2009)

58. Prodan, E., Schulz-Baldes, H.: Bulk and Boundary Invariants for Complex Topological Insulators. Mathematical Physics Studies. Springer, Berlin (2016)

59. Raghu, S., Haldane, F.D.M.: Analogs of quantum-Hall-effect edge states in photonic crystals. Phys. Rev. A **78**, 033,834 (2008). https://doi.org/10.1103/PhysRevA.78.033834

60. Rechtsman, M.C., Plotnik, Y., Zeuner, J.M., Song, D., Chen, Z., Szameit, A., Segev, M.: Topological creation and destruction of edge states in photonic graphene. Phys. Rev. Lett. **111**, 103,901 (2013). https://doi.org/10.1103/PhysRevLett.111.103901

61. Rechtsman, M.C., Zeuner, J.M., Plotnik, Y., Lumer, Y., Podolsky, D., Dreisow, F., Nolte, S., Segev, M., Szameit, A.: Photonic Floquet topological insulators. Nature **496**, 196–200 (2013). https://doi.org/10.1038/nature12066

62. Reed, M., Simon, B.: The scattering of classical waves from inhomogeneous media. J. Funct. Anal. **155**, 163–180 (1977). https://doi.org/10.1007/BF01214216

63. Safavi-Naeini, A.H., Hill, J.T., Meenehan, S., Chan, J., Gröblacher, S., Painter, O.: Two-dimensional phononic-photonic band gap optomechanical crystal cavity. Phys. Rev. Lett. **112**, 153,603 (2014). https://doi.org/10.1103/PhysRevLett.112.153603

64. Schneider, T., Serga, A.A., Neumann, T., Hillebrands, B., Kostylev, M.P.: Phase reciprocity of spin-wave excitation by a microstrip antenna. Phys. Rev. B **77**, 214,411 (2008). https://doi.org/10.1103/PhysRevB.77.214411

65. Schulenberger, J.R., Wilcox, C.H.: Completeness of the wave operators for perturbations of uniformly propagative systems. J. Funct. Anal. **7**, 447–474 (1971). https://doi.org/10.1016/0022-1236(71)90028-0

66. Shindou, R., Matsumoto, R., Murakami, S., Ichiro Ohe, J.: Topological chiral magnonic edge mode in a magnonic crystal. Phys. Rev. B **87**, 174,427 (2013). https://doi.org/10.1103/PhysRevB.87.174427

67. Silveirinha, M.G.: $\mathbb{Z}_2$ topological index for continuous photonic materials. Phys. Rev. B **93**, 075,110 (2016). https://doi.org/10.1103/PhysRevB.93.075110

68. Silveirinha, M.G.: Topological classification of Chern-type insulators with the photonic Green function. 1–46 (2018). arXiv:1801.09908

69. Soukoulis, C.M.: The history and a review of the modeling and fabrication of photonic crystals. Nanotechnology **13**(3), 420–423 (2002). https://doi.org/10.1088/0957-4484/13/3/335

70. Spohn, H.: Kinetic equations from Hamiltonian dynamics: Markovian limits. Rev. Mod. Phys. **52**(3), 569–615 (1980). https://doi.org/10.1103/RevModPhys.52.569

71. Spohn, H.: Large Scale Dynamics of Interacting Particles. Texts and Monographs in Physics. Springer, Berlin (1991). https://doi.org/10.1007/978-3-642-84371-6

72. Süsstrunk, R., Huber, S.D.: Observation of phononic helical edge states in a mechanical topological insulator. Science **349**, 47–50 (2015). https://doi.org/10.1126/science.aab0239

73. Süsstrunk, R., Huber, S.D.: Classification of topological phonons in linear mechanical metamaterials. Proc. Natl. Acad. Sci. U.S.A. **113**(33):E4767–E4775 (2016). https://doi.org/10.1073/pnas.1605462113

74. Thouless, D.J., Kohmoto, M., Nightingale, M.P., Den Nijs, M.: Quantized Hall conductance in a two-dimensional periodic potential. Phys. Rev. Lett. **49**, 405–408 (1982). https://doi.org/10.1103/PhysRevLett.49.405

75. Vorobets, M.: On the Bethe-Sommerfeld conjecture for certain periodic Maxwell operators. J. Math. Anal. Appl. **377**, 370–383 (2011). https://doi.org/10.1016/j.jmaa.2010.10.067
76. Wang, Z., Chong, Y.D., Joannopoulos, J.D., Soljačić, M.: Reflection-free one-way edge modes in a gyromagnetic photonic crystal. Phys. Rev. Lett. **100**(1), 013,905 (2008). https://doi.org/10.1103/PhysRevLett.100.013905
77. Wang, Z., Chong, Y.D., Joannopoulos, J.D., Soljačić, M.: Observation of unidirectional backscattering-immune topological electromagnetic states. Nature **461**(7265), 772–775 (2009). https://doi.org/10.1038/nature08293
78. Wilcox, C.: Theory of Bloch waves. J. Anal. Math. **33**(1), 146–167 (1978)
79. Wilcox, C.H.: Wave operators and asymptotic solutions of wave propagation problems of classical physics. Arch. Ration. Mech. Anal. **22**, 37–78 (1966). https://doi.org/10.1007/BF00281244
80. Wong, C.W., Yang, X., Rakich, P.T., Johnson, S.G., Qi, M., Jeon, Y., Barbastathis, G., Kim, S.G.: Strain-tunable photonic bandgap microcavity waveguides in silicon at 1.55 μm. In: Fauchet, P.M., Braun, P.V. (eds.) Tuning the Optical Response of Photonic Bandgap Structures, vol. 5511, pp. 156–164 (2004). https://doi.org/10.1117/12.560927
81. Wu, L.H., Hu, X.: Scheme for achieving a topological photonic crystal by using dielectric material. Phys. Rev. Lett. **114**, 223,901 (2015). https://doi.org/10.1103/PhysRevLett.114.223901
82. Wu, X., Meng, Y., Tian, J., Huang, Y., Xiang, H., Han, D., Wen, W.: Direct observation of valley-polarized topological edge states in designer surface plasmon crystals. Nat. Commun. **8**, 1–9 (2017). https://doi.org/10.1038/s41467-017-01515-2
83. Xu, S.Y., Belopolski, I., Alidoust, N., Neupane, M., Bian, G., Zhang, C., Sankar, R., Chang, G., Yuan, Z., Lee, C.C., Huang, S.M., Zheng, H., Ma, J., Sanchez, D.S., Wang, B., Bansil, A., Chou, F., Shibayev, P.P., Lin, H., Jia, S., Hasan, M.Z.: Discovery of a Weyl fermion semimetal and topological Fermi arcs. Science **349**, 613–617 (2015). https://doi.org/10.1126/science.aaa9297
84. Yablonovitch, E.: Inhibited spontaneous emission in solid-state physics and electronics. Phys. Rev. Lett **58**(20), 2059–2062 (1987). https://doi.org/10.1103/PhysRevLett.58.2059
85. Yablonovitch, E., Gmitter, T.J., Leung, K.M.: Photonic band structure: the face-centered cubic case employing nonspherical atoms. Phys. Rev. Lett **67**(17), 2295–2298 (1991). https://doi.org/10.1103/PhysRevLett.67.2295
86. Ye, L., Qiu, Y.Y.Z.H.H.C., Liu, Z.: Observation of valley-selective microwave transport in photonic crystals. Appl. Phys. Lett. **111**, 251,107 (2017). https://doi.org/10.1063/1.5009597
87. Zilberberg, O., Huang, S., Guglielmon, J., Wang, M., Chen, K., Kraus, Y.E., Rechtsman, M.C.: Photonic topological pumping through the edges of a dynamical four-dimensional quantum Hall system. Nature **553**:59–62 (2018). https://doi.org/10.1038/nature25011

The Localization Dichotomy for Gapped Periodic Systems and Its Relevance for Macroscopic Transport

Gianluca Panati

Abstract We review recent results concerning the localization of gapped periodic systems of independent fermions, as, e.g., electrons in Chern and Quantum Hall insulators. We show that there is a *"localization dichotomy"* which shows some analogies with phase transitions in Statistical Mechanics: either there exists a system of exponentially localized composite Wannier functions for the Fermi projector, or any possible system of composite Wannier functions yields a diverging expectation value for the squared position operator. This fact is largely model-independent, covering both tight-binding and continuous models. The results are discussed with emphasis on the main ideas and the broader context, avoiding most of the technical details.

Keywords Gapped periodic quantum systems · Chern insulators
Quantum Hall insulators · Haldane model · Hofstadter model
Periodic Schrödinger operators · Composite Wannier functions

1 Introduction

A prominent idea in Mathematical Physics, greatly boosted by the pioneering intuitions and the mathematical work of Herbert Spohn, is the fact that physical theories are organized within a hierarchical structure [31]. A theory which holds true at the microscopic level leads, in a suitable large-scale limit, to an effective mesoscopic or macroscopic theory, as, e.g., the kinetic description provided by the Boltzmann equation emerges, in the Boltzmann–Grad limit, from the Newton theory (or,

G. Panati (✉)
Dipartimento di Matematica, La Sapienza Università di Roma, Piazzale Aldo Moro 2,
00185 Rome, Italy
e-mail: panati@mat.uniroma1.it

© Springer Nature Switzerland AG 2018
D. Cadamuro et al. (eds.), *Macroscopic Limits of Quantum Systems*,
Springer Proceedings in Mathematics & Statistics 270,
https://doi.org/10.1007/978-3-030-01602-9_11

243

conjecturally, the Schrödinger theory) of N interacting particles, see, e.g., [29, 31] and some other contributions in this volume.

Quantum physics is not exceptional and, as clearly foreseen by Herbert Spohn, a hierarchical structure appears whenever there is a separation of scales in the physical system under investigation. From the fundamental quantum theory of the whole system, a simpler effective quantum theory governing the dynamics of the slow degrees of freedom alone emerges—in the appropriate time- or space-adiabatic limit. This idea is ubiquitous in atomic, molecular, solid-state, and high-energy quantum physics, and has been mathematically implemented in a series of papers of the Munich's group [13, 22–27, 32, 33].

Symmetries and other special invariants of the microscopic dynamics may be lost in the large-scale or adiabatic limit. For example, the reversible Newton dynamics leads to the irreversible Boltzmann evolution, and the corresponding breaking of time-reversal symmetry has been widely used by Boltzmann's opponents, under the name of *Umkehreinwand* (reversibility paradox), to criticize his kinetic theory of gases. As we know today, Boltzmann's opponents underestimated the power of probability and missed the idea of large-scale limit [31]. On the other hand, there are remarkable physical situations (often close to zero temperature) which escape this general paradigm, in the sense that the macroscopic theory reproduces some invariants of the microscopic one. The archetypical example is provided by the Quantum Hall effect (QHE). This effect appears in quasi-two-dimensional systems, and it is worth to notice that in $2d$ the *charge conductance* (a transport coefficient obtained as the ratio of two macroscopic quantities, namely, the current intensity and the voltage drop) has the same physical dimension and the same value of the *charge conductivity*, which is the corresponding microscopic transport coefficient.[1] Hence, the quantization of the Hall conductivity—a microscopic quantity—yields the quantization of the Hall conductance, which is measured by our macroscopic apparatus.

This contribution to the volume focuses on Quantum Hall systems and Chern insulators, two systems which exhibit an exceptional manifestation—at the macroscopic scale—of an invariant of the microscopic dynamics. As realized in a breakthrough paper by Thouless et al. [35], a nonvanishing Hall conductivity corresponds to a nontrivial topology of the space of occupied states, decomposed with respect to the crystal momentum—a space which is called *Bloch bundle* in the recent literature, following [24]. Besides this celebrated *Transport–Topology Correspondence*, a related **Localization–Topology Correspondence** has been recently noticed and proved for d-dimensional gapped periodic quantum systems, provided that $d \leq 3$ [17, 18]. Surprisingly, this correspondence was largely unnoticed by physicists, with two remarkable exceptions in the case $d = 2$ concerning the Landau Hamiltonian [36] and the Haldane model [34]. In fact, the Localization–Topology Correspondence is a general, model-independent fact, which holds true for any gapped periodic quantum system, and presumably also for gapped ergodic random systems. Since topology is

---

[1] The analogous claim for *spin transport* does not hold in general, since the vanishing of spin torque response is a necessary condition to obtain the equality of spin conductance and spin conductivity for two-dimensional systems, see [14] and references therein.

a dichotomy (either the Bloch bundle is trivial, or is not), an intriguing dichotomy for the localization of quantum states follows. The aim of this contribution is to explain it in a nontechnical style, focusing on essential ideas.

2 The Localization–Topology Correspondence

As a general paradigm, the localization properties of a quantum system are related to the spectral type of the Hamiltonian and to the (de)localization of the corresponding (generalized) eigenstates. However, when *periodic systems* are considered, the Hamiltonian operator has generically purely absolutely continuous spectrum.[2] Therefore, one needs a finer notion of localization, which allows to predict when a crystal, in the absence of any external magnetic field, exhibits a zero Hall conductivity, as it happens for ordinary insulators, and when a nonvanishing one, as in the case of the recently realized *Chern insulators* [2, 5], predicted by Haldane three decades ago [9, 11]. The main message of [18], reviewed in this contribution, is that such a finer notion of localization is provided by the rate of decay of *composite Wannier functions* (CWF) associated to the gapped periodic Hamiltonian. While a formal definition of CWFs will be provided in Sect. 4, we anticipate that composite Wannier functions provide a basis of the space of occupied states of the system (the states energetically concentrated below the spectral gap), which consists of m functions $(w_1, ..., w_m)$ together with their (magnetic) translates.

Equipped with this notion of localization, one is able to identify, for $d \leq 3$, two different regimes (illustrated in Table 1):

(i) whenever the system is time-reversal (TR) symmetric, there exist exponentially localized composite Wannier functions which are associated to the Bloch bands below the Fermi energy, assuming that the latter is in a spectral gap [3, 6, 8, 16, 21]; correspondingly, the Hall conductivity vanishes;

(ii) vice versa, as soon as the Hall conductivity is nonzero, as it happens for Chern insulators, the composite Wannier functions are delocalized. In such a case, any possible system of CWFs $w = (w_1, ..., w_m)$, associated to the Fermi projector, yields a diverging second moment of the position operator, namely,

$$\langle X^2 \rangle_w = \sum_{a=1}^{m} \int_{\mathbb{R}^d} |\mathbf{x}|^2 |w_a(\mathbf{x})|^2 d\mathbf{x} = +\infty. \tag{2.1}$$

The crucial point is that intermediate regimes are forbidden: either there exist exponentially localized CWFs, or any system of CWFs must satisfy (2.1).

---

[2]An exception is provided by the well-known Landau Hamiltonian. Notice, however, that if a periodic background potential is included in the model, one is generically back to the absolutely continuous setting.

Table 1 Synoptic table from [18] which illustrates the main concepts discussed in the text. The arrow $\Rightarrow$ symbolizes a general mathematical implication, while $\rightarrow$ refers to results which have been proved, so far, for specific models or under specific hypotheses, as, e.g., the Linear Response Ansatz [1]

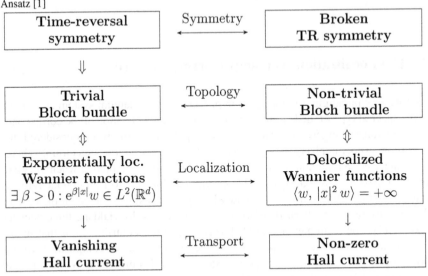

The analogy with phase transitions in Statistical Mechanics is striking, and indeed the therm "*topological phase transition*" is sometimes used. The role of the order parameter is here played by the variance of the position operator $\langle X^2 \rangle_w$.

As anticipated, the result is largely model-independent, as it relies only on the *periodicity* of the Hamiltonian, and on the existence of a *spectral gap* at the Fermi energy. It holds for a large class of models, both continuous and discrete, which will be specified in Sect. 3.

A more precise statement is the following (a precise definition of CWFs will be provided in Sect. 4).

Localization–Topology Correspondence: Consider a gapped periodic quantum system in dimension $d \leq 3$. Then it is always possible to construct a system $w = (w_1, \ldots, w_m)$ of CWFs for the occupied states such that

$$\sum_{a=1}^{m} \int_{\mathbb{R}^d} |\mathbf{x}|^{2s} \, |w_a(\mathbf{x})|^2 \mathrm{d}\mathbf{x} < +\infty \qquad \text{for every } s < 1. \tag{2.2}$$

Moreover, the following statements are equivalent:

(a) **Finite second moment**: there exists a system of CWFs $w = (w_1, \ldots, w_m)$ satisfying

$$\langle X^2 \rangle_w = \sum_{a=1}^{m} \int_{\mathbb{R}^d} |\mathbf{x}|^2 \, |w_a(\mathbf{x})|^2 d\mathbf{x} < +\infty;$$

(b) **Exponential localization**: there exists $\alpha > 0$ and a system of CWFs $\widetilde{w} = (\widetilde{w}_1, \ldots, \widetilde{w}_m)$ satisfying

$$\sum_{a=1}^{m} \int_{\mathbb{R}^d} e^{2\beta|\mathbf{x}|} \, |\widetilde{w}_a(\mathbf{x})|^2 d\mathbf{x} < +\infty \qquad \text{for every } \beta \in [0, \alpha);$$

(c) **Trivial topology**: the Bloch bundle associated to the occupied states is trivial.

The previous result can also be reformulated in terms of the localization functional introduced by Marzari and Vanderbilt [15], which with our notation reads

$$F_{MV}(w) = \sum_{a=1}^{m} \int_{\mathbb{R}^d} |\mathbf{x}|^2 |w_a(\mathbf{x})|^2 \, d\mathbf{x} - \sum_{a=1}^{m} \sum_{j=1}^{d} \left(\int_{\mathbb{R}^d} x_j |w_a(\mathbf{x})|^2 \, d\mathbf{x} \right)^2$$

$$=: \langle X^2 \rangle_w - \langle X \rangle_w^2. \tag{2.3}$$

Since there always exists a system of CWFs satisfying (2.2) for fixed $s = 1/2$, the first moment $\langle X \rangle_w$ is always finite. Hence, the Marzari–Vanderbilt functional is finite if and only if $\langle X^2 \rangle_w$ is finite. By the second part of the Localization–Topology Correspondence, the latter condition is equivalent to the triviality of the Bloch bundle. This result is in agreement with previous numerical and analytic investigations on the Haldane model [34] and with more recent results [4]. As a consequence, the minimization of F_{MV} is possible only in the topologically trivial case, and numerical simulations in the topologically nontrivial regime should be handled with care: one expects that the numerics become unstable when the mesh in k-space becomes finer and finer [4].

3   The Models

In this section, we introduce the class of models, both continuous and discrete, for which the main result holds true.

3.1   Continuos Models

First, the result holds true for magnetic periodic Schrödinger operators, acting in $L^2(\mathbb{R}^d)$ for $d \leq 3$, in the form[3]

---

[3]We use Hartree atomic units, and moreover we reabsorb the reciprocal of the speed of light $1/c$ and the sign of the charge carriers in the definition of the function A_Γ.

$$H_\Gamma = \frac{1}{2}\big(-i\nabla_{\mathbf{x}} - A_\Gamma(\mathbf{x})\big)^2 + V_\Gamma(\mathbf{x}).$$

Specific conditions on the magnetic and scalar potentials A_Γ and V_Γ are assumed, which guarantee in particular that H_Γ defines a self-adjoint operator on a suitable domain [18]. "Periodicity" of the Hamiltonian means that H_Γ should commute with translations by vectors in the Bravais lattice Γ of the solid under consideration, which is generated by a basis $\{\mathbf{a}_1, \ldots, \mathbf{a}_d\} \subset \mathbb{R}^d$ as $\Gamma = \mathrm{Span}_{\mathbb{Z}}\{\mathbf{a}_1, \ldots, \mathbf{a}_d\} \simeq \mathbb{Z}^d \subset \mathbb{R}^d$. The operator H_Γ is then required to commute with the lattice translation operators

$$(T_\gamma \psi)(\mathbf{x}) := \psi(\mathbf{x} - \gamma), \quad \gamma \in \Gamma, \ \psi \in L^2(\mathbb{R}^d), \tag{3.1}$$

as is the case when A_Γ and V_Γ are Γ-periodic functions. In particular, in two dimensions the magnetic flux per unit cell Φ_B should be zero.

The case of nonzero magnetic flux per unit cell, which appears when a **uniform magnetic field** is considered, can also be recast in our framework under a commensurability assumption. In this case, let $A_b(\mathbf{x})$ be a vector potential in $\mathbb{R}^d$ for a magnetic field of uniform strength $b \in \mathbb{R}$, e.g., $A_b(\mathbf{x}) = -\frac{1}{2c}\mathbf{x} \wedge \mathbf{B}$ when $d = 3$, where c is the speed of light and $\mathbf{B} = b\hat{\mathbf{B}}$ is the applied magnetic field. Consider the *Bloch-Landau Hamiltonian*

$$\widetilde{H}_{\Gamma,b} = \frac{1}{2}\big(-i\nabla_{\mathbf{x}} - A_b(\mathbf{x})\big)^2 + V_\Gamma(\mathbf{x}).$$

The role of the natural translations (3.1), which commute with $\widetilde{H}_{\Gamma,b}$ and among themselves, is now played by the **magnetic translations** [37, 38]

$$(T^b_\gamma \psi)(\mathbf{x}) := e^{i\gamma \cdot A_b(\mathbf{x})} \psi(\mathbf{x} - \gamma), \quad \gamma \in \Gamma.$$

These commute with $H_{\Gamma,b}$, but satisfy the pseudo-Weyl relations

$$T^b_\gamma T^b_\mu = e^{\frac{i}{c}\mathbf{B}\cdot(\gamma \wedge \mu)} T^b_\mu T^b_\gamma, \quad \gamma, \mu \in \Gamma.$$

If we assume the **commensurability condition**

$$\tfrac{1}{c}\mathbf{B}\cdot(\gamma \wedge \mu) \in 2\pi\mathbb{Q} \quad \text{for all} \ \gamma, \mu \in \Gamma, \tag{3.2}$$

then the magnetic translations provide a true unitary representation on $L^2(\mathbb{R}^d)$ at the price of choosing a smaller Bravais lattice Γ_b, i.e., of choosing a larger periodicity cell Y_b, depending on the strength of the magnetic field.

It turns out that the two effects can be superimposed without affecting the validity of our main result. Hence, in summary, we consider the Hamiltonian

$$\boxed{H_{\Gamma,b} = \frac{1}{2}\big(-i\nabla_{\mathbf{x}} - A_b(\mathbf{x}) - A_\Gamma(\mathbf{x})\big)^2 + V_\Gamma(\mathbf{x})} \tag{3.3}$$

acting in $L^2(\mathbb{R}^d)$, for $d \leq 3$, where A_b is a linear potential (as defined above) and A_Γ and V_Γ are Γ-periodic and satisfy the technical assumptions in [18]. Moreover, the commensurability condition (3.2) is always assumed. We also assume that the Hamiltonian has a spectral gap $\Delta \subset \mathbb{R}$, and for $\mu \in \Delta$ we define the Fermi projector $P_{\mu,b}$ as

$$\boxed{P_{\mu,b} = \chi_{(-\infty,\mu)}(H_{\Gamma,b})}.$$

3.2 Discrete Models

The dynamics of independent electrons in periodic crystals is often modeled by the so-called *tight-binding models*. In such models, the electron wavefunction is defined over a discrete set $\mathcal{C} \subset \mathbb{R}^d$ as, e.g., a triangular or a honeycomb structure.[4] In order to implement periodicity, $\mathcal{C}$ is equipped with a free action of a Bravais lattice $\Gamma = \mathrm{Span}_{\mathbb{Z}}\{\mathbf{a}_1, \ldots, \mathbf{a}_d\} \simeq \mathbb{Z}^d$, so that the quotient $F = \mathcal{C}/\Gamma$ is a finite set, representing the different "sublattices" or "substructures" in $\mathcal{C}$. For a hexagonal structure, F contains just two points.

After a specific choice of a periodicity cell Y (which is clearly not unique), one has that $\mathcal{C} \simeq \Gamma \times Y$. Hence, the Hilbert space of the model becomes

$$\ell^2(\mathcal{C}) \xrightarrow{\ \mathcal{S}\ } \ell^2(\Gamma) \otimes \mathbb{C}^N \simeq \ell^2(\mathbb{Z}^d, \mathbb{C}^N) \tag{3.4}$$

where $N = |Y| = |F|$ is the number of points in the periodicity cell, and the isomorphism $\mathcal{S}$ depends on the choice of the cell Y.

In view of the identification (3.4), we directly assume that the Hamiltonian is a bounded self-adjoint operator acting on $\mathcal{H}_{\mathrm{disc}} = \ell^2(\mathbb{Z}^d, \mathbb{C}^N)$. Every bounded operator A acting on $\mathcal{H}_{\mathrm{disc}}$ is represented by a family of matrices $\{A_{\mathbf{m},\mathbf{n}}\}_{\mathbf{n},\mathbf{m} \in \mathbb{Z}^d} \subset \mathrm{Mat}_N(\mathbb{C})$. In particular,

$$(A\psi)_{\mathbf{n}} = \sum_{\mathbf{m} \in \mathbb{Z}^2} A_{\mathbf{n},\mathbf{m}} \psi_{\mathbf{m}} \qquad \forall \psi \in \ell^2(\mathbb{Z}^d, \mathbb{C}^N).$$

The kernel $A_{\mathbf{m},\mathbf{n}}$ is a $N \times N$ matrix, and $|A_{\mathbf{m},\mathbf{n}}|$ denotes its matrix norm, while the operator norm on the full Hilbert space is denoted by $\|A\|$. We focus on a class of *short range* Hamiltonians, which are selected by the following condition.

Definition 1. *An operator A acting on $\mathcal{H}$ is called **nearsighted** if and only if there exist constants $C, \zeta > 0$ such that*

---

[4]We carefully avoid the use of the word "lattice", since the latter has a different meaning in physics and in mathematics. In particular, a honeycomb or a triangular structure is not a lattice in the mathematical sense (i.e., a discrete subgroup of $(\mathbb{R}^d, +)$ with maximal rank). As far as the Bravais lattice Γ is concerned, no ambiguity raises, since it is a lattice in both senses.

$$\left|A_{\mathbf{m},\mathbf{n}}\right| \leq C e^{-\frac{1}{\zeta}\|\mathbf{m}-\mathbf{n}\|_1} \quad \textit{for all } \mathbf{m}, \mathbf{n} \in \mathbb{Z}^2,$$

where $\|\mathbf{n}\|_1 := |n_1| + |n_2|$. *The constant* ζ *is called the **range** of A.*

Assumption 2. *The Hamiltonian operator* H *is a bounded self-adjoint operator acting on* $\mathcal{H}_{\mathrm{disc}}$ *which moreover*

(H_1) *is nearsighted with range* ζ_H;
(H_2) *is periodic, namely,* $H_{\mathbf{m},\mathbf{n}} = H_{\mathbf{m}-\mathbf{p},\mathbf{n}-\mathbf{p}}$ *for all* $\mathbf{m}, \mathbf{n}, \mathbf{p} \in \mathbb{Z}^2$;
(H_3) *has a spectral gap* $\Delta \subset \mathbb{R}$.

For $\mu \in \Delta$, we denote by $P_\mu := \chi_{(-\infty,\mu)}(H)$ the Fermi projection.

The above assumptions are satisfied by the most popular tight-binding models in solid-state physics, including the models proposed by Hofstadter [12] and Haldane [9]. Actually, for the latter models the matrix $H_{\mathbf{m},\mathbf{n}} = H_{\mathbf{m}-\mathbf{n},0}$ is nonzero only for finitely many values of $\mathbf{q} = \mathbf{m} - \mathbf{n}$ (finite range).

4 Periodicity, Bloch Functions, and Wannier Functions

4.1 Discrete Models

The periodicity of the Hamiltonian implies that H and the translation operators can be simultaneously diagonalized. As far as discrete models are concerned, the simultaneous diagonalization is simply achieved by choosing a unit cell ("dimerization", cf. (3.4)) and then using the Fourier transform $\mathcal{F} : \ell^2(\mathbb{Z}^d, \mathbb{C}^N) \to L^2(\mathbb{T}^d, \mathbb{C}^N)$, so that the Hamiltonian is decomposed as

$$\left(\mathcal{F} H \mathcal{F}^{-1} \phi\right)(\mathbf{k}) = H^{\mathcal{F}}(\mathbf{k}) \phi(\mathbf{k}) \qquad \forall \phi \in L^2(\mathbb{T}^d, \mathbb{C}^N), \tag{4.1}$$

for $\mathbb{T}^d = \mathbb{R}^d / 2\pi\mathbb{Z}^d$ a flat torus. Here, $H^{\mathcal{F}}(\mathbf{k})$ is an $N \times N$ matrix for each fixed $\mathbf{k} \in \mathbb{R}^d$, which is $\mathbf{k}$-periodic: $H^{\mathcal{F}}(\mathbf{k}+\mathbf{g}) = H^{\mathcal{F}}(\mathbf{k})$ for every $\mathbf{k} \in \mathbb{R}^d$ and every $\mathbf{g} \in 2\pi\mathbb{Z}^d$.

The construction above depends on the choice of the cell Y via (3.4). While such a dependence is often harmless, in some situations it is not. As remarked in [7], the **Berry curvature**—which is in principle physically observable—does depend on the choice of the cell! Hence, both physical consistency and mathematical elegance strive for a transform which is intrinsic, as it happens for the **modified Bloch–Floquet transform**, already used, e.g., in [21, 24] for its technical advantages when dealing with continuous models. This transform acts on a wavefunction $\Psi \in \ell^2(\mathcal{C})$ as

$$(\mathcal{U}\Psi)(\mathbf{k}, \mathbf{x}) := \sum_{\gamma \in \Gamma} e^{-i\mathbf{k} \cdot (\mathbf{x}-\gamma)} \Psi_{\mathbf{x}-\gamma}, \tag{4.2}$$

with the convention that $\Psi_{\mathbf{x}-\gamma} = 0$ whenever $\mathbf{x} - \gamma \notin \mathcal{C}$. Here $|\mathbb{B}|$ is the volume of the periodicity cell corresponding to the reciprocal lattice[5]

$$\Gamma^* = \left\{ \mathbf{k} \in \mathbb{R}^d : \mathbf{k} \cdot \gamma \in 2\pi\mathbb{Z} \text{ for all } \gamma \in \Gamma \right\}.$$

Since $(\mathcal{U}\Psi)(\mathbf{k}, \mathbf{x} + \gamma) = (\mathcal{U}\Psi)(\mathbf{k}, \mathbf{x})$ for every $\gamma \in \Gamma$, $(\mathcal{U}\Psi)(\mathbf{k}, \cdot)$ is an element of $\ell^2(F) \simeq \mathbb{C}^N$, which—in contrast to $\ell^2(Y)$—does not depend on the choice of the cell. The price to pay is that $\mathbf{k} \mapsto (\mathcal{U}\Psi)(\mathbf{k}, \cdot)$ is not periodic, but satisfies instead

$$(\mathcal{U}\Psi)(\mathbf{k} + \boldsymbol{\lambda}, \cdot) = \tau_\lambda (\mathcal{U}\Psi)(\mathbf{k}, \cdot) \tag{4.3}$$

where τ_λ is a unitary matrix. The transform $\mathcal{U}$ provides a decomposition analogous to (4.1), namely[6,7]

$$\left(\mathcal{U} H \mathcal{U}^{-1} \Phi\right)(\mathbf{k}) = H_{\mathbf{k}} \Phi(\mathbf{k}). \tag{4.4}$$

In this representation, periodicity is replaced by pseudo-periodicity or τ-covariance, namely

$$H_{\mathbf{k}+\lambda} = \tau_\lambda H_{\mathbf{k}} \tau_\lambda^{-1} \quad \text{for all } \lambda \in \Gamma^*, \mathbf{k} \in \mathbb{R}^d. \tag{4.5}$$

By spectral theorem, the Fermi projector $P \equiv P_\mu$ can be decomposed accordingly, and one gets $P_{\mathbf{k}+\lambda} = \tau_\lambda P_{\mathbf{k}} \tau_\lambda^{-1}$ for all $\lambda \in \Gamma^*, \mathbf{k} \in \mathbb{R}^d$.
Any normalized solution $u_{\mathbf{k},n}$ to the eigenvalue problem

$$H_{\mathbf{k}} u_{\mathbf{k},n} = E_{\mathbf{k},n} u_{\mathbf{k},n} \qquad u_{\mathbf{k},n} \in \ell^2(F) \simeq \mathbb{C}^N, n \in \{1, \ldots, N\}$$

is dubbed (periodic part of the) **Bloch function** for the nth-eigenvalue. Similarly, a **quasi-Bloch function** is a normalized solution to the equation

$$P_{\mathbf{k}} v_{\mathbf{k},a} = v_{\mathbf{k},a} \qquad v_{\mathbf{k},a} \in \ell^2(F), a \in \{1, \ldots, m\}$$

where $m = \dim P_{\mathbf{k}}$ (which is independent of $\mathbf{k}$ in view of the gap condition).

---

[5]It is easy to check that, for every lattice $\Gamma \subset \mathbb{R}^d$, one has $|Y||\mathbb{B}| = (2\pi)^d$.
[6]To be precise, one should introduce the Hilbert space

$$\mathcal{H}_\tau = \left\{ \Phi \in L^2_{\text{loc}}(\mathbb{R}^d, \ell^2(F)) : \Phi_{\mathbf{k}+\lambda} = \tau_\lambda \Phi_{\mathbf{k}} \text{ for a.e. } \mathbf{k} \in \mathbb{R}^d, \forall \lambda \in \Gamma^* \right\}.$$

Then the transform $\mathcal{U}$, defined by (4.2), extends to a unitary operator from $\ell^2(\mathcal{C})$ to $\mathcal{H}_\tau$, and formula (4.4) holds true for every $\Phi \in \mathcal{H}_\tau$. The subtle point is that the components of an element of $\ell^2(F) \simeq \mathbb{C}^N$ are intrinsic, while the components of an element of $\ell^2(Y) \simeq \mathbb{C}^N$ change according to the choice of the periodicity cell. Not all the $\mathbb{C}^N$s are equal, *some are more equal than others.*
[7]To agree with the physics literature, we use the notation $H_{\mathbf{k}}$, $P_{\mathbf{k}}$ and $u_{n,\mathbf{k}}$ although in mathematics the subscript usually refers to labels varying in a discrete set, and clearly $\mathbf{k} \in \mathbb{R}^d / \Gamma^* \simeq \mathbb{T}^d$ varies in a continuum set.

Definition 3. *A **Bloch frame** is a map* $\mathbf{k} \mapsto (v_{\mathbf{k},1}, \dots, v_{\mathbf{k},m})$ *that assigns to every point* $\mathbf{k} \in \mathbb{R}^d$ *an orthonormal frame spanning* Ran $P_{\mathbf{k}}$, *i.e., an orthonormal frame of* m *quasi-Bloch functions.*

As already discussed in the literature (see, e.g., [8, 16, 19, 20] and references therein) the existence of a Bloch frame that is both continuos and pseudoperiodic (i.e., satisfies (4.3)) might be topologically obstructed. If such an obstruction vanishes, we say that the *Bloch bundle* (corresponding to P_μ) is trivial. If $d \leq 3$, triviality holds true whenever the Hamiltonian is time-reversal symmetric [3, 6, 8, 21].

One might think that there is no topological obstruction to the existence of a Bloch frame with some prescribed discontinuities. Remarkably, and perhaps surprisingly, this is not the case: Bloch frames with "mild" discontinuities are forbidden. As proved in [18], a Bloch frame with Sobolev regularity H^1, i.e., that satisfies

$$\sum_{a=1}^{m} \int_{\mathbb{B}} \left\| \nabla_{\mathbf{k}} v_{\mathbf{k},a} \right\|^2 d\mathbf{k} < +\infty, \tag{4.6}$$

may exists if and only if the Bloch bundle is trivial. The Sobolev regularity H^s implies continuity only if $s > d/2$, by Sobolev embedding theorem. Hence, Bloch frames in the Sobolev space H^1 are in general discontinuous for $2 \leq d \leq 3$, but are nevertheless topologically obstructed if the Bloch bundle is not trivial. The last claims, proved in [18], are deeply rooted in the theory of Sobolev mappings with values in nonlinear targets, see [10] and references therein.

We now come back to the "position representation", by inverting the modified Bloch–Floquet transform.

Definition 4. *Given a Bloch frame* $\mathbf{k} \mapsto (v_{\mathbf{k},1}, \dots, v_{\mathbf{k},m})$, *the corresponding system of **composite Wannier functions** $w = (w_1, \dots, w_m)$, with $w_a \in \ell^2(\mathcal{C})$, is defined by*

$$w_a(\mathbf{x}) := \left(\mathcal{U}^{-1} v_{\mathbf{k},a} \right)(\mathbf{x}) = \frac{1}{|\mathbb{B}|} \int_{\mathbb{B}} e^{i\mathbf{k}\cdot\mathbf{x}} \, v_{\mathbf{k},a} \, d\mathbf{k}. \tag{4.7}$$

Since the modified Bloch–Floquet transform intertwines the operator $i\nabla_{\mathbf{k}}$ with the position operator X acting in $\ell^2(\mathcal{C})$, one concludes that the existence of a system of CWFs which are 1-localized (in the sense that (2.2) holds true for $s = 1$) implies the existence of a Bloch frame of Sobolev regularity H^1. The latter implies that the Bloch bundle is trivial, thus explaining the implication (a) $\Rightarrow$ (c) in the main result. The implication (c) $\Rightarrow$ (b) traces back to previous work, see [21] and reference therein. Clearly, claim (b) is stronger than (a), so the chain of implications is complete.

One might wonder whether is it possible to define Bloch frames by using the representation (4.1) of the Hamiltonian, namely, by defining Bloch functions $\psi_{\mathbf{k},n}$ as eigenfunctions of the Hamiltonian $H^{\mathcal{F}}(\mathbf{k})$, and correspondingly all the related concepts. Notice that this yields $\psi_{\mathbf{k}+\mathbf{g},n} = \psi_{\mathbf{k},n}$ for all $\mathbf{g} \in 2\pi\mathbb{Z}^d$. The answer is that one can certainly work with the representation (4.1) but one has to come back to the original representation by inverting the "dimerization" (3.4). For example, Wannier functions are given by $w_{\mathbf{k},a} = \mathcal{S}^{-1}\mathcal{F}^{-1}\psi_{\mathbf{k},a}$, where $\mathcal{S}$ is defined in (3.4). If this last step is forgotten, several properties will depend on the choice of the periodicity cell, yielding sometimes unphysical results.

4.2 Continuous Models

When dealing with continuous models, the same ideas and concepts used for discrete ones appear. Technicalities are however more involved, so that the use of the modified Bloch–Floquet transform becomes very advantageous, not only for the independence from the choice of the cell but for several technical reasons. We outline here the basic concepts of the continuous case, while we refer to [18] for assumptions and details.

We recall that, to recover periodicity when using magnetic translations, in view of (3.2) one has to consider a smaller lattice $\Gamma_b \subset \Gamma$, and hence a larger cell Y_b, where b is the strength of the magnetic field. To begin with, choose a basis $\{\mathbf{b}_1, \ldots, \mathbf{b}_d\}$ such that $\Gamma_b^* = \mathrm{Span}_{\mathbb{Z}}\{\mathbf{b}_1, \ldots, \mathbf{b}_d\}$ and consider the corresponding centered fundamental cell

$$\mathbb{B}_b := \left\{ \mathbf{k} = \sum_{j=1}^d k_j \mathbf{b}_j : -\frac{1}{2} \le k_j \le \frac{1}{2}, \ j \in \{1, \ldots, d\} \right\}.$$

The **magnetic Bloch–Floquet transform** is defined[8] on suitable functions $w \in C_0(\mathbb{R}^d) \subset L^2(\mathbb{R}^d)$ as

$$(\mathcal{U}_b w)(\mathbf{k}, \mathbf{x}) := \sum_{\gamma \in \Gamma_b} e^{-i\mathbf{k}\cdot(\mathbf{x}-\gamma)} (T_\gamma^b w)(\mathbf{x}), \quad \mathbf{x} \in \mathbb{R}^d, \ \mathbf{k} \in \mathbb{R}^d. \tag{4.8}$$

From (4.8), one immediately reads the (pseudo-)periodicity properties

$$\begin{aligned} T_\gamma^b (\mathcal{U}_b w)(\mathbf{k}, \mathbf{x}) &= (\mathcal{U}_b w)(\mathbf{k}, \mathbf{x}) && \text{for all} \ \gamma \in \Gamma_b, \\ (\mathcal{U}_b w)(\mathbf{k}+\lambda, \mathbf{x}) &= e^{-i\lambda\cdot\mathbf{x}} (\mathcal{U}_b w)(\mathbf{k}, \mathbf{x}) && \text{for all} \ \lambda \in \Gamma_b^*. \end{aligned} \tag{4.9}$$

Following [24], we reinterpret (4.9) in order to emphasize the role of covariance with respect to the action of the relevant symmetry group. Define the Hilbert space

$$\mathcal{H}^b_{\mathrm{f}} := \left\{ \psi \in L^2_{\mathrm{loc}}(\mathbb{R}^d) : T_\gamma^b \psi = \psi, \ \text{for all} \ \gamma \in \Gamma_b \right\} \simeq L^2(Y_b),$$

---

[8]The normalization here differs from the one used in [28] but agrees with the one used in [24], which is also the most common convention among solid-state and computational physicists. The latter is more convenient when a numerical grid in $\mathbf{k}$-space is considered, which becomes finer and finer in the thermodynamic limit.

with scalar product given by integration over the cell Y_b, namely

$$\langle \psi_1, \psi_2 \rangle_{\mathcal{H}^b_f} := \int_{Y_b} \overline{\psi_1(\mathbf{x})} \, \psi_2(\mathbf{x}) \, d\mathbf{x}.$$

Setting

$$\big(\tau(\lambda)\psi\big)(\mathbf{x}) := e^{-i\lambda \cdot \mathbf{x}} \psi(\mathbf{x}), \qquad \text{for } \psi \in \mathcal{H}^b_f,$$

one obtains a unitary representation $\tau : \Gamma^*_b \to \mathcal{U}(\mathcal{H}^b_f)$ of the group of translations by vectors of the dual lattice. One can then argue that $\mathcal{U}_b$ establishes a unitary transformation $\mathcal{U}_b : L^2(\mathbb{R}^d) \to \mathcal{H}^b_\tau$, where $\mathcal{H}^b_\tau$ is the Hilbert space

$$\mathcal{H}^b_\tau := \Big\{ \phi \in L^2_{\mathrm{loc}}(\mathbb{R}^d, \mathcal{H}^b_f) : \ \phi(k + \lambda, \cdot) = \tau(\lambda)\,\phi(k, \cdot) \ \forall \lambda \in \Gamma^*, \text{ for a.e. } k \in \mathbb{R}^d \Big\}$$

equipped with the inner product

$$\langle \phi_1, \phi_2 \rangle_{\mathcal{H}^b_\tau} = \frac{1}{|\mathbb{B}_b|} \int_{\mathbb{B}_b} \langle \phi_1(\mathbf{k}), \phi_2(\mathbf{k}) \rangle_{\mathcal{H}^b_f} \, d\mathbf{k}.$$

Clearly, functions in $\mathcal{H}^b_\tau$ are determined by the values they attain on the fundamental unit cell $\mathbb{B}_b$. Moreover, the inverse transformation $\mathcal{U}^{-1}_b : \mathcal{H}^b_\tau \to L^2(\mathbb{R}^d)$ is explicitly given by

$$\big(\mathcal{U}^{-1}_b \phi\big)(x) = \frac{1}{|\mathbb{B}_b|} \int_{\mathbb{B}_b} e^{i\mathbf{k} \cdot \mathbf{x}} \phi(\mathbf{k}, \mathbf{x}) \, d\mathbf{k},$$

which is sometimes called the **Wannier transform**.

Upon the identification of $\mathcal{H}^b_\tau$ with the direct integral [30, Section XIII.16]

$$\mathcal{H}^b_\tau \simeq \int_{\mathbb{B}_b}^{\oplus} \mathcal{H}^b_f \, d\mathbf{k},$$

we can reach a partial diagonalization of the magnetic Schrödinger Hamiltonian (3.3). Indeed, $H_{\Gamma,b}$ becomes a fibered operator in the Bloch–Floquet representation, i.e.,

$$\mathcal{U}_b \, H_{\Gamma,b} \, \mathcal{U}^{-1}_b = \int_{\mathbb{B}_b}^{\oplus} H(\mathbf{k}) d\mathbf{k},$$

where

$$H(\mathbf{k}) = \frac{1}{2}\big(-i\nabla_y - A_b(\mathbf{x}) - A_\Gamma(\mathbf{x}) + \mathbf{k}\big)^2 + V_\Gamma(\mathbf{x}). \tag{4.10}$$

Under suitable assumptions [18], the fiber operator $H(\mathbf{k})$, $\mathbf{k} \in \mathbb{R}^d$, acts on a $\mathbf{k}$-independent domain $\mathcal{D}^b_f \subset \mathcal{H}^b_f$, where it defines a self-adjoint operator. Moreover, the compactness of the resolvent implies that the spectrum of $H(\mathbf{k})$ is pure point: we label its eigenvalues as

$$E_0(\mathbf{k}) \leq E_1(\mathbf{k}) \leq \cdots \leq E_n(\mathbf{k}) \leq E_{n+1}(\mathbf{k}) \leq \cdots,$$

repeating them according to their multiplicities. The functions $\mathbb{R}^d \ni \mathbf{k} \mapsto E_n(\mathbf{k}) \in \mathbb{R}$ are called (*magnetic*) *Bloch bands*: these functions are Γ_b^*-periodic in view of the property of τ-*covariance* of the fiber Hamiltonian $H(\mathbf{k})$, namely

$$H(\mathbf{k} + \lambda) = \tau(\lambda) \, H(\mathbf{k}) \, \tau(\lambda)^{-1}, \qquad \lambda \in \Gamma_b^*,$$

which is formally identical to (4.5). A solution $u_n(\mathbf{k})$ to the eigenvalue problem

$$H(\mathbf{k})u_n(\mathbf{k}) = E_n(\mathbf{k})u_n(\mathbf{k}), \qquad u_n(\mathbf{k}) \in \mathcal{H}^b{}_f, \qquad \|u_n(\mathbf{k})\|_{\mathcal{H}^b{}_f} = 1,$$

constitutes the (periodic part of the) n-**th magnetic Bloch function**, in the physics terminology.

As in the discrete case, one can read properties of localization of the particle moving in the crystal from the Bloch functions, by going back to the position representation. To do so, one considers the rate of decay at infinity of the **Wannier function** w_n corresponding to the Bloch function $u_n \in \mathcal{H}_\tau^b$, defined—in analogy with (4.7)—as the preimage, via magnetic Bloch–Floquet transform, of the Bloch function, i.e.,

$$w_n(\mathbf{x}) := \left(\mathcal{U}_b^{-1} u_n\right)(\mathbf{x}) = \frac{1}{|\mathbb{B}_b|} \int_{\mathbb{B}_b} e^{i\mathbf{k}\cdot\mathbf{x}} u_n(\mathbf{k}, \mathbf{x})d\mathbf{k}. \tag{4.11}$$

In analogy with the discrete case, one introduces the concept of quasi-Bloch function and Bloch frame, and defines a system of *composite Wannier functions* as the set $w = (w_1, \ldots, w_m)$ of the preimages via $\mathcal{U}_b$ of a Bloch frame $(\chi_1, \ldots, \chi_m)$, with $\chi_a \in \mathcal{H}_\tau$.

5 Conclusions

Composite Wannier functions provide an excellent tool to measure the localization of fermions in extended systems, as, e.g., those appearing in solid-state physics. By measuring localization or delocalization in terms of CWFs, the authors of [18] were able to state and prove a localization dichotomy which is largely model-independent, since it applies to all d-dimensional gapped periodic quantum systems of independent fermions, provided $d \leq 3$. Conceptually, this result corresponds to a topological phase transition where the role of the order parameter is played, *mutatis mutandis*, by the expectation value of the squared position operator with respect to a system of CWFs, compare Eq. (2.1).

As far as the future is concerned, several developments are on the agenda. First of all, one might consider the extension of the localization dichotomy to (egodic) random systems. In this case, the lack of periodicity will force us to abandon the geometric concepts, including the idea of Bloch bundle and Bloch frame, and to replace them with the corresponding generalized concepts of noncommutative geometry.

Second, the result might be generalized to systems of moderately interacting fermions. The main difficulty along this route is the fact that the interacting one-body density matrix, in the thermodynamic limit, will presumably not be a projector. Moreover, a spontaneous breaking of periodicity in the thermodynamic limit might appear and is indeed expected for strongly correlated electrons (density waves).

Despite some nonnegligible technical difficulties, we expect the localization dichotomy to hold true for ergodic random operators and for moderately interacting electrons, and hence to become a new paradigm in the theory of localization of states in extended quantum systems.

Acknowledgements I am indebted to Herbert Spohn for his precious guidance in my early scientific steps. The opportunity to work with him has been invaluable to me. This survey would not have been possible without the fruitful collaboration with many colleagues, including (in chronological order) S. Teufel, Ch. Brouder, N. Marzari, A. Pisante, D. Fiorenza, D. Monaco, É. Cancès, A. Levitt, and G. Stolz.

References

1. Aizenman, M., Warzel, S.: Random Operators. Graduate Studies in Mathematics, vol. 168. Spinger, Berlin (2015)
2. Bestwick, A.J., Fox, E.J., Kou, X., Pan, L., Wang, K.L., Goldhaber-Gordon, D.: Precise quantization of the anomalous Hall effect near zero magnetic field. Phys. Rev. Lett. **114**, 187201 (2015)
3. Brouder, Ch., Panati, G., Calandra, M., Mourougane, Ch., Marzari, N.: Exponential localization of Wannier functions in insulators. Phys. Rev. Lett. **98**, 046402 (2007)
4. Cancès, É., Levitt, A., Panati, G., Stoltz, G.: Robust determination of maximally-localized Wannier functions. Phys. Rev. B **95**, 075114 (2017)
5. Chang, C.Z. et al.: High-precision realization of robust quantum anomalous Hall state in a hard ferromagnetic topological insulator. Nat. Mater. **14**, 473 (2015)
6. Cornean, H.D., Herbst, I., Nenciu, G.: On the construction of composite Wannier functions. Ann. Henri Poincaré **17**, 3361–3398 (2016)
7. Fruchart, M., Carpentier, D., Gawedzki, K.: Parallel transport and band theory in crystals. EPL **106**, 60002 (2014)
8. Fiorenza, D., Monaco, D., Panati, G.: Construction of real-valued localized composite Wannier functions for insulators. Ann. Henri Poincaré **17**, 63–97 (2016)
9. Haldane, F.D.M.: Model for a quantum Hall effect without Landau levels: condensed-matter realization of the parity anomaly. Phys. Rev. Lett. **61**, 2017 (1988)
10. Hang, F., Lin, F.H.: Topology of Sobolev mappings II. Acta Math. **191**, 55–107 (2003)
11. Hasan, M.Z., Kane, C.L.: Colloquium: topological insulators. Rev. Mod. Phys. **82**, 3045–3067 (2010)
12. Hofstadter, D.R.: Energy levels and wave functions of Bloch electrons in rational and irrational magnetic fields. Phys. Rev. B **14**, 2239–2249 (1976)

13. Hövermann, F., Spohn, H., Teufel, S.: Semiclassical limit for the Schrödinger equation with a short scale periodic potential. Commun. Math. Phys. **215**, 609–629 (2001)
14. Marcelli, G.: A Mathematical Analysis of Spin and Charge Transport in Topological Insulators, Ph.D. thesis in Mathematics. La Sapienza Università di Roma, Rome (2017)
15. Marzari, N., Vanderbilt, D.: Maximally localized generalized Wannier functions for composite energy bands. Phys. Rev. B **56**, 12847–12865 (1997)
16. Monaco, D., Panati, G.: Symmetry and localization in periodic crystals: triviality of Bloch bundles with a fermionic time-reversal symmetry. Acta App. Math. **137**, 185–203 (2015)
17. Monaco, D., Panati, G., Pisante, A., Teufel, S.: The localization dichotomy for gapped periodic quantum systems. arXiv:1612.09557
18. Monaco, D., Panati, G., Pisante, A., Teufel, S.: Optimal decay of Wannier functions in Chern and quantum Hall insulators. Commun. Math. Phys. **359**, 61–100 (2018)
19. Nenciu, G.: Existence of the exponentially localised Wannier functions. Commun. Math. Phys. **91**, 81–85 (1983)
20. Nenciu, G.: Dynamics of band electrons in electric and magnetic fields: rigorous justification of the effective Hamiltonians. Rev. Mod. Phys. **63**, 91–127 (1991)
21. Panati, G.: Triviality of Bloch and Bloch-Dirac bundles. Ann. Henri Poincaré **8**, 995–1011 (2007)
22. Panati, G., Spohn, H., Teufel, S.: Space-adiabatic perturbation theory in quantum dynamics. Phys. Rev. Lett. **88**, 250405 (2002)
23. Panati, G., Spohn, H., Teufel, S.: Space-adiabatic perturbation theory. Adv. Theor. Math. Phys. **7**, 145–204 (2003)
24. Panati, G., Spohn, H., Teufel, S.: Effective dynamics for Bloch electrons: Peierls substitution and beyond. Commun. Math. Phys. **242**, 547–578 (2003)
25. Panati, G., Spohn, H., Teufel, S.: Motion of electrons in adiabatically perturbed periodic structures. In: Mielke, A. (ed.) Analysis, Modeling and Simulation of Multiscale Problems. Springer, Berlin (2006)
26. Panati, G., Spohn, H., Teufel, S.: The time-dependent Born-Oppenheimer approximation. Math. Model. Numer. Anal. **41**, 297–314 (2007)
27. Panati, G., Sparber, Ch., Teufel, S.: Geometric currents in piezoelectricity. Arch. Rat. Mech. Anal. **191**, 387–422 (2009)
28. Panati, G., Pisante, A.: Bloch bundles, Marzari-Vanderbilt functional and maximally localized Wannier functions. Commun. Math. Phys. **322**, 835–875 (2013)
29. Pulvirenti, M., Simonella, S.: Propagation of chaos and effective equations in kinetic theory: a brief survey. Math. Mech. Complex Syst. **4**, 255–274 (2016)
30. Reed, M., Simon, B.: Methods of Modern Mathematical Physics. Vol. IV: Analysis of Operators. Academic Press, New York (1978)
31. Spohn, H.: Large Scale Dynamics of Interacting Particles. Springer, Berlin (1991)
32. Spohn, H., Teufel, S.: Adiabatic decoupling and time-dependent Born-Oppenheimer theory. Commun. Math. Phys. **224**, 113–132 (2001)
33. Teufel, S.: Adiabatic Perturbation Theory in Quantum Dynamics. Springer, Berlin (2003)
34. Thonhauser, T., Vanderbilt, D.: Insulator/Chern-insulator transition in the Haldane model. Phys. Rev. B **74**, 235111 (2006)
35. Thouless, D.J., Kohmoto, M., Nightingale, M.P., den Nijs, M.: Quantized Hall conductance in a two-dimensional periodic potential. Phys. Rev. Lett. **49**, 405–408 (1982)
36. Thouless, D.J.: Wannier functions for magnetic sub-bands. J. Phys. C **17**, L325–L327 (1984)
37. Zak, J.: Magnetic translation group. Phys. Rev. **134**, A1602 (1964)
38. Zak, J.: Identities for Landau level orbitals. Europhys. Lett. **17**, 443 (1992)

Printed in the United States
By Bookmasters